Stefan Sinzinger, Jürgen Jahns

Microoptics

WILEY-VCH

Stefan Sinzinger, Jürgen Jahns

Microoptics

WILEY-VCH

Weinheim · New York · Chichester
Brisbane · Singapore · Toronto

Dr. Stefan Sinzinger
Prof. Dr. Jürgen Jahns
University of Hagen
Optical Information Technology
Feithstraße 140
D-58084 Hagen
Germany

The cover image combines a photograph of a technical GRIN microlens array (courtesy of Dr. M. Oikawa of Nippon Sheet Glass Inc, Japan) and an REM photography of the facetted eyes of an insect (Calliphora spec.) (courtesy of Dr. C. Chr. Meinecke and Dr. J. Rosenberg, Institut für Tierphysiologie, Ruhr-Universität Bochum).

Library of Congress Card No.: applied for

A catalogue record for this book is available from the British Library.

Deutsche Bibliothek Cataloguing-in-Publication Data:
Sinzinger, Stefan:
Microoptics / Stefan Sinzinger/Jürgen Jahns. - Weinheim ; New York ;
Chichester ; Brisbane ; Singapore ; Toronto : Wiley-VCH, 1999
ISBN 3-527-29428-7

Printing: betz-druck gmbh, D-64291 Darmstadt.
Bookbinding: Wilhelm Osswald & Co., D-67433 Neustadt.
Printed in the Federal Republic of Germany.

Contents

Preface

It is a great honour and pleasure to have the opportunity to write the Preface to the book on "Microoptics" by Dr. Stefan Sinzinger and Prof. Jürgen Jahns.

As the authors stated in their book, the concept of microoptics can be thought of in analogy to microelectronics and more widely, microtechnologies. Moreover, readers may discover some different aspects in "Microoptics" after reading this book.

The word "microoptics" was presented by Dr. Teiji Uchida and Dr. Ichiro Kitano in the late 1960's for forming practical optical components based on gradient index fibers and lenses. By adding some other miniature optical elements, microoptics has been really playing an important role to provide various optical subsystems in the optoelectronics field.

Along with the development of optical fiber communication, the concept of "integrated optics" was proposed by Dr. S. E. Miller in 1969. This concept is based upon planar waveguides which can be prepared by a monolithic fabrication process to deal with lightwaves. Fortunately, we can use now some practical components, such as semiconductor integrated optics based upon semiconductor lasers integrated with modulators and amplifiers, silica-based optical circuits, ultrafast lithium niobate-based modulators, and so on. At that time, I tried to use the new wording "microlens", but this was not accepted by optical societies. But now, it is registered in the standard keywords.

When I wrote a book in this technical field: "Fundamentals of Microoptics", published by Academic Press in 1984, I felt that these two concepts were considered separately and should meet some innovative integration consideration to match the development of rapidly growing optoelectronics field such as optical fiber communication, optical disks, optoelectronics equipments, and so on.

Therefore, I think that modern microoptics should involve so-called integrated optics and classical microoptics to provide solutions for responding to the new demand of optoelectronics which we may meet in the 21st century, such as terabit networks and terabyte optical memories, advanced displays, and so on.

This book is beautifully organized and covers important and attractive topics in this field. I found in this book many descriptions which are expected by a lot of readers, i.e., smart pixel including surface emitting lasers, array illuminators, information processing, and so on.

I believe that this book may be read with the highest favour not only by experts in this technical area but also beginners who are going to start research in microoptics.

Congratulations on the publication of "Microoptics"!!

Kenichi Iga
Professor, Tokyo Institute of Technology
Autumn 1998 in Tokyo

1 From macrooptics to microoptics — an overview

Microoptics has emerged as a new branch of science during the past 10–20 years and is gradually making its way towards commercialization in a number of fields. The term microoptics indicates a relationship with microelectronics. Such a relationship does exist as far as the fabrication techniques are concerned. Like microelectronics (and micromechanics, for that matter), microoptics uses planar, lithographic fabrication techniques. Hence, it seems appropriate to define the term "microoptics" based on the fabrication aspect: *microoptics is fabricated by microtechnology.* As we shall see later, this is not a clearcut definition in the mathematical sense. There exist boundaries and transitions between what we consider as microoptics and other areas of optics.

1.1 Optics technology

If we distinguish between different areas of optics based on the fabrication technique, we may identify the following areas: classical optics, fiber optics and microoptics (Table 1.1). The history of optics started with the fabrication of glass. This tradition has existed for several thousand years. Artificial glass was discovered accidentally in fired earthenware, through the combination of arenaceous limestone, containing sand, with soda. The clay tablet library of the Assyrian king Assubanipal (7 BC) contains the oldest remaining glass recipe: "Take 60 parts sand, 180 parts ash from sea plants, 5 parts chalk — and you get glass." The traditional methods for the processing of glass are grinding and polishing. Grinding is a mechanical process used to remove material. It provides a surface shape as close as possible to the desired structure. Polishing is based on mechanical as well as chemical processes. By polishing, the final optical surface may be obtained with tolerances well below the wavelength of the light. An overview of these techniques can, for example, be found in [1]. In the 70s, diamond turning was added to the list of fabrication tools in an effort to generate "arbitrary" surface shapes. Dimensions for classical optics are in the range from millimetres up to metres (for astronomical telescopes). This is why it may be justified to speak of "macrooptics". Macrooptics is closely connected with mechanical mounting hence the term "optomechanics" . The precision of the fine-mechanical parts is typically on the order of 0.1 mm.

technology	processing techniques	mounting techniques
classical optics ("macrooptics")	grinding, polishing, diamond turning	fine mechanics
"miniature optics"	grinding, polishing, gradient index optics, LIGA process (components), fiber pulling	miniaturized mechanics, micromechanics
"microoptics"	lithographic: optical, electron beam, X-ray, LIGA non-lithographic: diamond turning, microjet printing	micromechanics, integration on single substrate bonding techniques

Tab. 1.1: Areas of optics defined by their fabrication and mounting techniques.

The advent of fiber optics for communication purposes as well as for illumination and image transmission systems (e.g., endoscopy) brought with it a trend to miniaturization. Also, several new techniques were developed such as the pulling of fibers in combination with the generation of the preform (see, for example, [2]) and the fabrication of gradient index optics [3] (such as the SELFOC^{TM} lenses by ion diffusion). The dominant fabrication techniques for miniaturized optics, however, continued to be the classical ones. Ball lenses used for laser-to-fiber coupling or for endoscopic systems are one example of "miniature optics". More recently, new fabrication techniques have also been used to make miniature optics. An example is the LIGA technique [4, 5] which was developed in the 80s. Tolerance requirements for the alignment of miniature optics shifted to the micrometre regime or even below, for example, for the coupling of optical fibers. Dimensions of miniature optical components are typically on the order of 0.1–1 mm.

In the 1970s planar lithographic fabrication techniques were adapted from semiconductor processing to the fabrication of optical components, for example, to fabricate special beam splitters [6] and lenslet arrays [7]. The use of these techniques allows one to generate optical components with dimensions in the micrometre range (however, all the way up to 1 m or more) and (sub-)micron features. Various lithographic techniques have been developed for microoptic fabrication (see Table 1.1). The LIGA technique mentioned above as well as diamond turning have also been demonstrated for the manufacture of microoptical elements. More recently, in an effort to move to low cost fabrication, non-lithographic techniques like microjet printing of lenses using polymer materials have been investigated.

The use of lithographic fabrication techniques allows for a large amount of flexibility in the design of the microoptics. In addition, a large variety of materials is available, ranging from glass to semiconductor materials and on to plastic. The possibility to process materials like, for example, silicon and gallium arsenide is of interest since it allows one to put microoptics directly onto optoelectronic devices. This indicates a special feature of micro-technology, i.e., the trend towards the integration of components and systems. This trend is exemplified by names like MST (micro systems technology) and MOEMS (micro-opto-electro-mechanical systems) which means the combination of different functions by the use of lithographic fabrication.

1.2 Classification of optical hardware

So far, we have not distinguished between waveguide and free-space optics. In waveguide optics, a light wave is confined by a lateral variation of the refractive index (either using a step profile or a gradient profile) (Fig. 1.1a). In the longitudinal direction, the propagation medium is usually homogeneous. Lateral dimensions vary from a few micrometres for single mode waveguides to the order of 1 mm for multimode plastic fibers, for example. In free-space optics, a light wave is not confined laterally. Rather, it is "guided" by lenses (as the key elements in free-space optics), beam splitters and mirrors which are positioned at discrete positions in a longitudinal direction (i.e., along the optical axis). Between these components, the propagation medium (air, glass, ...) is homogeneous in the lateral direction (Fig. 1.1b). In integrated optics one can find "hybrid" structures that combine waveguide and free-space optics as shown in Fig. 1.1c.

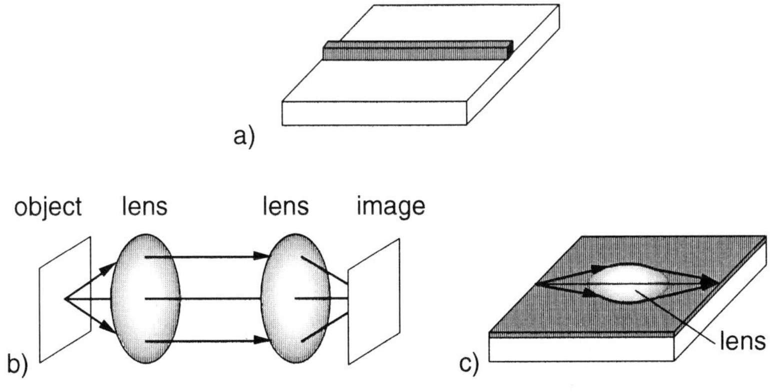

Fig. 1.1: Schematic representation of a) waveguide optics, b) free-space optics, and c) a combination of waveguide and free-space optics.

Another distinction can be made between discrete and integrated optics (Table 1.2). Discrete means that a system consists of individual components which have to be mounted together mechanically. Classical optics and fiber optics/miniaturized optics belong to the class of "discrete optics". Mounting is mostly achieved by using fine mechanics. For fiber-optical applications, micro-mechanical components like V-grooves etched into silicon have recently been used since they provide better precision than fine mechanical parts. The difficulty of alignment and mechanical stability have been the motivation for trying to integrate optics. This attempt was certainly also motivated by the success of microelectronic integration (VLSI) which is the cause for the high functionality, low cost and reliablity of electronic systems. In 1969, Miller [8] proposed to build integrated waveguide-optical circuits that combine several functions on a single optoelectronic chip.

In the 1980s, when free-space optics was heavily investigated for interconnection and computing purposes, the integration of free-space optics was suggested. Two approaches were put forward: the "stacked planar microoptics" [9] and the "planar integrated free-space optics" [10]. Common to the different approaches for integrated optics is the absence of mechanics, the stability and small size. By using hybrid integration techniques (such as flip-chip bonding and thermo-anodic bonding) the passive optics can be combined with other types of components or sub-modules.

	waveguide optics	**free-space optics**
discrete mounting	fiber optics lenses	beam splitters etc.
integrated optics	IGWO	IFSO

Tab. 1.2: Classification of optics. IGWO - integrated waveguide optics, IFSO - integrated free-space optics.

Yet another distinction has to be made between "passive" and "active" optics. By passive optics, we mean optical elements for light propagation, such as waveguides, lenses, lens arrays, beam splitters etc. By active optics, we mean optoelectronic devices for light generation, modulation, amplification and detection.

1.3 Optical functions and their implementation

The implementation of a free-space optical system requires two basic operations: imaging (or focusing and collimation, respectively) and beam deflection (or $1 \times N$ beam splitting). For the first task, one uses lenses, for the latter one uses prisms, gratings and mirrors. Both functions can be implemented by using refraction, diffraction and reflection, as well as combinations thereof (Fig. 1.2).

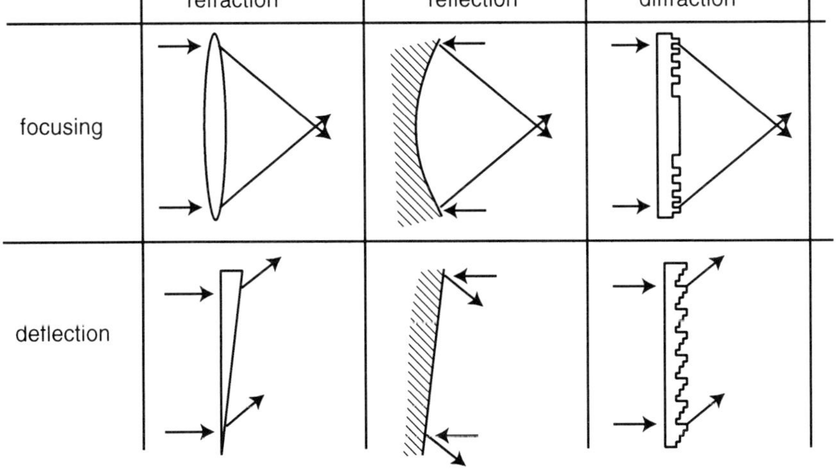

Fig. 1.2: Optical functions and their implementation.

In this section, we will briefly survey refractive, diffractive and reflective optics. On the one hand, the purpose of this section is to give the reader an overview of these classes of devices and the technology used for their fabrication. Furthermore, we wish to introduce the terminology we are going to use in this book.

Most classical macrooptical elements are based on refraction at an optical interface, for example, between air and glass, as described by Snell's law. More recently, in the seventies and eighties, elements with a gradient-index structure were developed, so that now we can distinguish between refractive surface relief elements and GRIN-type elements (Fig. 1.3).

For refractive optical elements (ROEs), diffraction only occurs at finite apertures. This means that diffraction is not usually utilized for the functionality but rather limits the performance. This is of importance especially for optical arrays, when, for example, "crosstalk" becomes an issue. In a few cases, refractive array components may be used as diffractive elements. An example is the implementation of an optical array illuminator (the term will be explained later) based on Fraunhofer diffraction at a lenslet array.

The example just mentioned already implies the fundamental distinguishing feature between refractive and diffractive optics. A diffractive optical element (DOE) is a *periodic structure*. The classical diffraction grating is an example. Its action is described by the grating equation which one may consider as the analogue to Snell's

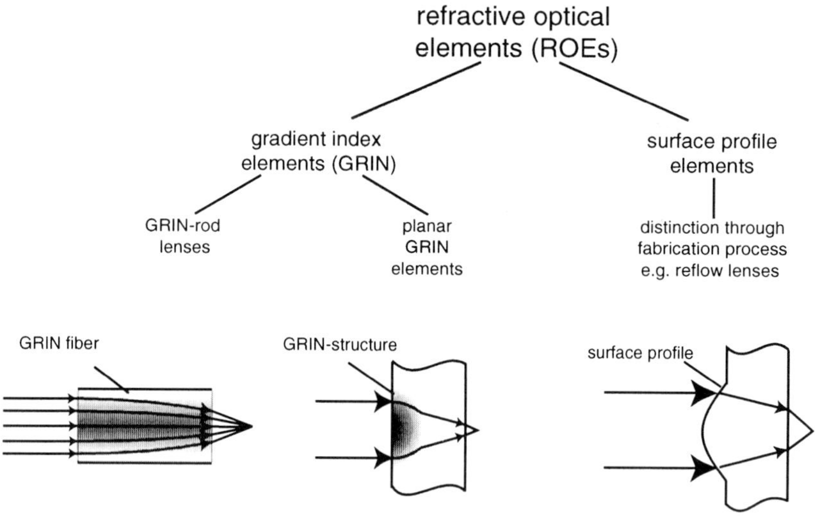

Fig. 1.3: Classification of refractive optical elements.

equation for refraction. In microoptics, a whole variety of DOEs has been developed including diffractive lenses, lenslet arrays and special types of beam splitters.

There exists a large variety of techniques for fabricating diffractive optics. We are going to distinguish between amplitude and phase gratings on the one hand and blazed and quantized phase profiles on the other (Fig. 1.4).

In the early 19th century, Joseph von Fraunhofer measured the wavelength of light by using grating diffraction. Initially, he performed his experiments with gratings consisting of a set of thin stretched wires. Soon after that, ruling was developed as a technique for grating fabrication. Fraunhofer already achieved periods of a few micrometres by ruling with a metallic "comb" over a glass plate coated with soot. This type of grating would now be called a (binary) amplitude grating since it only influences the amplitude of a light wave, not its phase.

In accordance with this definition, a phase grating acts only upon the phase of a light wave. An early example of a phase grating is the blazed grating (or echellette grating) introduced by R. W. Wood in 1910 [11]. This type of grating has a continuous sawtooth profile. In a sense, a blazed grating represents a diffractive-reflective element, since to fully understand its operation, one has to take diffraction and reflection into account. The diffracted energy will be maximum in the direction corresponding to a reflection. With blazed gratings, very high diffraction efficiencies are obtained that are close to the theoretical value of 1. A practical problem is the high cost associated with the fabrication of blazed metallic gratings if mechanical ruling is used. Therefore, many blazed gratings are made by replicating from a master.

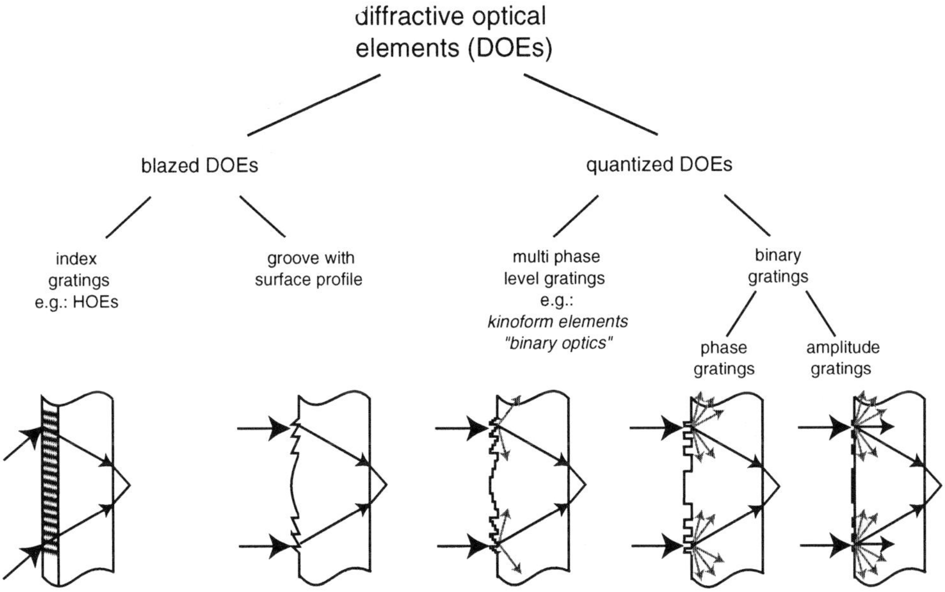

Fig. 1.4: Classification of diffractive optical elements.

In the 60s and 70s, the advent of the laser caused much interest in areas such as optical image processing using spatial filtering. Furthermore, holography was developed as a major tool for optics (Gabor [12]). Analog and computer-generated holograms were added to the hardware catalogue of optics. In analog holography, an optical setup (two interfering waves) is used to generate an interferogram in a (thin or thick) photographic emulsion holography. Holography has also been used as a technique to fabricate microoptical elements (beam splitters and lenslets) in materials such as dichromated gelatin and photopolymers.

Computer-generated holograms (CGHs) were invented in order to be able to implement "arbitrary" wavefronts without the need for optical recording (Lohmann and Brown [13]). Rather, the elements were designed by computer and fabricated using digital plotters. Almost simultaneously with CGHs, kinoform elements were introduced by Lesem et al. [14]. However, whereas CGHs are usually based on the detour-phase principle, kinoforms are phase-only elements where the phase modulation was originally realized by a dielectric layer of variable thickness. Despite certain limitations of both the CGH and the kinoform, which are mostly due to limited capabilities of the technology existing at the time, they can be considered as a new paradigm introduced into the world of optical fabrication. This implies the use of computer design techniques in combination with digital or analog processing tools.

This was perpetuated by the adaptation of lithographic fabrication for the manufacture of optics in the seventies and eighties (Dammann [6], D'Auria [7]). Lithographic fabrication includes the structuring of a photosensitive layer (photoresist) and the transfer of the structure into some substrate material (usually some glass or semiconductor). Binary and multi-level phase technology was developed which allowed one to implement elements with high diffraction efficiencies. "Binary optics" (Veldkamp [15]) (where binary is reminiscent of the digital approach to fabrication rather than the number of phase levels) can be considered as a continuation of what started with CGHs, only based on improved fabrication technology. More recently, during the nineties, analog lithographic techniques using, for example, direct-writing with laser and electron beams allowed one to realize continuous or stepped phase profiles with very high precision, thus finally realizing the kinoform concept with precision that could not be achieved in the sixties.

As already mentioned in the context of blazed gratings, reflection can play an important role for optical elements. In principle, any optical element can be made reflective by some metallic or dielectric coating. Purely reflective elements are of importance for a number of purposes. In macrooptics, examples are telescope mirrors or spectroscopic components. For microoptics, reflective elements may also be of relevance, for example, for integrated systems, and therefore deserve mentioning.

1.4 Scope of this book

For an overview like this it is necessary for practical and intellectual reasons to confine oneself to a certain area. Most of this book will deal with passive free-space optics based on microfabrication techniques. This means, we will talk about the fabrication techniques themselves, about individual components (like lenses, beamsplitters, ...), about microoptic integration, about systems aspects and applications. Microoptics can be refractive or diffractive — depending on the physics of the elements. They can also be "hybrid" in the sense that diffraction and refraction play an equally important role for the performance of the element.

However, for a number of reasons, we have decided also to include chapters on active devices for free-space optics, on waveguide optics and miniature optical elements like the SELFOC^{TM} lenses. This is supposed to help giving the reader a better overview and understanding of the whole field. Literature about related topics like a comprehensive treatment of waveguide optics or micromechanics can be found in [16, 17].

1.5 Organization of the book

After this introduction to the subject, in **Chapter 2**, "Optical components with small dimensions", we approach the question of how the lateral dimensions affect the performance of optical components. We define quality criteria ("figures of merit") for the performance of optical components and observe how they develop under scaling of the elements. This also gives us a chance to introduce some basic optical parameters which will be used throughout the book.

In **Chapter 3** we focus on "lithographic fabrication technology". The main goal here is to categorize the basic fabrication steps and discuss the critical parameters. Although we have the application to microoptics fabrication in mind, the chapter gives a general overview of lithographic processing.

The topic of **Chapter 4** is "refractive microoptics". The functionality of surface profile microoptics is well known from conventional optics. For this group of microoptical elements we thus focus on fabrication techniques, rather than recalling the optical basics. Gradient index (GRIN) optical elements, on the other hand, are more "exotic" elements, specifically interesting for miniature and microoptics. Therefore we are also addressing, e.g., the laws of light propagation in GRIN media.

Of specific importance for microoptics is the topic of **Chapter 5**, namely "diffractive microoptics". Diffractive optics is perfectly adjusted to binary fabrication by lithographic means. The chapter is devoted to an overview of the field, addressing the basic rules of phase quantization, fabrication techniques, a variety of approaches to the theoretical modelling of diffractive optics, as well as design issues.

Chapter 6 is devoted to "integrated waveguide optics", which, according to the previous definitions, is a detour from our main topic. Nevertheless, we think it is helpful, in this context, to give an overview of waveguide optical components and integrated waveguide optics as an approach to optical systems integration. The chapter contains sections on the development of discrete waveguide modes, waveguide couplers and the physical aspects of light modulation in waveguides. In addition we introduce some typical application areas of waveguide optics.

In **Chapter 7** we focus on microoptical systems integration or integrated free-space optics. This is one of the most important issues in microoptics, since systems integration is the key to "real world" applications. Three different approaches are discussed in this chapter, i.e., "micro-opto-electro-mechanical systems (MOEMs)", "stacked optics" and "planar optics". The second half of the chapter is devoted to microoptical imaging systems specifically adjusted to interconnect applications.

Systems aspects are strongly influenced by the available optoelectronic components. This is the reason for addressing the basics of optoelectronics in **Chapter 8**. We discuss the physics of "superlattices and multiple quantum wells" which are fundamental for optoelectronic devices like "SEEDs (self electro-optic-effect devices)" and "VCSELs (vertical-cavity surface emitting lasers)". Finally we introduce the concept of "smart pixel arrays".

The remainder of the book is devoted to applications of microoptical components and systems. The topic of **Chapter 9** is "array illumination". A variety of different approaches are discussed for splitting an incoming laser beam into a 1D or 2D array of beamlets. This task is interesting, e.g., for optical scanners or copying machines as well as optical interconnects.

In **Chapter 10** we present a more general discussion of "microoptical elements for beam shaping", of which the beam-splitting components discussed in chapter 9 are a specific subgroup. Here we focus on the shaping of coherent laser beams. The chapter is subdivided into sections on lateral, axial, temporal beam shaping as well as on multiple aperture beam shaping such as beam combination and aperture filling. The most elegant solution to the beam-shaping problem is "intracavity" beam shaping which also can be applied to single lasers or to laser arrays.

Chapter 11 is devoted to another field of applications, where microoptical techniques are gaining more and more impact. "Optical information technology" can be subdivided into information processing, optical interconnects as well as optical data storage. We address aspects where microoptical techniques are already applied or where microoptics might be useful in the near future.

In the final **Chapter 12** a variety of further applications is discussed, several of which have already been mentioned throughout the book in a different context. We focus on "imaging with microlenses", "beam steering", "microoptical sensors", and "microoptics for optical design". The main intention of this collection is to point out the large variety of application areas which might help the reader to develop own ideas about where to apply microoptical techniques.

Each of the chapters contains an extensive **List of references** on related publications. We tried to be as comprehensive as possible but are fully aware that such a list cannot possibly be complete. Since it is impossible not to omit a significant number of references, we tried to select some of the authentic pioneering work and supplement it through references to some of the most recent publications in the field.

A **Glossary** and a **List of symbols** provided for each chapter is supposed to help the reader find the way through the complex subject. Frequently used **Abbreviations** are listed separately at the end of the book. **Exercises** and **Solutions to exercises** are meant to test and support the understanding of specific issues discussed in the respective chapter.

1.6 Further reading

One purpose of this book is to serve as a reference for the field of microoptics. For this reason, we have tried to add as many references to the chapters as possible (and reasonable). In a rapidly growing field, however, this effort is always like the work of sisiphos. The reader may find interest in related books on the topic. During the past years, a few have been published that should be mentioned [18, 19, 20, 21].

1.7 Acknowledgment

For the successful work on a book like this, the support of a large number of people is indispensible. First we would like to thank all actual and former members of the institute for "Optische Nachrichtentechnik" at the FernUniversität Hagen for their help during our work on the manuscript. They provided several of the pictures and figures and contributed on numerous occasions through discussions in group meetings. Specifically we owe many thanks to Susanne Kinne for her help with formatting issues.

It is a great honour to us that Prof. K. Iga (Tokyo Institute of Technology), one of the pioneers in the field of microoptics, provided his support by writing the Preface to this book.

Prof. A. W. Lohmann (Universität Erlangen-Nürnberg) deserves our special thanks. He initiated the idea to write a book on microoptics. Although he was not directly involved in the writing, he contributed on several occasions with continuous encouragement and by us referring to notes from many of his lectures.

We are especially grateful to J. Leger, University of Minnesota, R. A. Morgan, Honeywell Corp., E. Meusel, Universität Dresden, E. B. Kley, Friedrich Schiller Universität Jena, J. Rosenberg, Ruhr-Universität Bochum, M. Oikawa, Nippon Sheet Glass, Inc. and W. Däschner, Aereal Imaging Corp. for their supportiveness and for providing results and pictures used in the book.

We also thank Prof. H. G. Schuster (Universität Kiel) who decided to include this book in the series "Modern Trends in Physics".

The logistic support from the publisher Wiley-VCH, in person through R. Wengenmayr, Dr. M. Bär and B. Pauli has been very important for the successful completion of this book. Dr. A. J. Owen deserves many thanks for numerous suggestions for language and style improvements.

References

[1] H. Bach and N. Neuroth (eds), "The properties of optical glass", Springer Verlag, Berlin (1995).

[2] J. M. Senior, "Optical Fiber Communications", Prentice Hall, Englewood Cliffs (1985).

[3] T. Uchida, M. Furukawa, I. Kitano, K. Koizumi and H. Matsamura, "Optical characteristics of a light-focusing fiber guide and its applications", *IEEE J. Quant. El.* **QE-6** (1970), 606–612.

[4] A. Heuberger, "X-ray Lithography", *Solid State Technol.* **28** (1986), 93–101.

[5] E. W. Becker, W. Ehrfeld, P. Hagmann, A. Maner and D. Münchmeyer, "Fabrication of microstructures with high aspect rations and great structural heights by synchrotron radiation, lithography, galvanoforming, and plastic moulding", *Microelectronic Engineering* **4** (1986), 35–56.

[6] H. Dammann, "Blazed Synthetic Phase-Only Holograms", *Optik* **31** (1970), 95–104.

[7] L. D'Auria, J. P. Huignard, A. M. Roy and E. Spitz, "Photolithographic fabrication of thin film lenses", *Opt. Comm.* **5** (1972), 232–235.

[8] S. E. Miller, "Integrated optics: an introduction", *Bell Systems Techn. J.* **48** (1969), 2059–2068.

[9] K. Iga, M. Oikawa, S. Misawa, J. Banno and Y. Kokubun, "Stacked planar optics: an application of the planar microlens", *Appl. Opt.* **21** (1982), 3456–3460.

[10] J. Jahns and A. Huang, "Planar Integration of Free-Space Optical Components", *Appl. Opt.* **28** (1989), 1602–1605.

[11] M. Born and E. Wolf, "Principles of Optics", Pergamon Press, London (1980), 6th ed.

[12] D. Gabor, "A new microscope principle", *Nature* **161** (1948), 177.

[13] B. R. Brown and A. W. Lohmann, "Complex spatial filtering with binary masks", *Appl. Opt.* **5** (1966), 967–969.

[14] L. B. Lesem, P. M. Hirsch and J. Jordan, "The kinoform: a new wavefront reconstruction device", *IBM J. Res. Dev.* **13** (1969), 150–155.

[15] W. B. Veldkamp and T. J. McHugh, "Binary Optics", *Scientific American* (1992), 50–55.

[16] H. Nishihara, M. Haruna and T. Suhara, "Optical integrated circuits", McGraw-Hill, New York (1989).

[17] W. Menz and P. Bley, "Mikrosystemtechnik für Ingenieure", VCH Verlagsgesellschaft, Weinheim (1993).

[18] K. Iga, Y. Kokubun and M. Oikawa, "Fundamentals of microoptics", Academic Press, Tokyo (1984).

[19] H.-P. Herzig (ed.), "Micro-optics: Elements, systems, and applications", Taylor & Francis, London (1997).

[20] J. Turunen and F. Wyrowski (eds), "Diffractive optics for Industrial and Commercial Applications", Akademie Verlag, Berlin (1997).

[21] M. Kufner and S. Kufner, "Micro-optics and lithography", VUB Press, Brussels, Belgium (1997).

2 Optical components with small dimensions

In this chapter we analyse the scaling behaviour of optical components, which means, we take a look at how their performance is affected by the physical dimensions. We will concentrate on the performance of lenses since they are undoubtedly the most important components of the optical system. At the end of the chapter we will briefly discuss prisms which represent the group of deflecting optical elements.

Before we can address the scaling behaviour of the various types of optical elements, it is necessary to establish some criteria for the quality of optical elements. However, it is beyond the scope of this book to give a detailed analysis of quality criteria for lenses. We will rather give an overview of some of the aspects which have to be considered when talking about the quality of a lens. This is meant to make the reader aware of the problems related to this issue. The definition of the quality of a lens is one of the most important problems in technical optics and lens design. It is not possible to establish one single figure of merit, which satisfactorily allows one to evaluate the quality of a specific lens. The main reason for this is the fact that the performance of the lens depends highly on the specific situation. In order to be able to say whether a lens is "good" or "bad", we need specific information about the imaging geometry, the necessary imaging contrast, the detectors and the light sources which are used in the system. In the following sections we focus on parameters which are often used in the effort to evaluate the quality of lenses. We continue by discussing how these parameters are influenced by different scaling of the lenses.

2.1 Microlens performance

2.1.1 Diffraction limit

It is well known that light is diffracted at the apertures of optical elements. This effect occurs at any optical component (e.g., lens or prism) and affects its performance. Let us, for example, consider a lens with an aperture D (Fig. 2.1).

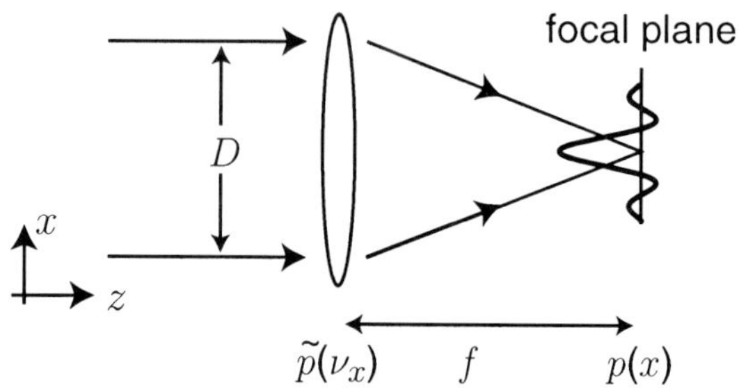

Fig. 2.1: Focusing of a collimated beam by a lens.

Ideally, the focus of the plane wave should be infinitely small, being the image of a point source located at infinity. In our example the lens is supposed to have an ideal geometrical shape, e.g., to focus an incoming collimated plane wave at a distance f from the lens (Fig. 2.1). Although in this ideal case no other aberrations are introduced, the focus will have a finite extension. Diffraction at the lens aperture (D) causes a blur of the focus. The light distribution in the focus is determined by the Fourier transform of the pupil function of the lens. The 1D pupil function ($\tilde{p}(\nu_x)$) of this ideal lens is described by a rect-function ($\mathrm{rect}(\frac{\nu_x}{\Delta\nu})$). The frequency coordinate ν_x is related to the physical coordinate x in the Fourier domain by $\nu_x = \frac{x}{\lambda f}$ and $\Delta\nu = \frac{D}{\lambda f}$, where λ denotes the wavelength of the illuminating light beam. The point spread function (psf, $p(x)$), i.e., the image of a point source generated by the lens, is calculated as the Fourier transform of the pupil function $\tilde{p}(\nu_x)$:

$$p(x) \quad \propto \quad \int \mathrm{rect}\left(\frac{\nu_x}{\Delta\nu}\right) e^{-2\pi i \nu_x x}\, d\nu_x \quad \propto \quad \mathrm{sinc}\left(x \cdot \frac{D}{\lambda f}\right) \tag{2.1.1}$$

Here we used the following definitions for $\mathrm{rect}(x)$ and $\mathrm{sinc}(x)$:

$$\mathrm{rect}(x) = \begin{cases} 1 & : \quad |x| < \frac{1}{2} \\ 0 & : \quad \text{else} \end{cases} \quad ; \qquad \mathrm{sinc}(x) = \frac{\sin(\pi x)}{\pi x} \tag{2.1.2}$$

For the more common case of lenses with circular apertures, the ideal psf is calculated as the Fourier transform of the $\mathrm{circ}(\frac{r}{D})$ function. This yields the so-called Airy pattern $\frac{J_1(r \cdot \frac{D}{\lambda f})}{r \cdot \frac{D}{\lambda f}}$, where $J_1(x)$ is the first order Bessel function [1].

The psf corresponds to the shape of the point image formed by the lens. In the absence of aberrations a lens is called ideal or *diffraction-limited.* This means that the psf is determined by the sinc-function resulting from the diffraction at the lens pupil. Fig. 2.2 shows the shape of the psf of such an ideal lens. When the lens aperture D decreases, the extension of the psf increases proportionally. This has important effects on the scaling behaviour of lenses. With a reduction of the lens

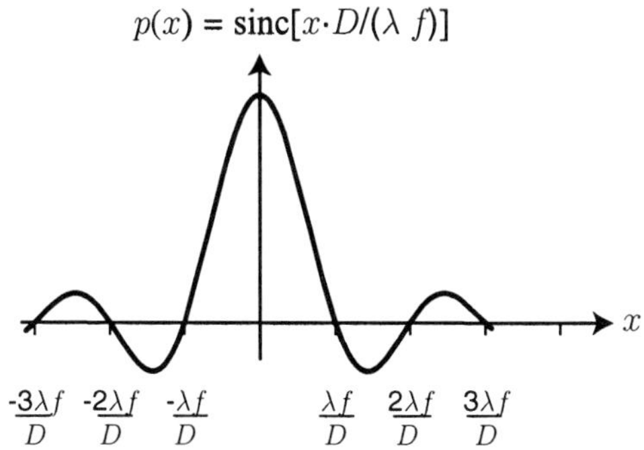

Fig. 2.2: The 1D point spread function (psf) of an ideal lens.

diameter it becomes easier to achieve diffraction-limited performance. Since the extension of the diffraction-limited psf becomes larger the constraints on the shape of the phase profile become less stringent. The effects of deviations in the phase profile on the point image are hidden under the wide diffraction-limited psf. In the extreme case the microlens diameter becomes very small, e.g., the size of a pinhole [2]. Here diffraction-limited imaging is possible without any specific phase profile in the lens pupil (e.g., pinhole camera). Phase errors due to different optical path lengths of the rays passing through the pupil vanish, because of the small extension of the pinhole diameter. This is also one of the reasons why small droplets of a transparent liquid often form relatively good lenslets (cf. water droplets on the shower curtain or a window; see also Chapter 4 on the fabrication of dispensed microlenses). Due to surface tension, the droplets assume an approximately spherical shape. Aberrations due to deviations from the ideal profile are covered by the wide diffraction-limited psf.

From these considerations we learn that diffraction-limited performance alone does not guarantee good lens quality. Especially if we are dealing with microoptics, we have to find better criteria to be able to evaluate the optical components. Before we go on with our considerations about the scaling of the lens properties, we need to consider the non-ideal lens, i.e., the lens with aberrations.

2.1.2 Aberrations

Wavefront aberrations

Besides the influence of diffraction, the performance of a lens is affected by aberrations. They are caused by deviations of the shape of the lens from the ideal profile. It has to be taken into account, that the ideal shape of the lens strongly depends on the specific imaging situation. Depending on the lens type, aberrations are caused, e.g., by deformations of the surface or the refractive index profile of the lens. For a brief discussion of aberrations, we consider again the case of a lens focusing a collimated light beam. The job of the lens is to transform the incoming plane wave into a spherical wave (or, in the paraxial approximation, into a parabolic wave), also called the Gaussian reference sphere. For a real lens with aberrations, however, the real wavefront $A \cdot e^{i\boldsymbol{k}R(x,y)}$ behind the lens shows deviations from the ideal spherical shape $A \cdot e^{i\boldsymbol{k}W(x,y)}$ (Fig. 2.3). These deviations are called the wavefront aberrations $\Psi(x,y)$ and can be calculated as:

$$\Psi(x,y) = W(x,y) - R(x,y) \tag{2.1.3}$$

For a classification, the wavefront aberrations Ψ are expanded over a set of suitable polynomials. For the classification of the primary aberrations Kingslake [3] suggested the following representation [4]:

$$\Psi(x,y) = \mathcal{A}(x^2+y^2)^2+\mathcal{B}y(x^2+y^2)+\mathcal{C}(x^2+3y^2)+\mathcal{D}(x^2+y^2)+\mathcal{E}y+\mathcal{F}x+\mathcal{G} \tag{2.1.4}$$

In this expansion the commonly known ray aberrations are represented by the coefficients as follows:

$\mathcal{A}$: spherical aberration

$\mathcal{B}$: coma

$\mathcal{C}$: astigmatism

$\mathcal{D}$: defocusing

$\mathcal{E}$: tilt about x-axis

$\mathcal{F}$: tilt about y-axis

$\mathcal{G}$: constant term

These aberrations are sometimes also called the primary, Seidel or the third order aberrations.

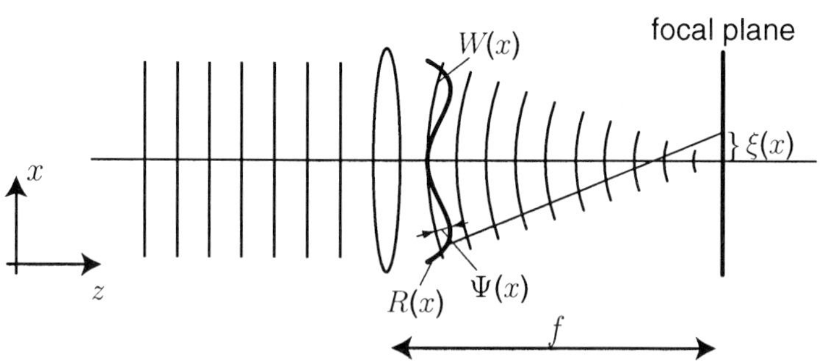

Fig. 2.3: Relation between wavefront and ray aberrations.

There are several other sets of suitable polynomials over which the wavefront aberrations can be expanded. For example, Zernike polynomials are often used for such an expansion. Each of these expansions results in its own representation of the primary and higher order aberrations. We do not go further into detail with the mathematical representation of the aberrations. For a comprehensive treatment of the theory of aberrations the reader is referred to the literature (e.g., H. H. Hopkins "Wave theory of aberrations" [5] and D. Malacara (ed.) "Optical Shop testing" [6]). We continue with a discussion of some types of aberrations from a more phenomenological point of view. Later on we address suitable ways to measure the quality of microlenses.

Ray aberrations and ray tracing

From geometrical optics we know that the directions of optical rays are defined by the $\boldsymbol{k}$-vectors of an optical wavefront. These $\boldsymbol{k}$-vectors are the vectors directed perpendicularly to the planes of constant phase (W(x,y) = const.). They can be defined mathematically by the gradient of the wavefront.

$$\boldsymbol{k}(x,y) = \left(\frac{\partial W(x,y)}{\partial x}, \frac{\partial W(x,y)}{\partial y}\right) \tag{2.1.5}$$

Wavefront aberrations cause deformations of the real wavefronts behind the optical element. Consequently, the optical rays are misdirected and hit the focal plane at locations different from the ideal position (Fig. 2.3). The distance between the ideal and the real ray position in the focal plane is called the transverse geometrical or ray aberration $\xi(x,y)$ (Eq. 2.1.6).

$$\begin{aligned} \xi(x,y) &= f \cdot \left[\frac{\partial(W(x,y)-R(x,y))}{\partial x}, \frac{\partial(W(x,y)-R(x,y))}{\partial y}\right] \\ &= f \cdot \left[\frac{\partial\Psi(x,y)}{\partial x}, \frac{\partial\Psi(x,y)}{\partial y}\right] \end{aligned} \tag{2.1.6}$$

The quality of an optical system can be evaluated by observing a set of optical rays travelling through the system at a variety of different angles. This is the principle of the ray tracing approach. The calculations in the first order neglect the influence of diffraction. For an ideal imaging system all rays should hit the focal plane at the same position. Due to the transverse ray aberrations, however, there is a spot distribution around this ideal image point. An analysis of the shape of the spot distribution provides information about the types and amount of aberrations in the system.

Ray tracing is the most widespread technique for the numerical modelling and the design of optical systems. The starting point for the analysis of an optical system is a set of optical rays represented by the $\boldsymbol{k}$-vectors. According to Eq. 2.1.5 these rays represent the incident wavefront $W(x,y)$. In homogeneous media the rays propagate linearly, maintaining their directions. Thus, by basic geometrical considerations, the propagation of those rays to the planes containing optical surfaces (e.g., the principal planes of a lens) can be simulated. The coordinates of the spots where the rays hit the surfaces are calculated (Fig. 2.4) from:

$$\boldsymbol{r}_1 = \boldsymbol{r}_0 + |\boldsymbol{r}_1 - \boldsymbol{r}_0| \frac{\boldsymbol{k}}{|\boldsymbol{k}|} \tag{2.1.7}$$

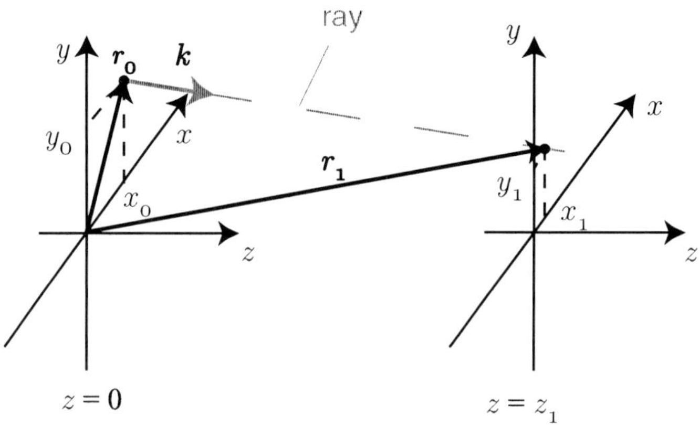

Fig. 2.4: Geometrical considerations for ray tracing.

An optical element located in the plane z_1 influences the $\boldsymbol{k}$-vectors of the rays. Different methods can be used to calculate the new $\boldsymbol{k}$-vectors behind the component. If the curvature of the surface and the refractive index of the optical component is known at all locations, Snell's law of refraction yields the new propagation directions (i.e., the $\boldsymbol{k}$-vectors $\boldsymbol{k}_{\text{out}}$) of the rays. Alternatively the $\boldsymbol{k}$-vectors ($\boldsymbol{k}_{\text{comp}}$) due to the component can be calculated from the phase profile. These vectors $\mathbf{k}_{\text{comp}}$ are added to the $\boldsymbol{k}$-vectors of the incident rays ($\boldsymbol{k}_{\text{in}}$) to yield the new propagation directions ($\boldsymbol{k}_{\text{out}}$).

$$\boldsymbol{k}_{\text{out}} = \boldsymbol{k}_{\text{in}} + \boldsymbol{k}_{\text{comp}} \tag{2.1.8}$$

After a series of propagation steps and planes with optical components, we find the points where the optical rays are incident on the observation plane. An evaluation of the resulting pattern yields the performance of the optical system.

In the first approximation, the ray tracing analysis is a geometrical method, which does not consider diffractive effects. Nevertheless, it is possible to analyse systems which consist of or contain diffractive optical elements. This is possible by calculating the $\boldsymbol{k}$-vectors, i.e., the angular spectrum of the diffractive element in a Fourier expansion. These $\boldsymbol{k}$-vectors can be used for the calculation of the ray propagation according to Eqs 2.1.7 and 2.1.8.

The ray tracing algorithm can also be used for systems design. Using Eq. 2.1.8 we calculate the $\boldsymbol{k}$-vectors of the optical component ($\boldsymbol{k}_{\text{comp}}$) from the $\boldsymbol{k}$-vectors of the incident ($\boldsymbol{k}_{\text{in}}$) and the desired wavefronts ($\boldsymbol{k}_{\text{out}}$). Since the $\boldsymbol{k}$-vectors are defined as the gradient of the wavefield (Eq. 2.1.5), the inverse calculation yields the necessary phase distribution ($\varphi(x, y)$) of the optical component:

$$\varphi(x,y,z) = \varphi(x_0,y_0,z_0) + \int_{x_0}^{x} k_x(\rho,y_0,z_0)d\rho + \int_{y_0}^{y} k_y(x,\varsigma,z_0)d\varsigma + \int_{z_0}^{z} k_z(x,y,\tau)d\tau \tag{2.1.9}$$

Thus, knowledge of the distribution of the $\boldsymbol{k}$-vectors of the ideal optical system yields the necessary phase profile of the components. This phase function is the goal for the fabrication of the diffractive or refractive components of the system. In the next chapter we consider the fabrication of such elements. Now, however, we want to continue our discussion of the performance of optical components with an introduction of quality criteria for lenses.

2.1.3 Quality criteria for lens performance

In the previous section we discussed two seperate phenomena which influence the quality of an image generated by a lens system. Diffractive effects at the lens aperture determine the psf of the aberration free component. Aberrations are deviations of the wavefront behind the lens from the ideal shape. One of the most difficult problems in lens design as well as lens characterisation is the definition of a suitable figure of merit for the quality of a lens. In this section we address a variety of different figures of merit for lens performance as well as the problems related to them.

The second Gaussian moment of the ray aberrations G_{ray}

G_{ray} can be considered as a measure for the ray-optical quality of the image.

$$G_{\text{ray}}(x) = \int [\xi^2(x)]dx = \overline{\xi^2(x)} \tag{2.1.10}$$

G_{ray} is the variance of the distance between the real and the ideal rays in the focal plane of the lens. A possible criterion for the image quality based on G_{ray} could be: The smaller the second Gaussian moment of the ray aberrations G_{ray} the better the image quality. Fig. 2.5 shows a typical shape of the wavefront $\Psi(x)$ and ray $\xi(x)$ aberrations which increase towards the edges of the lens.

The problem of this figure of merit is that it does not take diffraction into account [7]. Due to the diffraction blur the performance of the optical element does not necessarily improve with the geometrical performance. According to Fig. 2.5, G_{ray} can be improved by simply reducing the aperture D of the lens, e.g., through an aperture stop. In this case, however, the diffraction blur is increased and reduces the image quality, even though our criterion G_{ray} is improved. Consequently, the second Gaussian moment of the ray aberrations is not an ideal criterion for the performance of a lens. This is even more true in microoptics where the diffraction-limited point spread function is relatively wide due to small lens diameters. Although the introduction of an aperture stop does not correspond to the scaling of the lens, it is important to keep this feature of G_{ray} in mind.

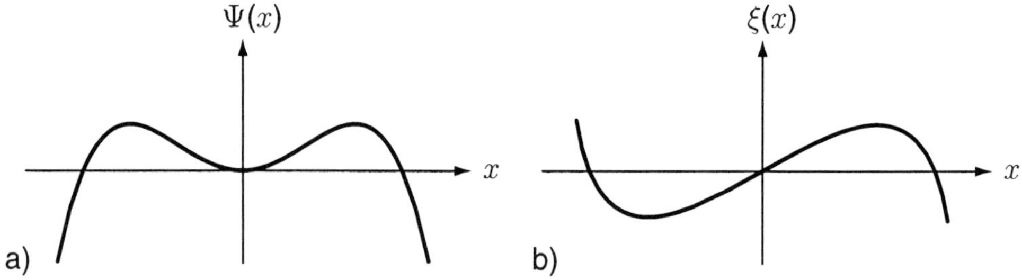

Fig. 2.5: The typical shape of a) wavefront and b) ray aberrations introduced by a lens.

The second Gaussian moment of the point spread function G_{psf}

G_{psf} is a criterion which contains both ray and wave aspects. It is an interesting figure of merit for lens performance. The most important aspect which makes G_{psf} suitable for lens evaluation is the separation into a wave-optical and a ray-optical part (for derivation, see Exercises no. 2) [8]:

$$G_{psf}(x) = \int x^2 \cdot |p(x)|^2 dx = \frac{1}{4\pi^2} \left[\overline{\left(\frac{dA(\nu_x)}{d\nu_x} \right)^2} + k^2 \overline{A^2 \, \xi^2} \right] = G_{wave} + G_{ray} \tag{2.1.11}$$

G_{psf} can be identified as a measure of the pixel size in the image generated by a lens with the pupil function:

$$\tilde{p}(\nu_x) = A(\nu_x) e^{ik\Psi(\nu_x)} \tag{2.1.12}$$

G_{psf} represents the extension of the psf of the optical system. It is a useful criterion, e.g., for the calculation of the space-bandwidth product.

Rayleigh criterion

Other figures of merit for lens performance directly relate the amount of wavefront aberration $\Psi(x, y)$ to the image quality. For example, the widely used Rayleigh criterion states that a system is ideal, i.e., diffraction-limited, as long as the maximum wavefront deviation $|\Psi_{max}(x)|$ is less than one quarter of a wavelength [9]. In other words to fulfil the Rayleigh criterion the real wavefront needs to be confined within two concentric spheres with a spacing of $2 \cdot \lambda/4$ (Fig. 2.6a):

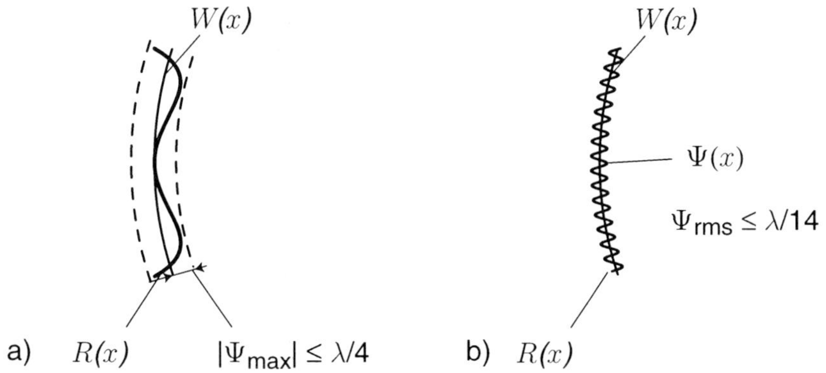

Fig. 2.6: Illustration of a) the Rayleigh criterion and b) the Maréchal criterion for the tolerable amount of wavefront aberrations.

$$\text{Rayleigh:} \qquad |\Psi_{\max}(x)| \leq \frac{\lambda}{4} \tag{2.1.13}$$

Lord Rayleigh derived this criterion (Eq. 2.1.13) for imaging configurations with low-order aberrations such as primary spherical aberration. In this case it provides a good quality measure. This can be understood with the help of Fig. 2.6a and by considering the relationship between the quarter-wave criterion and the Rayleigh resolution limit. For low-order aberrations up to the quarter wave criterion the shape of the psf is not affected (see Exercises no. 3). The resolution of the optical system can still be considered ideal or diffraction-limited. In an effort to find a measure for the resolution of an optical apparatus, Rayleigh suggested considering two image points to be resolvable if they are separated by at least half the width of the central lobe of the psf (Fig. 2.7). In this case one image point is located at precisely the position of the first zero transition of the psf of the adjacent image point. For an ideal system with rectangular pupils and no aberrations, the intensity between the spots drops to about 0.81 of the maximum intensity. This results in a contrast between the darkest and brightest image location of $K = (I_{\max} - I_{\min})/(I_{\max} + I_{\min}) = 0.1$. Whether two spots can be resolved depends on what contrast is enough to resolve the image points. This is influenced by the quality of the detecting system. In many cases the Rayleigh resolution limit can be overcome significantly with highly sensitive, low-noise detectors. On the other hand, aberrations, even if they remain within the quarter-wave limit, reduce the image contrast. Small aberrations merely cause a shift of light from the main lobe of the psf to the side lobes. Higher aberrations additionally cause a broadening of the psf. This directly affects the Rayleigh resolution. The correlation between the image quality and the maximum value of the wavefront aberration is lost in the presence of higher order aberration functions.

For low order aberration functions with only moderate modulations, the Rayleigh criterion, by restricting the maximum deviation, automatically limits the gradient

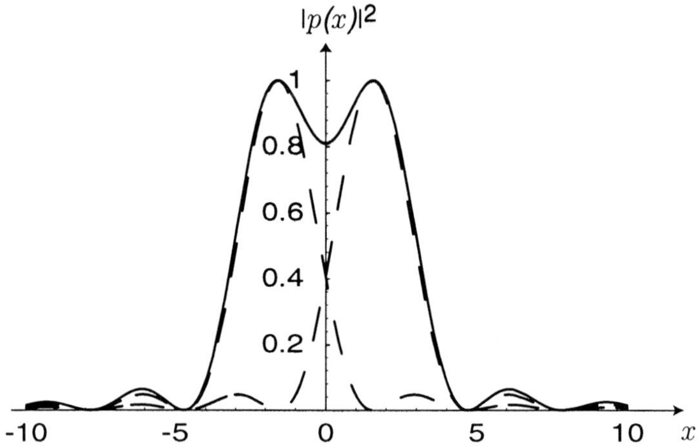

Fig. 2.7: The Rayleigh criterion for the resolution of adjacent image points.

$\frac{\partial\Psi(x)}{\partial x}$, i.e., the ray aberrations. For higher order modulations of the aberration function, the ray aberrations can become very significant, even if the maximum wavefront deviation remains within the Rayleigh limits [10]. In this case a criterion restricting the average amount of aberration which can be tolerated, is more useful .

Maréchal criterion

The Maréchal criterion uses the root-mean-square wavefront aberration Ψ_{rms} to define the lens quality (Fig. 2.6b) [9]:

$$\text{Maréchal:} \qquad |\Psi_{\mathbf{rms}}| = \sqrt{\int |\Psi(x)|^2 dx - \left[\int |\Psi(x)| dx\right]^2} = \sqrt{\int |\Psi(x)|^2 dx} \leq \frac{\lambda}{14} \tag{2.1.14}$$

This criterion is closely related to the Strehl ratio $\mathcal{S}$, which will be discussed next. One practical issue to be taken into account during measurements is that the Maréchal criterion is sensitive to statistical noise on the measured phase profile. A peak-to-valley criterion such as the Rayleigh criterion is little affected by statistical noise.

Strehl ratio

For some applications high resolution of the imaging system is the most essential feature. In this case the overall extension of the psf of the lens is of interest. For other applications, e.g., interconnection systems, the maximum energy which can be deposited at a detector of a certain size is of more concern. In such a situation the Strehl ratio $\mathcal{S}$, also called "Definitionshelligkeit", is a suitable figure of merit. $\mathcal{S}$ is defined as the normalized peak intensity of the point spread function of the lens.

$$\mathcal{S} = \frac{I_{\text{real}}(0,0)}{I_{\text{ideal}}(0,0)} = \left| \int\int e^{i\frac{2\pi}{\lambda}\Psi(x,y)} dxdy \right|^2 \tag{2.1.15}$$

Here $I_{\text{real}}(0,0)$ and $I_{\text{ideal}}(0,0)$ denote the intensities at the center of the real point image and the ideal psf without aberrations, respectively. Even for aberrations small enough not to affect the extension of the psf, the light intensity in the center peak can drop significantly. In optical interconnection systems this affects the signal-to-noise ratio of the detector signal. A Strehl ratio of $\mathcal{S} \geq 0.8$ is generally considered to correspond to diffraction-limited performance.

By a Taylor expansion of Eq. 2.1.15 we find the relation between the Strehl ratio and the Maréchal criterion [11]:

$$\mathcal{S} = 1 - \left(\frac{2\pi}{\lambda} \Psi_{\text{rms}} \right)^2 \tag{2.1.16}$$

If only low-order aberrations are taken into account, Rayleigh's $\frac{\lambda}{4}$-criterion also results in a Strehl ratio of $\mathcal{S} = 0.8$ [12].

Due to interference effects the behaviour of the Strehl ratio becomes uncorrelated to image quality for large aberrations, i.e., for $\mathcal{S} \leq 0.3$. Like many other figures of merit, $\mathcal{S}$ defines the image quality for one specific object point imaged through the optical system. Generally the image quality varies over extended image fields. For a more comprehensive definition of the quality of an optical system the Strehl ratio would have to be determined for every image point separately.

Space-bandwidth product (SBP)

A phenomenological measure for the quality of a lens is the space-bandwidth product (SBP) [1]. It indicates the number of pixels (or channels) which can be transmitted through an imaging system. We can define:

$$\text{SBP} = \frac{\Delta x \Delta y}{\delta x \delta y} \tag{2.1.17}$$

Here, Δx and Δy denote the extensions of the image field, whereas δx and δy represent the size of one resolution cell in the image plane.

The significance and the problems connected to the SBP can be seen when relating it to the well known sampling theorem (Fig. 2.8) [1]. The pupil function $\tilde{p}(\nu_x, \nu_y)$ in the Fourier plane of an imaging system (e.g., a 4f-setup) determines

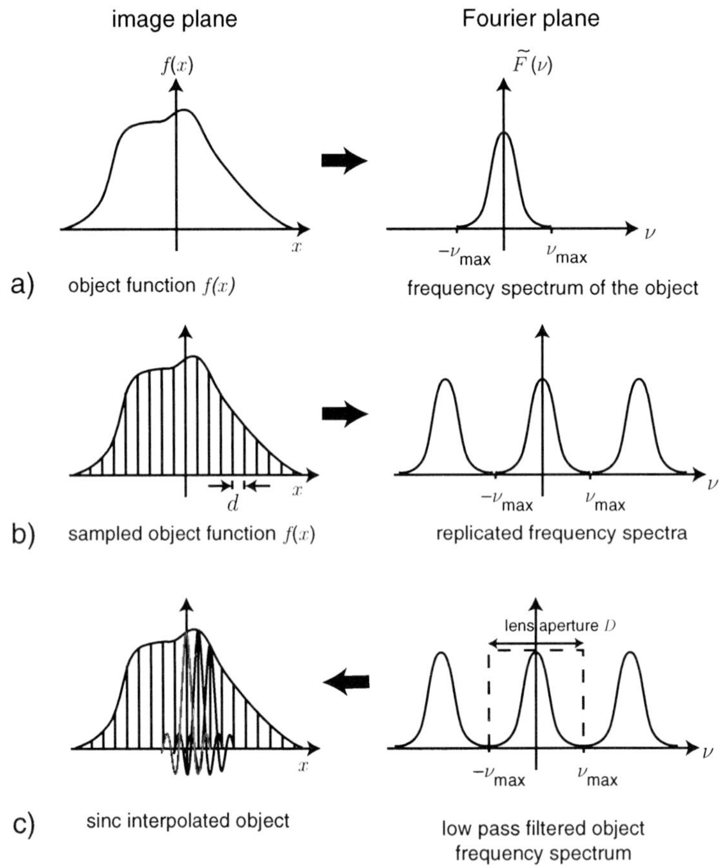

Fig. 2.8: The effect of sampling and spatial filtering on the image of an object function ($f(x)$). Further explanation in the text.

the frequency bandwidth transmitted through the system. In a coherent imaging system an ideal lens transmits a frequency range $[-\nu_{\max}, +\nu_{\max}]$, where $\nu_{\max}$ is determined by the numerical aperture of the lens system:

$$\nu_{\max} = \frac{D}{2\lambda f} = \frac{\mathrm{NA}}{\lambda} \tag{2.1.18}$$

D denotes the diameter of the pupil function and f the focal length. The maximum bandwidth $\Delta\nu$ of a signal, transmitted through this ideal lens system, in the coherent case is given by:

$$\Delta\nu = 2\nu_{\max} = \frac{2\mathrm{NA}}{\lambda} \tag{2.1.19}$$

A signal with a bandwidth of $\Delta\nu$, according to Shannon's sampling theorem, can be sampled in the spatial domain at distances $\delta x \leq \frac{1}{2\Delta\nu}$ without information loss [13]. This determines the maximum information density carried by such a bandlimited function. Sampling at distances smaller than δx does not increase the amount of information in the signal.

In the frequency domain the sampling of the signal results in a replication of the signal frequency spectrum at distances $\Delta\nu$ (Fig. 2.8b). The ideal pupil function $\tilde{p}(\nu_x, \nu_y)$ transmits one of these multiple spectra. This spatial filtering process causes the sampling function (a so-called δ-comb) to be convoluted with the psf of the lens system (Fig. 2.8c). In the ideal case the sampling points in the image plane are thus sinc-interpolated. In the presence of aberrations, however, the sinc-functions are distorted or blurred. As aberrations get worse for off-axis object points, eventually the sampling points cannot be resolved anymore. This marks the edge of the image field and determines the SBP.

For calculating the SBP, the size of the resolution cell as well as the size of the image field have to be determined. For the former we can apply Rayleigh's resolution criterion. For the latter we need information about the psf of the lens for all object points within the image field. We need to check the resolution criterion for any two neighbouring image points with increasing distance from the optical axis, until the points are no longer resolved. Noise in the system is an additional aspect which has to be taken into account.

These considerations illustrate that with the SBP we have the same problem as with most of the other quality criteria for imaging systems. Without specific knowledge about the experimental configuration, the detecting system and the application any quality criterion remains somewhat arbitrary. It is generally not possible to give a single number which answers the question of how good or bad a lens is. For these reasons, it is especially important to adjust the measurement techniques for optical elements to the intended applications.

2.2 Scaling — from macro- to micro-components

2.2.1 Scaling of diffractive and refractive lenses

The scaling of lenses can be performed in various ways (Fig. 2.9) [14]. In order to illustrate the effect of the different scaling techniques, we consider how they would influence the lithographic masks if the lenses were fabricated as diffractive lenses. These masks consist of a series of concentric rings. The period and thickness of the rings decrease with increasing radii. The numerical aperture or the $f/\#$ of the lens is determined by the largest diffraction angle, i.e., by the period of the outermost ring system. A closer discussion of the physical background of diffractive optical elements will be given in Chapter 5.

- **Photographic scaling** results in a scaling of the diameter D as well as the minimum feature size $w_{\min}$ (Fig. 2.9a). Consequently, the maximum deflec-

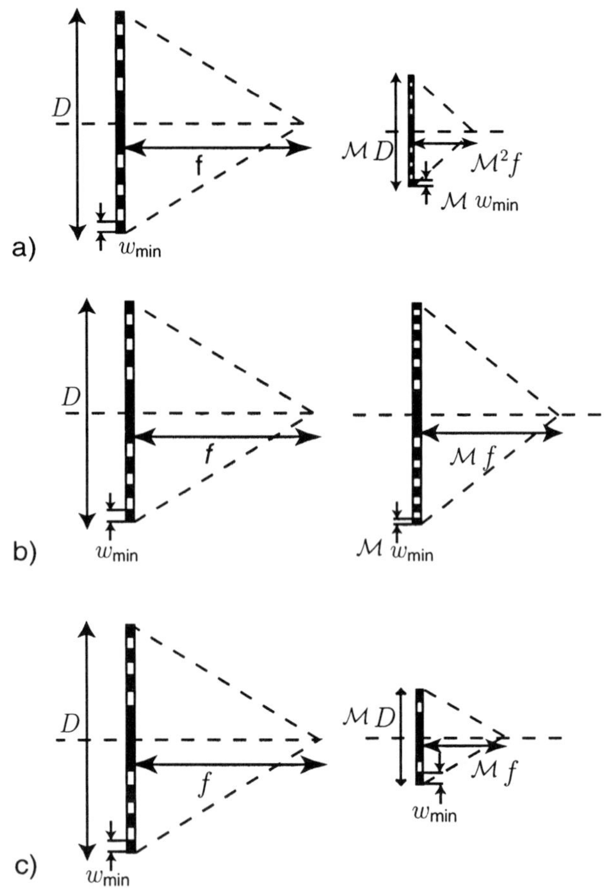

Fig. 2.9: Illustration of the different ways to scale lenses (using the example of a diffractive lens).

tion angle, i.e., the numerical aperture is scaled with the magnification $\mathcal{M}$:

$$\begin{array}{rccc} \text{diameter:} & D & \longrightarrow & \mathcal{M}\cdot D \\ \text{minimum feature size:} & w_{\min} & \longrightarrow & \mathcal{M}\cdot w_{\min} \\ \text{focal length:} & f & \longrightarrow & \mathcal{M}^2\cdot f \end{array} \qquad (2.2.1)$$

- In **microscopic scaling** the microscopic structure of the lens is scaled, while the diameter D is kept constant (Fig. 2.9b). The focal length is scaled with $\mathcal{M}$, due to the change in the minimum feature size.

$$\text{diameter:} \quad D \quad \longrightarrow \quad \text{const.}$$

$$\begin{array}{rccc} \text{minimum feature size:} & w_{\min} & \longrightarrow & \mathcal{M} \cdot w_{min} \\ \text{focal length:} & f & \longrightarrow & \mathcal{M} \cdot f \end{array} \tag{2.2.2}$$

- **Constant $f/\#$ scaling** is yet another type of scaling to be applied to lenses (Fig. 2.9c) [15]. To this end, all angles are kept constant, while the diameter and the focal length of the element are scaled. For diffractive lenses this means that the microscopic features of the lithographic mask need to be kept constant, while the lens diameter is scaled.

$$\begin{array}{rccc} \text{diameter:} & D & \longrightarrow & \mathcal{M} \cdot D \\ \text{minimum feature size:} & w_{\min} \propto \frac{f}{D} & \longrightarrow & \text{const.} \\ \text{focal length:} & f & \longrightarrow & \mathcal{M} \cdot f \end{array} \tag{2.2.3}$$

For diffractive lenses, constant $f/\#$ scaling means that the technological requirements on the lithographic fabrication (i.e., the minimum feature size, $w_{\min}$) remain constant. We want to consider the effect of this scaling concept on some of the quality criteria.

$$\begin{array}{rccc} \text{numerical aperture:} & \text{NA} = \frac{D}{2f} & \longrightarrow & \text{const.} \\ \text{wave aberrations:} & \Psi & \longrightarrow & \mathcal{M} \cdot \Psi \\ \text{ray aberrations:} & \xi & \longrightarrow & \mathcal{M} \cdot \xi \end{array} \tag{2.2.4}$$

For the behaviour of the other quality criteria due to scaling we obtain:

$$\begin{array}{rccc} G_{\text{ray}}: & G_{\text{ray}}(x) = \int\int[\xi^2(x)]dxdy & \longrightarrow & \mathcal{M}^2 \cdot G_{\text{ray}}(x) \\ G_{\text{psf}}: & G_{\text{psf}}(x) = G_{\text{wave}}(x) + G_{\text{ray}}(x) & \longrightarrow & \text{const.} + \mathcal{M}^2 \cdot G_{\text{ray}}(x) \\ \text{Strehl ratio:} & S = 1 - (\frac{2\pi}{\lambda}\Psi_{\text{rms}})^2 & \longrightarrow & S = 1 - \mathcal{M}^2 \left(\frac{2\pi}{\lambda}\right)^2 \Psi_{\text{rms}}^2 \\ \text{SBP:} & \text{SBP} = \frac{\Delta x \Delta y}{\delta x \delta y} = \frac{\Delta x \Delta y}{G_{\text{psf}}(x,y)} & \longrightarrow & \mathcal{M}^2 \frac{\Delta x \Delta y}{\delta x \delta y + \mathcal{M}^2 \bar{\xi}^2} \end{array} \tag{2.2.5}$$

As can be seen from the relations in Eq. 2.2.5 the ray optical aberrations of a lens are significantly reduced under constant $f/\#$ scaling. At the same time the wave optical psf, which is proportional to the numerical aperture, remains constant. The Strehl ratio is improved with reduced lens diameter since it is proportional to the root-mean-square wavefront aberration. The amount of information transmitted through the lens is calculated from the space-bandwidth product. With a reduction of the lens diameter the image field of the lens is reduced, as is the ray-optical psf.

Since the wave-optical psf remains constant the overall SBP is reduced. Thus, as expected, a reduction of the lens diameter results in a reduced SBP. This is illustrated in Fig. 2.10. From the figure we see that the SBP curve saturates for large lenses if the effect of aberrations is taken into account. This is the case as long as aberration effects dominate the psf of the lens. For interconnection applications it is interesting to consider the information density ρ_{inf} transmitted through the optical system. ρ_{inf} can be calculated from the SBP and the image field size as:

$$\rho_{\text{inf}} = \frac{\text{SBP}}{\Delta x \Delta y} = \frac{1}{G_{\text{psf}}} \tag{2.2.6}$$

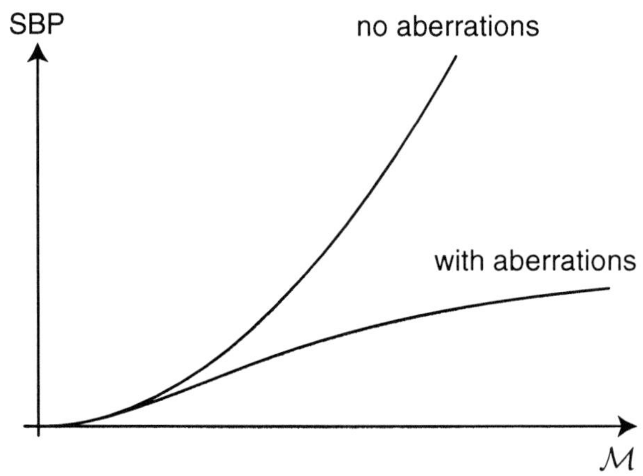

Fig. 2.10: The behaviour of the space-bandwidth product of a lens under constant $f/\#$-scaling.

In spite of the reduced SBP of single microlenses the information density of an optical system can be maximized by using a large number of microlenses in parallel rather than one single lens with larger dimensions. In Chapter 7 on array optics we will see that, especially for interconnection applications, combinations of microlens arrays with single imaging lenses offer optimum performance.

2.2.2 Scaling of prisms

Besides collimation and focusing performed by lenses, light deflection is another important task in optical systems. This job can be performed using diffraction gratings or prisms.

The performance of a (micro)prism is determined by the wedge angle, the smoothness of the phase slope and the sharpness of the edges. The edge quality can be characterised by its radius of curvature r (Fig. 2.11). Generally r is determined by the fabrication process. We want to assume that the quality of the edges, i.e., the radius of curvature, remains constant for the compared prisms. Due to the curvature at the edge, within a distance r from the edge, the light will be deflected into undesired directions. If the prism baseline Δx is scaled by a factor $\mathcal{M}$, the portion of the light amplitude $\frac{A}{A_0}$ which experiences wrong deflection angles is scaled inversely proportional to $\mathcal{M}$.

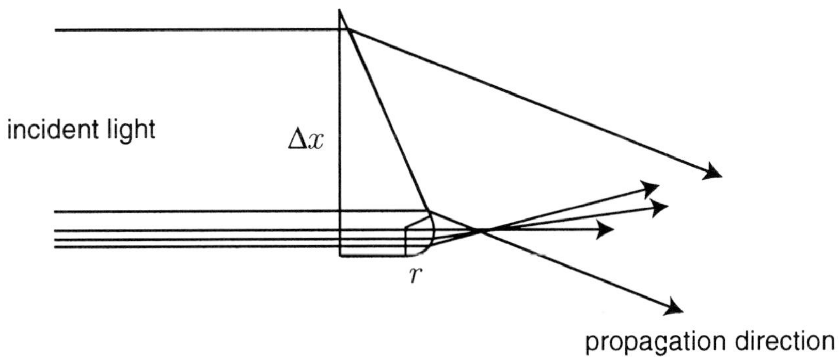

Fig. 2.11: The influence of the edges on the performance of (micro) prisms.

$$\frac{A}{A_0} \propto \frac{r}{\Delta x} = \frac{1}{\mathcal{M}} \cdot \frac{r}{\Delta x'} \tag{2.2.7}$$

The efficiency η of a prism can be defined as the portion of the incoming light which is deflected into the right direction. η thus scales as:

$$\eta = 1 - \frac{A}{A_0} \propto 1 - \frac{1}{\mathcal{M}} \cdot \frac{r}{\Delta x'} \tag{2.2.8}$$

As with microlenses, diffraction also gains importance when miniaturising prisms. It causes an increase in the extension of the psf in the image plane which is proportional to the scaling factor $\mathcal{M}$. The quality of microprisms is mostly determined by the fabrication process. In fact, because of the need for sharp edges and continuous profiles in the same component, the fabrication of microprisms is even more difficult. To be able to fabricate refractive prisms with reasonably big wedge angles, fabrication processes which allow high aspect ratios are necessary. This problem can be alleviated by blazing the prisms so that smaller phase depths suffice to achieve similar performance (Chapter 5). The blazing has some interesting effects on the performance of the gratings under polychromatic illumination [16]. Phase quantization of the blazed gratings leads to multi-phase-step gratings with similar behaviour.

2.3 Glossary

- **aberrations**
  The deviations of a wavefront from the ideal shape ("wave" aberrations) or of the position of an optical ray from the ideal position ("ray aberrations").

- **diffraction-limited**
  The performance of a lens is called diffraction-limited, if its resolution is as good as can be expected from the physical size of the lens aperture, i.e., no effect of aberrations is noticed.

- **point spread function (psf)**
  of an optical system: The shape of the image of an ideal point source formed by the optical system; calculated from the Fourier transform of the pupil function.

- **Rayleigh criterion**
  A criterion defining two image points as being resolved, if they are separated by more than half the width of the central lobe of the psf. For low-order aberrations this is equivalent to Rayleigh's wavefront criterion, stating that an optical system is diffraction-limited as long as the maximum wavefront aberration remains smaller than $\frac{\lambda}{4}$.

- **space-bandwidth product (SBP)**
  of an optical system: the total number of resolved points transmitted through the system; calculated from the ratio of the maximum size of the image field divided by the minimum size of a resolution cell.

- **Strehl ratio $\mathcal{S}$**
  also called "Definitionshelligkeit": measure for the amount of light contained in the main lobe of the psf of an optical system; for low-order aberration functions $\mathcal{S} = 0.8$ corresponds to the Rayleigh limit.

2.4 List of symbols

A:	amplitude of a wavefront
$\mathcal{A}; \mathcal{B}; \mathcal{C}; \mathcal{D}; \mathcal{E}; \mathcal{F}; \mathcal{G}$:	aberration coefficients
$\mathrm{circ}(x)$:	circular function defining a circle with radius $x = 1$
D:	diameter of the physical aperture of a lens or the optical beam
η:	light efficiency
f:	focal length of the optical system
$f/\#$:	f-number of a lens
G_{psf}:	second Gaussian moment (i.e., standard deviation) of the point spread function
$G_{\mathrm{ray}}(x)$:	second Gaussian moment of the ray aberrations
$J_n(x)$:	n-th order Bessel function
$\boldsymbol{k} = (k_x, k_y, k_z)$:	k-vector of a wave
λ:	wavelength of the light source
$\mathcal{M}$:	scaling factor
NA:	numerical aperture
$\nu_x;\ \nu_y$:	spatial frequency components defined as $\nu_x = \frac{x}{\lambda f}$
$\Delta\nu$:	spatial frequency bandwidth of a signal
$\Psi(x, y)$:	wavefront aberrations
Ψ_{rms}:	root-mean-square value of the wavefront aberrations
$p(x)$:	point spread function (psf) of a lens
$\tilde{p}(\nu_x)$:	pupil function
$\boldsymbol{r}; r$:	spatial vector and its length
r:	radius of curvature at the edge of a prism
$R(x, y)$:	phase profile of the real wavefront (including aberrations)
$\mathrm{rect}(x)$:	rectangle function, defining a rectangle with width $x = 1$
ρ_{inf}:	information density transmitted through an optical system
rms:	root-mean-square
SBP:	space-bandwidth product
$\mathcal{S}$:	Strehl ratio
$\mathrm{sinc}(x)$:	sinc-function defined as $\frac{\sin(\pi x)}{\pi x}$
$\varphi(x, y, z)$:	phase distribution, e.g., of an optical element
$W(x, y)$:	phase profile of the ideal wavefront
w_{min}:	minimum feature size of a lithographic mask
x, y, z:	spatial coordinates of the system
$\xi(x, y)$:	ray aberrations

2.5 Exercises

1. **Second Gaussian moment of the point spread function**

   The second Gaussian moment of the point spread function G_{psf} separates into a wave optical and a ray-optical part (Eq. 2.1.11). Derive Eq. 2.1.11 for the pupil function defined by Eq. 2.1.12.

   Note: Use the following Fourier relations:

$$\int |g(x)|^2 dx = \int |\tilde{g}(\nu_x)|^2 d\nu_x$$

$$\tilde{g}(\nu_x) = \frac{d\tilde{p}(\nu_x)}{d\nu_x} \quad \longrightarrow \quad g(x) = 2\pi i x p(x)$$

2. **Rayleigh criterion and geometrical aberrations:**

   For low-order aberrations the Rayleigh $\frac{\lambda}{4}$-criterion determines the amount of aberration which is acceptable without causing a broadening of the psf of the optical system. This is to be shown for the example of the primary spherical aberration ($\Psi(x) \propto Ax^4$).

3. **Strehl ratio and rms wavefront aberrations:**

   The Strehl ratio S is defined through Eq. 2.1.15. Use a Taylor expansion to derive the relationship between S and the rms wavefront aberrations (Eq. 2.1.16).

References

[1] J. W. Goodman, "Introduction to Fourier Optics", McGraw Hill, New York (1968).

[2] M. Young, "Pinhole optics", *Appl. Opt.* **10** (1971), 2763–2767.

[3] R. Kingslake, "Lens Design Fundamentals", Academic Press, New York (1978).

[4] D. Malacara and S. L. DeVore, "Interferogram evaluation and wavefront fitting", in Optical Shop Testing, D. Malacara (ed.), John Wiley & Sons, New York (1992), 455–499.

[5] H. H. Hopkins, "Wave theory of aberrations", Clarendon Press, Oxford (1950).

[6] D. Malacara, "Optical shop testing", John Wiley & Sons, New York (1992).

[7] V. Gerbig and A. W. Lohmann, "Is lens design legal?", *Appl. Opt.* **28** (1989), 5198–5199.

[8] N. Bareket, "Second moment of the point spread function as an image quality criterion", *JOSA* **69** (1979), 1311–1312.

[9] M. Born and E. Wolf, "Principles of Optics", Pergamon Press, London (1980), 6th ed.

[10] R. Barakat, "Rayleigh wavefront criterion", *JOSA* **55** (1965), 572–573.

[11] W. B. King, "Dependence of the Strehl ratio on the magnitude of the variance of the wave aberration", *JOSA* **58** (1968), 655–661.

[12] V. N. Mahajan, "Strehl ration for primary aberrations: some analytical results for circular and annular pupils", *JOSA* **72** (1982), 1258–1266.

[13] C. E. Shannon, "The mathematical theory of communication", Univ. of Illinois Press, Urbana, Ill, USA (1949).

[14] H. M. Ozaktas, H. Urey and A. W. Lohmann, "Scaling of Diffractive and Refractive Lenses for Optical Computing and Interconnections", *Appl. Opt.* **33** (1994), 3782–3789.

[15] A. W. Lohmann, "Scaling Laws for Lenses", *Appl. Opt.* **28** (1989), 4996–4998.

[16] S. Sinzinger and M. Testorf, "The transition between diffractive and refractive micro-optical components", *Appl. Opt.* **34** (1995), 5970–5976.

3 Lithographic fabrication technology

After our discussion of the quality criteria and the scaling behaviour of optical components, in this chapter we focus on fabrication techniques for microoptics. According to our definition in Chapter 1, lithography is the most important technological approach for microoptics fabrication. Here we discuss the fundamentals of the lithographic technology available for the structuring of planar substrates. The goal is to give the reader an overview of the variety of techniques available for lithographic fabrication. Several of the approaches are specified in more detail in the following chapters for the fabrication of specific microoptical components.

Lithography is the name for a sequence of processing steps for structuring the surfaces of planar substrates. Two types of lithographic fabrication procedures can be distinguished (Fig 3.1), "mask lithography" and "scanning lithography". In mask lithography the pattern of the component is encoded as an amplitude distribution in a lithographic mask. Uniform illumination of the mask is used to expose a photosensitive coating on the substrate. In scanning lithography no masks are used. Rather, local variation of the photoresist exposure is achieved in a so-called direct-writing process. To this end a laser or electron-beam is scanned over the substrate, while the beam intensity and exposure time (also called "dwell time") of the beam are modulated. After the exposure of the photoresist layer a development step converts the exposed photoresist into a surface profile. In a further processing step the surface profile of the photoresist pattern can be transferred into the substrate. We find four major processing steps in the lithographic fabrication procedure, which we call "pattern generation", "coating or thin layer deposition", "alignment and exposure" and "pattern transfer". In the following we discuss the possible technologies applied in each of these categories. The choice of which technology to use mostly depends on precision requirements, materials and other experimental boundary conditions.

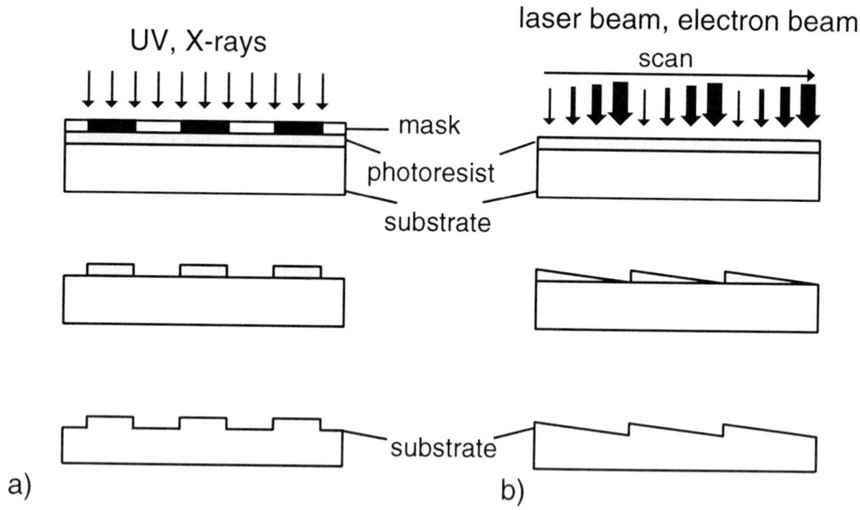

Fig. 3.1: Approaches to lithographic fabrication: a) mask lithography, b) scanning lithography.

3.1 Pattern generation

This paragraph is devoted to the variety of techniques available for the generation of the microscopic pattern of microoptical components. We will discuss approaches like "plotting and photoreduction", "laser beam writing", "x-ray and e-beam writing" and "grey-level masks". Except for the plotting and photoreduction technique all of these approaches can be exploited in both mask lithography for generation of the masks and in scanning lithography for directly writing the pattern into the substrate.

However, before discussing the techniques, we need to understand which are the most important parameters of lithographic masks for microoptics fabrication. A good way to do so is to consider one specific example. To this end and in anticipation of the detailed discussion of diffractive microoptics in Chapter 5, we briefly discuss the amplitude diffraction grating. Such a grating consists of a periodic structure. Parts of the period are either absorbing or reflecting the light while the rest is transmitting. Fig. 3.2 shows the transparency function $f(x)$ of an amplitude grating. $f(x)$ can be described by a Fourier series:

$$f(x) = \sum_m A_m \, e^{2\pi i m \frac{x}{p}} \tag{3.1.1}$$

where the A_m denote the diffraction amplitudes, which are calculated from the Fourier transform relationship:

$$A_m = \frac{1}{p} \int_{-\frac{p}{2}}^{+\frac{p}{2}} f(x)\, e^{-2\pi i m \frac{x}{p}}\, dx \tag{3.1.2}$$

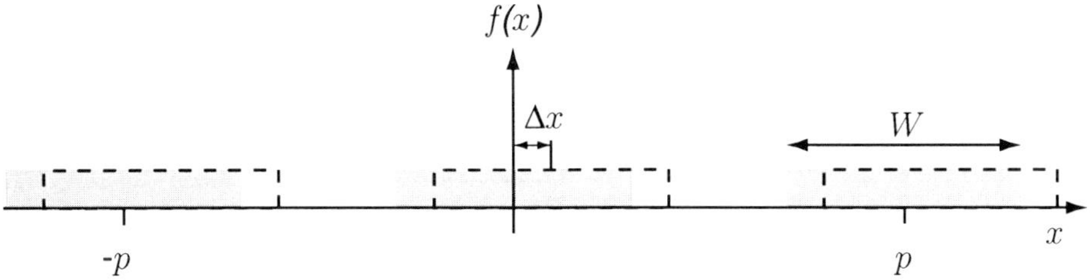

Fig. 3.2: The transmission function of a diffraction grating.

Thus, for the grating shown in Fig. 3.2 the significance of the various mask parameters can be found by an analysis of the diffraction orders A_m, which are calculated from:

$$A_m = \frac{1}{m\pi} \sin(m\pi \frac{W}{p})\; e^{2\pi i m \frac{\Delta x}{p}} \tag{3.1.3}$$

Here, W is the width of a grating bar, p is the period of the grating and Δx is a lateral shift of the grating. The diffraction angles α_m under which the A_m appear, are determined by:

$$\sin(\alpha_{\mathrm{m}}) = m\; \frac{\lambda}{p} \tag{3.1.4}$$

The minimum feature size $w_{\min}$ achieved in the mask writing process limits the period to values $p \geq 2w_{\min}$. We conclude that the minimum feature size limits the diffraction angles α_m of the achievable pattern. The limited resolution might also cause errors in the ratio $\frac{W}{p}$ (the so-called duty cycle of the grating) or on the position of the grating bar within the period. According to Eq. 3.1.3 the duty cycle $\frac{W}{p}$ determines the amplitudes in the diffraction orders, whereas the lateral position Δx determines the phase of the diffracted light. More sophisticated diffraction gratings, e.g., Dammann gratings or Fresnel zone lenses, consist of complex periodic patterns. Multiple beam interference of the light diffracted from these structures forms the desired light distribution in the diffraction plane. The correct phase and amplitude relation between these interfering orders is very critical for the resulting diffraction pattern. According to the example of the amplitude grating, the phase corresponds to the position of the transition points of the grating, whereas the amplitude is determined by the size of the grating elements.

After this discussion we can define the minimum feature size $w_{\rm min}$ and the space-bandwidth product (SBP, i.e., the number of resolution points) as the most important features of lithograpic patterns for microoptics fabrication. An additional important feature of lithographic masks is the contrast provided for the illumination radiation.

The first step to generating the pattern of a microoptical component is the numerical design. Since the design process is closely related to the physics of the specific microoptical component, we will come back to the design problem in Chapters 4 and 5. In the following paragraphs we address the various possibilities for a technological realisation of the respective structure.

3.1.1 Plotting and photoreduction

The cheapest way to generate photolithographic masks from the design data involves the use of plotters as output media. High resolution laser printers or drafting plotters can be used for this purpose. In order to enhance the resolution an enlarged copy of the mask pattern is plotted first. In a photoreduction step the mask of the desired lateral extension is imaged onto a photographic high resolution plate. With this technique generally a maximum resolution of 5–10 μm can be achieved. More stringent, however, is the limitation with respect to the SBP of the masks. Generally the size of the plotter output is limited to A1 or A0 formats, i.e., to 84.1 cm × 59.4 cm or 118.9 cm × 84.1 cm. A typical characteristic for almost any mask writing device is the fact that the minimum feature size $w_{\rm min}$ is considerably larger (e.g., by a factor of 10) than the positioning precision $\delta x_{\rm p}$ of the scanning system. In the case of a drafting plotter used for mask writing the typical minimum feature size is determined by the width of the writing pen. A typical value is $w_{\rm min} = 250\ \mu$m. For an output format of A1 this would yield a SBP of:

$$\mathrm{SBP}_{\rm plotter} = \frac{\Delta x \Delta y}{\delta x \delta y} \approx 8 \cdot 10^6 \tag{3.1.5}$$

With the use of step and repeat cameras for the photoreduction step, the plotted pattern can be imaged several times onto the photographic plate, which increases the SBP of the generated masks.

3.1.2 Laser beam writing

A more sophisticated technology for pattern generation is laser beam writing. The pattern is written into a light sensitive layer of photoresist by a focused laser beam.

Fig. 3.3 schematically shows the setup of a typical laser scanner. A laser with an emission wavelength near the UV region (e.g., HeCd laser at λ = 442 nm) is modulated (e.g., by an acousto-optical modulator AOM) and focused onto the wafer coated with a layer of photoresist. The focus of the beam is scanned over the whole wafer area and thus writes the mask pattern. Reasonable scanning speed can be achieved in a complex scanning process which uses a combination of direct deflection of the laser beam (e.g., by an acousto-optical deflector AOD) and scanning of the wafer using an x-y stage.

Critical parameters of the process are the synchronisation between the modulator and the deflector, the positioning of the x/y stage, and the precise focusing of the laser beam. The acousto-optical modulation can be realised by using a deflector and a spatial filtering system. When deflected the laser beam misses a pinhole (spatial filter) and is thus blocked. The modulation is electronically synchronised with the beam deflection and the x-y scanning of the substrate. The position of the x-y stage is controlled interferometrically. The positioning precision δx_{p} of such a laser scanner is typically $\delta x_{\mathrm{p}} = 0.5\ \mu$m.

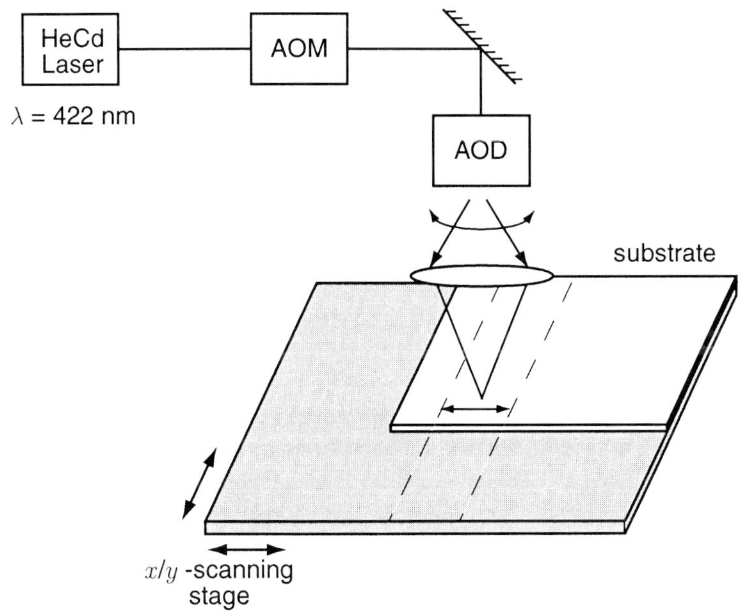

Fig. 3.3: Schematic setup for laser beam writing of lithographic masks [1]; AOM: acousto-optical modulator; AOD: acousto-optical deflector.

The most critical issue for the minimum feature size $w_{\min}$ is the exact focusing of the laser beam. This must not be affected by the roughness of the photoresist layer which causes a variation of the distance between the focusing lens and the layer to write in (defocusing). In a common solution to this problem the focusing objective is floating on an air cushion of constant pressure. With this trick it is possible to achieve a constant minimum spot size depending on the illumination wavelength (typical value: 0.8 μm at λ = 442 nm). The maximum image field of such a laser scanning system is generally about 150 mm $\times$ 150 mm. With a minimum size of $w_{\min} = 1.5\ \mu$m the SBP of this system can be calculated:

$$\mathrm{SBP}_{\mathrm{laser}} \approx 10^{10} \tag{3.1.6}$$

3.1.3 X-ray and e-beam writing

The minimum diameter of the focus spot of a writing beam is determined by the wavelength of the beam (Chapter 2). For optical lithography typically UV light with wavelengths $\lambda \geq 200$ nm is used. Such systems are limited to a resolution of about 0.5–1 μm. For a further reduction of the minimum feature size, radiation of shorter wavelength such as x-rays or electron-beams (e-beams) can be used. Although x-rays with wavelengths of about 4–8 Å could theoretically reach extremely high resolution, x-ray lithography is confronted with various problems such as [2]:

- x-ray radiation is photon radiation which cannot be influenced by electric or magnetic fields;
- x-rays are reflected from only a few materials, the reflectivities are low;
- there are no good lenses for x-rays;
- x-rays have high penetration depths.

As a consequence it is not possible to develop x-ray steering and modulating systems, which allow the writing of a lithographic pattern in a way similar to laser beam lithography (Section 3.1.2). However, parallel bundles of x-rays are available, e.g., from synchrotron radiation. These can be used in mask lithography for the "exposure" process to transfer the pattern from the lithographic mask to the substrate. The advantages of using x-rays for this purpose will be discussed in Section 3.3.3.

As compared to x-rays, e-beams have the advantage of consisting of charged particles, which can be influenced, i.e., focused and deflected, by electric and magnetic fields. The wavelength attributed to e-beams is even shorter than for x-rays and can be calculated from De Broglie's relation:

$$\lambda = \frac{\mathrm{h} \cdot c}{E[\mathrm{J}]} = \frac{12.3}{E[\mathrm{eV}]}\,\text{Å} \tag{3.1.7}$$

h: Planck's constant; E: energy of the electrons.

For a typical electron energy of 25 keV the corresponding wavelength is $\lambda = 0.078$ Å. Nevertheless, the minimum features generated with electron beams are significantly larger and range from sizes of about 20–30 nm in laboratory experiments to about 0.2 μm in standard machines. Reasons for this extended focus are the limited quality of the electrical lenses and deflecting devices as well as repulsive interaction of the electrons in the tightly focused beam.

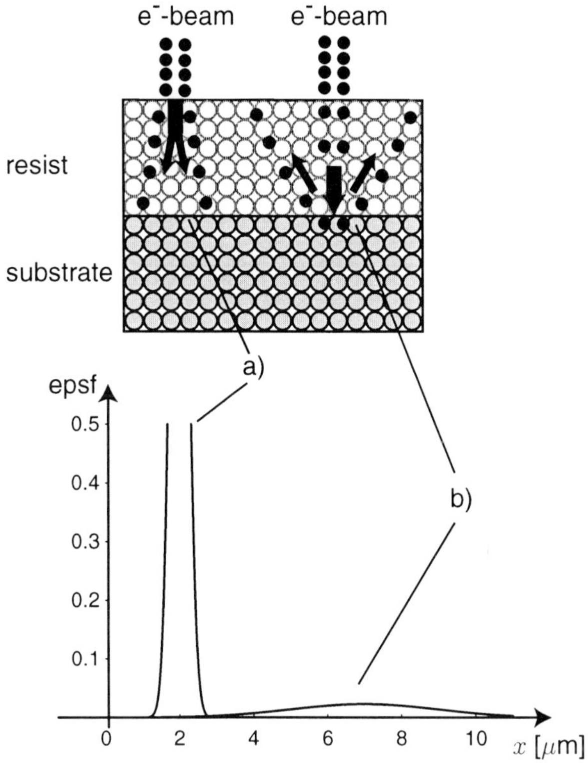

Fig. 3.4: Electron scatterings responsible for the proximity effect and their effect on the epsf for typical values of $\alpha = 0.3$ and $\beta = 2.8$; a) elastic forward scattering of the electrons in the resist layer; b) back-scattering of secondary electrons generated in the substrate.

When interacting with the resist material some of the electrons are scattered elastically (Fig. 3.4a), which is also called “forward scattering”. This causes the electron beam to diverge and increases the focus extension. A second scattering effect occurs upon interaction of the e-beam with the substrate. There the electrons generate secondary electrons, which are “backscattering” into the resist layer (Fig. 3.4b). Both of these scattering effects contribute to the “proximity effect”. The shape of the point spread function of the electron beam (epsf) is generally described as a superposition of two Gaussian functions which stem from the two scattering effects [3].

$$\mathrm{epsf}(r) = \frac{1}{\pi(1+\eta)}\left[\frac{1}{\alpha^2}e^{\left(-\frac{r^2}{\alpha^2}\right)} + \frac{\eta}{\beta^2}e^{\left(-\frac{r^2}{\beta^2}\right)}\right] \tag{3.1.8}$$

Here η stands for the ratio between forward and backward scattering, while α and β are the parameters of the backward and forward scattering, respectively.

Due to the proximity effect, electron energy is also deposited at locations outside the original focus or beam diameter. When a structure is written very close to an exposed area, the resist in the overlap area has already been partially exposed by the scattered electrons. In the normal writing scheme with constant beam intensity and exposure time, these regions tend to be overexposed [2]. The precise amount of broadening of the exposed resist area depends on a variety of factors such as photoresist type and thickness as well as the substrate material. The proximity effect has been the subject of intensive research throughout recent years. Several methods to reduce the negative effect on the pattern quality have been suggested, among them the precise control of the electron energy, the use of multilayer resist coatings as well as the calculation and precompensation of the energy deposition within and outside the pattern areas [4]. To this end, the parameters of the epsf have to be determined experimentally [5, 6]. With the dose adjustment technique it is possible to improve the minimum feature size of e-beam writing to values of $w_{\text{min}} \leq 0.1\ \mu\text{m}$. Problems, however, still remain if thick photoresist layers are to be structured. This is desirable for direct writing of blazed and continuous profile diffractive optical elements in scanning lithography (Chapter 5). Due to the proximity effect the minimum feature size increases with the thickness of the resist layer. The trade-offs with the direct writing approaches will be discussed in Chapter 4.

For a comparison of the performance of e-beam writing with the other pattern generation technologies we calculate the SBP for a typical e-beam machine (Jenoptik ZBA 23) [7]. This machine provides a minimum feature size of $w_{\text{min}} = 0.2\ \mu\text{m}$ over a maximum image field of $\Delta x^2 = 162^2\ \text{mm}^2$. For the SBP results:

$$\text{SBP}_{\text{e-beam}} = 6.5 \cdot 10^{11} \qquad (3.1.9)$$

Writing schemes for e-beam scanners:

The large SBP of e-beam writers illustrates the achievable pattern quality but also brings about some problems. With the increasing number of resolved pixels the writing time for the laser or e-beam writer increases dramatically. One factor which influences the writing speed is the data transfer to the e-beam machine. With the massive amount of data, i.e., tens of Gigabits per pattern, this data transfer has become an important issue. Additionally the response time of the systems which actively control the scanning process influence the writing speed. Electron beam scanners are equipped, e.g., with an automatic vibration damping and alignment control system. We find the average writing time for one pixel to be determined by:

- the average time necessary to transfer the pixel data;
- the exposure time per pixel, which can be calculated from the resist sensitivity and the intensity of the writing beam;
- the moving time of the scanning units;

- the vibration damping of the deflecting stage;
- the time necessary for alignment corrections, e.g., to respond to the interferometer control.

A typical value for the writing time per pixel is 1.2 μs. In this case the e-beam machine takes about $1\frac{1}{2}$ days to expose the whole pattern. For comparison, with the laser writing system an illumination of the whole image field consisting of only 10^{10} pixels takes about $3\frac{1}{2}$–$4\frac{1}{2}$ hours.

In order to apply e-beam writing in mask or scanning lithography efficiently it is essential to reduce the writing time for the patterns. The approaches to achieve this goal are twofold: New photoresists are developed which are more sensitive to the e-beam radiation and allow shorter illumination times. In a second approach the scanning process is optimized [2].

The scanning scheme often used in e-beam systems is the so-called "raster scan", which is adapted from scanning beam microscopes or TV-like scanning. The sharply focused electron beam with a circular cross-section and a gaussian intensity profile is scanned over the substrate and is modulated between the "on" and "off" state as it passes over structured areas. This writing scheme is very time consuming since areas with no structures are scanned by the e-beam as well as structured areas. Additionally the raster scan puts severe requirements on the data transfer rate.

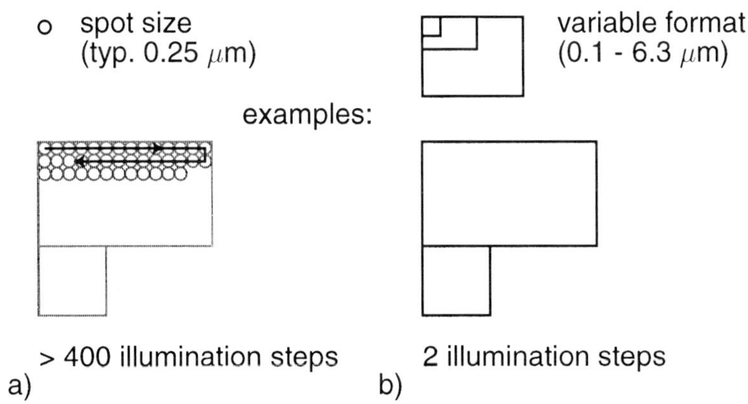

Fig. 3.5: Writing schemes for e-beam scanners: a) vector scanning, b) variable-shaped-beam illumination [7].

"Vector scanning" is an alternative scanning scheme, where time and data transfer requirements are reduced significantly. The electron beam in this case is directed only to scan the areas which contain structures. Thus, the amount of data to be transferred is reduced as well as the writing time, since only a portion of the addressable area needs to be scanned. This scheme is widely used nowadays in commercially available e-beam systems (Fig. 3.5a).

The "variable-shaped-beam" writing scheme is an approach for further cutting back e-beam writing time without sacrificing precision. To this end, the size of the e-beam is adjusted to the extension of the currently written structures. If coarse structures need to be written, a wide e-beam is used, whereas fine structures are written with the finest possible beam focus. Fig. 3.5 shows a simple example to illustrate the idea of the variable-beam-shape technique. In this example only two illumination steps with adjusted beam shape are enough to write the pattern for which, in the vector scanning mode, more than 400 illumination steps are necessary. However, for several reasons the real savings in writing time are significantly lower. First of all, the exposure time with the wider beam shapes needs to be longer to deposit enough energy in the resist layer. Secondly, time is lost when changing the beam size.

The variable-shaped-beam method requires a significant amount of numerical preparation of the structure data. The pattern needs to be subdivided into elements (e.g., square or rectangular) of different sizes. Additionally the dose deposition for each of the different beam sizes has to be calculated and optimized. It is important that the dose necessary to expose the photoresist is not deposited too quickly, in order not to heat the resist beyond the melting point. This would cause the resist pads to loose their exact shape. It has been estimated that the variable-shaped-beam method can speed up the e-beam writing process by at least a factor of 2–3.

Data formats for e-beam writing

An additional contribution to accelerate the e-beam writing process can be expected from suitable data formats. The conventional way to describe a lithographic pattern is through bitmap addressing of each individual pixel. Due to the increase of the number of pixels in high resolution lithography, this requires huge data storage and data transfer capacity.

Alternative data formats approximate the pattern with polygon tracks. Depending on the geometry this yields a considerable reduction in the amount of data. Reduction is highest for linear and rectangular patterns. However, for circular or arbitrarily shaped structures (e.g., for optimized diffractive lenses) a large number of data points is necessary to provide sufficiently high lateral resolution. Another set of curves, the so-called Bézier curves, can provide further data reduction. Bézier curves are popular in CAD systems since they allow one to determine arbitrary curves with the least possible number of parameters [8]. The coordinates of the starting and ending points as well as the gradients of the curves at these points are sufficient for fully describing the Bézier curves. Encoding the pattern using the Bézier curves helps to cut back significantly on the amount of data. In this case the data transfer no longer limits the writing time. Nevertheless, due to the high resolution, e-beam writing remains a relatively time consuming and expensive technology for pattern generation.

Long writing times and high cost are the main reason why nowadays high resolution e-beam scanners are mostly used for mask lithography rather than for the direct fabrication of components in scanning lithography. Lithographic masks when written once by e-beam writing can be used many times to be copied onto the substrate. Thus, the most expensive process is only needed once for mask fabrication.

However, in combination with replication techniques for surface profiles, direct e-beam writing is an interesting approach due to the achievable writing precision. Direct-writing techniques with variable dose deposition can also be applied to the fabrication of continuous surface relief-phase elements such as blazed components [7]. Replication techniques will be discussed later on in this chapter. In Chapters 4 and 5 we will come back to the potential of direct writing methods in the context of specific optical components.

3.1.4 Grey-level masks

Originally, for the fabrication of kinoform elements, it was suggested that one could use the phase effect occurring in photoemulsions after the bleaching of grey-scale masks [9]. Meanwhile, with lithographic means, more sophisticated techniques for the fabrication of phase elements have been developed. Nevertheless, with new techniques emerging, interest in grey-scale masks for lithographic exposure has been growing again recently. The fabrication of multilevel and quasi-continuous phase profiles in a single lithographic illumination step has been demonstrated using grey-level masks. The process is interesting especially for low cost mass fabrication of microoptical components.

In mask lithography the structured exposure required for the fabrication of surface-relief microoptics can be achieved with grey-level masks. Instead of scanning the substrate with a laser or electron beam of variable intensity, a one step exposure using a grey-level mask can be applied. This technique has great potential for reducing the cost of microoptics fabrication. With grey-level masks fabricated with similar technological effort as surface relief DOEs, the advantage for commercial fabrication lies in the easy single step replication possibilities. The grey-scale mask can be used to copy the pattern directly into a resist coating with analog response curve. After development, the surface relief of the DOE is generated. Resolution enhancement is possible if a reduction stepper is used to copy the mask onto the substrate. Several different approaches for the fabrication of grey-scale masks are feasible:

- In principle, grey-level masks can be fabricated with any of the techniques decribed for the writing of lithographic patterns. The straightforward approach is to use the analog response of a photoemulsion to generate a grey-level mask through **variable-intensity writing**, e.g., with a laser beam scanner [10]. Commercial slide imagers also have been used to generate grey-level slides from computer-generated patterns. Subsequent photoreduction on high resolution photographic plates yields reasonable lateral resolution of $w_{\min} < 5\ \mu$m [11]. In a slightly different approach the colour sensitivity of high resolution black and white photographic plates is used to generate the grey-level pattern [12]. The lateral resolution in these approaches is limited by the finite grain size of the photoemulsion. Typical grain sizes for this technique are of the order of several micrometres.

- Another approach to grey-scale mask fabrication is **halftoning**. Here, grey-levels are encoded through the density and diameter of binary structures. Main advantage of this technique is the easy control of the writing and further processing since only binary structures are written. Resolution is sacrificed since for every grey pixel several binary structures are necessary [13]. With a subsequent photoreduction process, including a spatial filtering step to remove the halftoning structures, good quality grey-scale masks have been fabricated [14]. The lateral resolution in these experiments was $w_{\min} \approx 8\ \mu$m.

- More sophisticated grey-level mask writing techniques have been developed recently using **thin layer coatings**. Here, a material with a precisely defined absorption coefficient is deposited with variable thickness on a substrate. This technique has been demonstrated using a nickel(75%)-chromium(16%)-iron (8%)-alloy called "Inconel" as absorbing material. Inconel is generally used as absorbing coating for neutral density filters. Thermal evaporation, which will be described later on in this chapter, is a suitable technique for deposition of thin layers of this material [15]. Variable thickness of the coating can be achieved through multiple binary lithographic steps. Thus, for the generation of grey-level masks with this technique a similar technological effort is necessary as for the fabrication of multilevel diffractive optical elements (see also Chapter 5). The advantage lies in the easy replication when using the element in mask lithography.

- In recent years a further high resolution grey-scale mask writing procedure has been developed [16]. The process is based on a glass substrate which is sensitive to high energy beams, called **HEBS glass** (acronym for high-energy beam sensitive). Based on a conventional borosilicate glass matrix the glass contains alkali ions as well as photoinhibitor materials. After production the glass substrates undergo an ion-exchange process. In a high temperature bath of approx. 320°C containing Ag^+ ions, the alkali ions (e.g., Na^+) from the glass are exchanged for silver ions. Generally these silver ions are bound in a silver-alkali-halide complex within the glass matrix so that no absorption occurs due to elementary silver. The purpose of the photoinhibitor is to avoid breaking down the silver complex (e.g., under UV illumination). This would immediately be followed by a reduction of the silver ions and an increasing absorption. In HEBS glass the breaking of the silver complex and subsequent reduction of the silver ions can be activated by high-energy radiation such as e-beams. When illuminated, the absorbing characteristics of the silver-halide complex depositions cause a variation of the optical density of the glass. Increasing dosage of the electron beam causes higher density of the silver-halide crystals within the glass, resulting in lower optical transmittance of the HEBS glass substrate. Although this working principle is very similar to the grey-scale picture formation in conventional photographic emulsions the HEBS glass has significant advantages for the fabrication of lithographic masks. In photographic emulsions the silver-halide crystals cluster together during the development to form grains with extensions of several microns. These macroscopic grains act as scatterers for the light beam and significantly reduce the dynamic range and the signal-to-noise ratio of the masks. In HEBS glass the silver-halides are constrained in the SiO_2 glass matrix. Thus, the formation of larger grains is avoided. The silver-halide crystals in the exposed HEBS glass typically measure 10–20 nm.

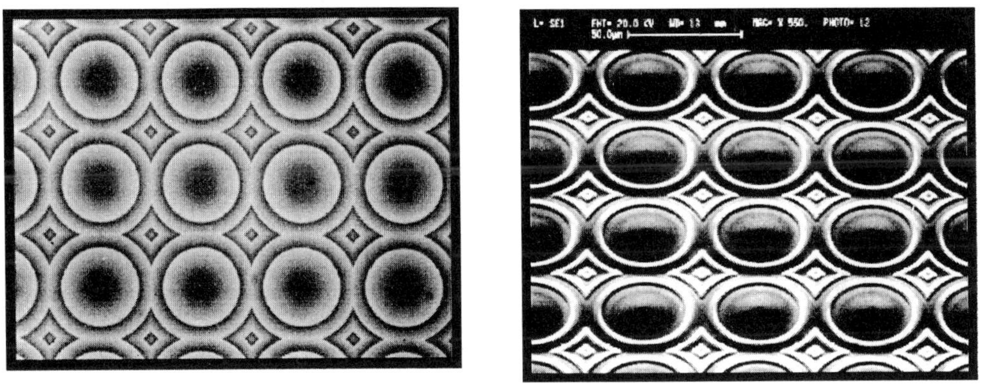

a) b)

Fig. 3.6: Fabrication of diffractive optical elements using grey-scale masks in HEBS glass: a) grey-scale mask; b) DOE after etching into fused silica glass substrate. (photographs courtesy of Dr. W. Däschner, Aereal Imaging Corp.).

The necessary dose variation of the writing e-beam is achieved by a variation of the scanning speed of the electron beam. Mask generation in the HEBS glass is straightforward. The only necessary processing step is the coating of a thin conductive chromium layer to avoid local charging of the glass during the e-beam writing. This chromium layer is removed after the scanning process. No further processing is necessary. The grey-scale mask is visible immediately after the writing process. Grey-level masks have been used for the fabrication of refractive as well as diffractive optical elements (see Chapters 4 and 5) (Fig. 3.6). For a 20 keV electron beam the response function of the HEBS glass was determined to be a 0.8 μm diameter gaussian function. Elements with periods of 5–8 μm have been fabricated with this technology. The use of a reduction stepper allows one to reduce the achievable minimum feature size even further.

3.1.5 Special masks

So far we have discussed several techniques for fabricating photolithographic masks generally consisting of alternating areas which are transparent or opaque to the illuminating radiation. As long as UV light is used for the exposure, chromium masks on quartz glass substrates can be used. The chromuim structures are highly absorbing while the quartz glass provides good transmission for the UV radiation.

In deep high-resolution lithography, x-ray or proton beams are often interesting for exposure. In this case the thin chromium or photomasks do not sufficiently absorb the illuminating radiation. In order to get high mask contrast it is necessary to use materials with high radiation transparency as the mask substrate (e.g., thin layers of Ti or Be) and highly absorbing materials (e.g., Au) as the structured mask layer. Other boundary conditions on the choice of materials are determined by the fabrication processes. Commonly, a sequence of e-beam lithography, galvanometric growth of the absorber structures and x-ray illumination is applied to mask fabrication [17]. An iterative process is necessary because thick absorber layers are needed which cannot be fabricated in a single etching step. A typical example for the fabrication of such masks is the LIGA (lithography, galvanic forming and molding) process. As an alternative technology deep structuring techniques, such as ion-beam milling, can be applied.

Here we finish our discussion of techniques for writing the electronically generated mask data in photosensitive layers of, e.g., photoresist. The following section is devoted to the fabrication of coatings or thin layers with little surface roughness and precisely defined thickness.

3.2 Coating or thin layer deposition

For mask generation as well as wafer structuring the fabrication of high quality layers of variable thickness and material is necessary. This is possible with a large variety of different processes. They vary in the precision achievable, the technological effort necessary and the kind of materials which can be processed. In the following paragraphs we will discuss approaches such as "spin coating" as well as "physical" and "chemical vapour deposition".

3.2.1 Spin coating

The cheapest and fastest way to generate thin layers of liquid or soluble materials is by spin coating. The substrate is fixed, e.g., on a vacuum chuck, which is rotated at a rate of typically 3000–6000 rotations per minute (rpm). When some of the coating material is deposited onto the rotating substrate, due to centrifugal forces a thin layer of the material is spread over the substrate. The rest of the material is thrown off the substrate. The thickness of the coating depends on various boundary conditions:

- the spinning speed;
- the viscosity of the coating material;
- the temperature of substrate and environment.

Although this coating method is fairly simple it yields good quality coatings with highly uniform layer thickness, provided the wafer is free of dust particles and the spinning speed is high enough. The spin coating technique is limited to coating materials which are dissolved before the coating process. By post baking after coating the solvent is removed and the coating, typically a photosensitive polymer (e.g., photoresist), remains on the substrate. For non-soluble layer materials more sophisticated coating techniques are necessary.

3.2.2 Physical vapour deposition (PVD)

In the basic process of PVD, the coating is deposited onto the substrate from the atmosphere in a reaction chamber. In order to achieve precisely defined layer thicknesses and compositions, vacuum conditions are required in the chamber. The coating material is introduced into the atmosphere from a source by physical processes. The PVD processes are categorized according to the mechanisms by which the material is introduced into the atmosphere.

- **Thermal Evaporation**

  One way to increase the partial pressure of the coating material in the atmosphere of the chamber is by heating the source material [2]. The Clausius-Clapeyron equation (Eq. 3.2.1) relates the vapour pressure $p(T)$ in the atmosphere to the absolute temperature T of the material (ΔE here is the latent heat of vaporization and V_g, V_l denote the molar volume of the gas and the liquid phase of the coating material, respectively).

  $$\frac{dp(T)}{dT} = \frac{\Delta E}{T(V_g - V_l)} \tag{3.2.1}$$

  The vapour pressure $p(T)$ directly determines the deposition rate $\mathcal{R}$ of the coating material on the substrate [18].

$$\mathcal{R} \propto \mathrm{C} \cdot \sqrt{\frac{M}{2\pi \mathrm{k}T}} \cdot p(T) \qquad (3.2.2)$$

Here M denotes the molecular mass of the coating material, k is Boltzmann's constant. For reasonable deposition rates a vapour pressure of $p(T) = 10^{-5}$ bar is necessary. In this case a typical deposition rate of approximately $\mathcal{R} \approx 10^{-3} \frac{\mathrm{kg}}{\mathrm{m}^2 \cdot \mathrm{s}}$ is achieved.

Eq. 3.2.1 indicates the nearly exponential dependence of $p(T)$ and $\mathcal{R}$ on the temperature T. Consequently, for reproducible coating thickness, a precise control of the temperature is indispensable. Several ways to heat the coating material are generally applied. Inductivity or I^2R heated boats, which hold the coating materials, form the most widespread sources for directly heated evaporation coating systems. Electron gun heated systems use an electron beam (electron energy $\approx$ 10 keV) to deposit the heating energy in the material. This process is necessary if materials with high melting points are to be used as coating materials. The energy is deposited directly into the source material without heating of the boat. The boat itself is only heated indirectly. Consequently, contamination of the layer by the boat material is significantly reduced. The uniformity of the coating layer over the substrate is determined by the shape of the source and the target-source distance. Long distances ensure good uniformity but low deposition rates.

- **Sputtering**

  Another technique for thin film deposition on substrate wafers is the sputtering process. In contrast to the evaporation technique the coating material is released from the source after a physical impact of plasma ions. Source and substrate are used as plane parallel cathode and anode, respectively (Fig. 3.7) [18]. In between these two electrode plates a plasma is ignited, e.g., an Argon plasma. Due to the dc voltage ("DC-sputtering") between the plates the positive Ar^+ cations are accelerated towards the cathode covered with the coating material. By momentum transfer they break the molecular and atomic bondings of the cathode material. Uncharged target atoms are released into the space between the electrodes. Due to their relatively high kinetic energy (1–10 eV) they reach the substrate. They form a thin layer of the same composition as the cathode material.

  If insulating materials are to be deposited the described dc-configuration cannot be used because the positive ions arriving at the cathode would neutralize the cathode and prevent new ions from being attracted. The sputtering of insulators is possible by high frequency (rf) ignition of the plasma. The positive ions are then removed from the cathode surface when the voltage polarisation is inverted.

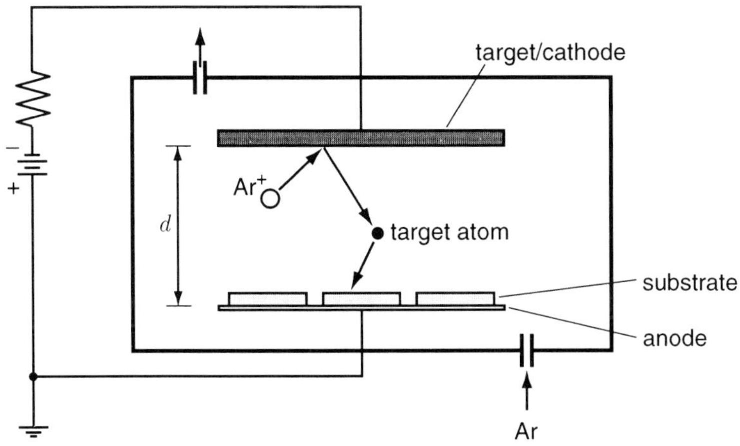

Fig. 3.7: Schematic of a plasma chamber for sputter coating [2].

Sputtering uses momentum transfer rather than thermal processes for the deposition of the coating material. This can be performed at much lower temperatures than, e.g., the evaporation technique. The only heating of the wafer is due to the plasma electrons which reach the substrate. The main advantages over thermal deposition techniques are:

- the sources can have a large area, so that very homogeneous layers over big areas can be generated;
- no dissociation of composite materials occurs due to the low temperatures;
- good film adhesion is provided;
- the deposition rate is independent of the vapour pressure of the material.

- **Molecular beam epitaxy (MBE)**

During the conventional evaporation process at high temperatures (e.g., 1200°C for Al evaporation) relatively high partial pressures of the evaporant occur in the chamber atmosphere. Thus, the material transport rate (or material flux density) as well as the deposition rate are relatively high (e.g., 4 $\frac{\mu m}{s}$). This deposition rate is significantly too high if epitaxial growth is to be achieved. Epitaxial layers are grown on monocrystalline substrates and assume the same crystalline lattice and the same orientation as the substrate. This can be achieved if the growth rate of the layer is reduced to very low values of about 1 $\frac{\mu m}{h}$, corresponding to less than one atom layer per second. Such low growth rates can be achieved with sophisticated source setups, so-called "effusion cells". These effusion cells release the source material at very small and precisely controlled doses. Due to the slow epitaxial growth of the layers in MBE it is possible to control the stochiometric composition as well as the doping of the layer. This technology is important for the fabrication of semiconductor devices such as VCSELs or multiple quantum well modulators (e.g., SEEDs) (see Chapter 5).

An important feature of the coating process is the absolute thickness of the layer as well as its uniformity over the substrate or over several substrates coated in parallel. To monitor the thickness of the coating online during the PVD process, the resonance detuning of a quartz crystal is observed. The quartz crystal is placed at a position where the same amount of material is deposited as on the substrate probe itself. It serves as a frequency-determining element in a resonant circuit [2]. When material is deposited on the quartz surface the mass change detunes the resonance frequency of the crystal. This allows a very precise measurement of the thickness which has been deposited. The thickness t of the coating is proportional to the frequency detuning Δf:

$$t = \mathcal{K}\Delta f \tag{3.2.3}$$

The constant $\mathcal{K}$ is determined empirically for different materials or compositions. Typical values of $\mathcal{K}$ are:

$$\begin{aligned} \text{Al:} \qquad & \mathcal{K} = 0.6\ \frac{\text{Å}}{\text{Hz}} \\ \text{SiO:} \qquad & \mathcal{K} = 0.3\ \frac{\text{Å}}{\text{Hz}} \end{aligned} \tag{3.2.4}$$

3.2.3 Chemical Vapour Deposition (CVD)

In PVD processes generally no chemical reactions are involved in achieving the deposition of the layer material. The material is physically transported from a source to the target where it forms the coating. In CVD the coating materials are generated in chemical reactions before they form the layer coating on the substrate. The compounds of the chemical reactions are introduced as reactive gases. The energy to activate the chemical reaction stems from external sources. These sources allow a classification of the various methods of CVD into thermal CVD, laser induced CVD and plasma-enhanced CVD (PECVD). Very stringent control of the conditions and chemical concentrations during CVD allows one to grow epitaxial films (e.g., silicon on silicon layers or silicon on sapphire layers).

3.3 Alignment and exposure

In scanning lithography the pattern is written directly into the coating on the substrate. This situation is different in mask lithography. The pattern of the lithographic mask needs to be transferred into the substrate coating with best possible alignment relative to existing structures in the substrate. The relative alignment of the lithographic mask and the substrate is controlled in "mask aligners" with high precision x/y stages. The substrate coating is exposed through the masks with uniform illumination. In this section we address this important processing step in mask lithography. We discuss the various possible "exposure geometries", different "light sources for lithography" and address the critical issues of "multimask" and "through-wafer alignment".

3.3.1 Exposure geometry

The transfer of the mask pattern to the photoresist coating on the substrate is performed in three possible geometries, referred to as "contact printing", "proximity printing" and "projection printing" (Fig. 3.8).

- In **contact printing** the mask is placed in physical contact with the surface of the coated wafer. In vacuum chucks the space between substrate and mask is evacuated, to optimize the uniformity of the contact pressure. In order not to damage mask or substrate coating during the alignment the mask is kept at a small distance of $\delta z \approx 100\ \mu m$ from the substrate surface. The direct contact during the exposure prevents any diffractive beam spread between the mask and the photosensitive layer. Sub-μm resolution can be achieved with this technique depending on the illuminating wavelength. Deviations from perfect flatness of the soft coatings are corrected by the pressure applied in contact printing. Problems in contact printing stem from a possible damaging of either the substrate coating or the mask patterns as a result of the physical contact. For industrial mass fabrication of microoptical components such mask damage significantly increases the production cost of components fabricated with contact printing.

Michel et al. of IBM research Laboratories recently suggested so-called light coupling masks also referred to as light stamp for contact lithography. Instead of a thin amplitude mask they use a binary surface profile mask during the lithographic illumination. The contact between mask and substrate only occurs at the ridges of the mask. Upon illumination from the backside, the ridges couple the light efficiently into the photoresist. In the non-contact areas photoresist exposure is much lower due to reflection losses at the material interfaces. For an increased contrast it is possible to apply an absorption coating additionally. With this technique researchers hope to reduce mask wear. This would justify the additional effort during the mask production, e.g., by replication from an expensive silicon master. The light stamp masks can be made of a flexible polymer which is not damaged by dust particles on the substrate. During the lithographic production, cheap replica rather than the expensive master masks are used. As a further advantage, due to the immersion effect it is possible to copy structures smaller than the vacuum wavelength of the illuminating light source. The fabrication of feature sizes of 100 nm has been demonstrated with a 248 nm light source [19].

- In order to speed up lithographic fabrication and reduce mask wear, in **proximity printing** direct contact between mask and wafer is avoided. For good resolution it is necessary to reduce the distance between the layers, also called the "proximity gap", as much as possible. Typical values for the proximity gap are $g_P = 2$–$20\ \mu$m. Due to diffraction effects at the mask, for proximity printing the minimum feature size w_{min} printed into a photoresist layer of thickness t_R can be estimated from [18]:

$$w_{min} \approx \sqrt{\lambda(g_P + \frac{t_R}{2})} \qquad (3.3.1)$$

For UV radiation (e.g., $\lambda = 400$ nm) a minimum feature size of $w_{min} \approx 3$–$4\ \mu$m results.

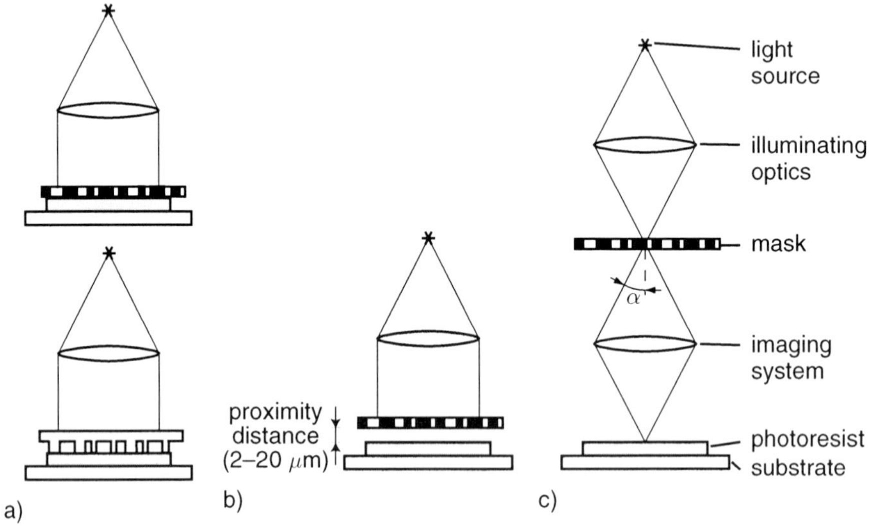

Fig. 3.8: The different geometries for lithographic illumination: a) contact printing using amplitude masks (top) or light coupling masks (bottom); b) proximity printing; c) projection printing. [18].

- In **projection printing** the mask pattern is imaged onto the wafer optically. The distance between mask and wafer is relatively large. Due to the huge space-bandwidth product of the lithographic masks, very sophisticated optical systems are necessary for this imaging step. In today's projection printing systems reflective lenses are used for the imaging. Compared to refractive lenses, mirror systems can be corrected very well over relatively large image fields. Nevertheless, to achieve the high SBP of lithographic masks a stepping or scanning process is used. Only the best section of the mirror system is used to image a small section of the mask. Mask and substrate are scanned with respect to the optical system in order to achieve optimum image quality over the whole substrate area. With this technique minimum features sizes below 1 μm can be achieved. From the extension of the diffraction-limited psf we can calculate the resolution limit of an imaging system:

$$w_{\min} \propto \frac{\lambda}{\mathrm{NA}} = \frac{2\lambda f}{D} \tag{3.3.2}$$

NA, D, and f denote the numerical aperture, diameter and focal length of the system, respectively. NA is defined as:

$$\mathrm{NA} = \frac{D}{2f} \tag{3.3.3}$$

Optical imaging systems for projection printing nowadays are an interesting area of application for microoptical systems. An array of microoptical projection systems can be used to image large mask areas onto the substrate. Details on this interesting application of microoptics will be discussed in Chapter 12.

3.3.2 Light sources for mask lithography

Eq. 3.3.2 suggests two possibilities for enhancing the resolution of the projection printing system. Large numerical apertures of the optical system or short illumination wavelengths allow better resolution. The fundamental problem of systems with large numerical apertures results from the reduction of the depth of focus with increasing NA. The depth of focus $\delta z_{\mathrm{focus}}$ can be calculated from:

$$\delta z_{\mathrm{focus}} = \pm \frac{\lambda}{2\mathrm{NA}^2} \tag{3.3.4}$$

For NA $= 0.38$ and $\lambda = 248.5$ nm the depth of focus is already as small as ± 0.86 μm. For best possible resolution it is necessary to avoid a decrease of the depth of focus to values smaller than the thickness of the photoresist coatings. Thus, for projection printing lithography it is not desirable to use optical systems with numerical apertures larger than 0.4–0.5. Nowadays, resolution enhancement of the lithographical systems is achieved by using light sources with shorter wavelengths. For optical lithography UV illumination is generally used. High pressure mercury short-arc lamps are the most important light sources in todays lithographic mask aligners. These mercury lamps emit spectral lines at wavelengths of 436, 405, 365 nm, respectively [20]. Other lines in the mercury spectrum occur at 334, 313 and 254 nm (called g-, h-, i-lines). However, it becomes increasingly difficult to achieve enough intensity in these lines. The latest generation of mask aligners uses pulsed KrF- and ArF-excimer lasers with emission lines at 248 and 193 nm. Further in the future even shorter wavelength lasers such as carbon plasma lasers may become feasible. A major disadvantage of excimer lasers is the toxic gas filling, which requires careful handling. Other technological problems for the deep UV lithography arise from the materials which can be used for the optical components. For wavelengths shorter than 200 nm, fused silica and quartz glass start to exhibit significant absorption. Thus, in the deep UV region it will become necessary to fabricate optical elements made of materials such as CaF_2 or MgF_2.

wavelength [nm]	**source type**
18.2 nm	pulsed carbon plasma laser
157 nm	pulsed F^2 excimer laser
193 nm	pulsed ArF excimer laser
248 nm	pulsed KrF excimern laser
254 nm	Hg/Xe short arc
313 nm	Hg/Xe short arc
326 nm	Cd/Xe short arc
335 nm	Hg/Xe short arc
365 nm	Hg short arc (Hg i-line)
405 nm	Hg short arc (Hg h-line)

Tab. 3.1: Light sources for UV lithography [21, 22, 23].

3.3.3 Illumination with x-ray (synchrotron) and proton radiation

Instead of using optical light sources for exposing the substrate it is possible to use other types of radiation, such as x-rays or proton beams. Due to the extremely short wavelengths this reduces diffraction effects and enhances the resolution. X-ray radiation and particle beams (electron or proton beams) offer interesting features for lithographic applications [2]. Neither x-rays nor proton radiation can be controlled (focused and deflected) well enough so far to build systems for scanning lithography. In mask lithography sufficiently collimated beams of both radiation types can be used in the wafer illumination process.

Due to the high penetration depths and small diffractive beam widening these lithographic techniques allow the fabrication of deep structures with high aspect ratios (Chapter 4). This potential is very interesting for fabrication of micromechanical systems as well as refractive microoptical components. X-ray and proton beam methods, e.g., have been applied to the fabrication of refractive microprisms [24].

3.3.4 Multimask alignment

For lithographic fabrication of efficient microoptical components or systems it is often necessary to perform a sequence of binary structuring steps. A typical example is the fabrication of multiple-phase level diffractive optical elements. We will learn more about these elements in Chapter 5. Here we want to discuss the alignment steps, which are necessary to fabricate these elements. In scanning or grey-scale mask lithography the multilevel pattern is transferred into the photoresist coating on the substrate in a single exposure step, e.g., during scanning through variable dose writing. In this case the alignment accuracy of the different phase levels is determined by the positioning accuracy of the scanning system. In mask lithography multilevel elements are often fabricated in several binary fabrication steps (Fig. 3.9). For the fabrication of a phase grating with N phase levels a minimum of $\log_2 N$ binary structuring steps are necessary. Each of the fabrication steps is performed using a binary mask. For good performance of the resulting DOE, good relative alignment of the masks is very important. Sub-μm repositioning accuracy can be achieved in mask aligners using suitable alignment marks.

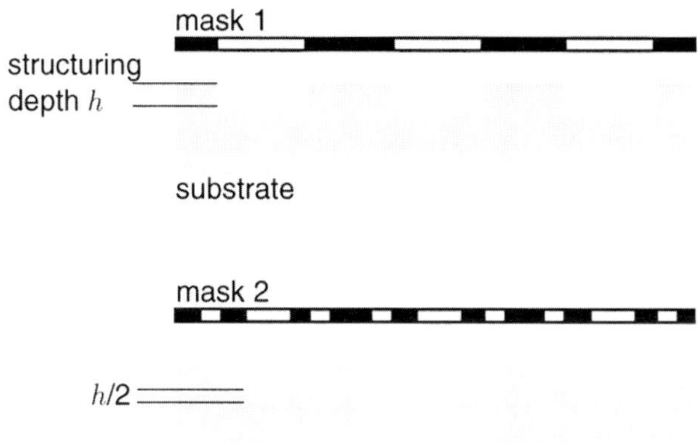

Fig. 3.9: Fabrication of a 4 phase level element using two binary lithographic steps.

The achievable precision with alignment marks for manual alignment is based on the ability of the human eye to register small deviations in symmetrical patterns. Suitable designs of alignment marks support the alignment process. Several aspects should be considered when designing the alignment marks:

- The resolution, field of view and depth of focus of the alignment microscope need to be taken into account. It is important that the alignment structures are designed such that they can be imaged with an appropriate magnification, good contrast and sufficient depth of focus. Suitable image size and contrast ease the observation, while a sufficient depth of focus is necessary, since two separate planes have to be aligned.

- Two alignment structures in different corners of the substrate wafer help to avoid errors due to a relative rotation of the substrates. Many mask aligners are equipped with a split-field microscope which allows one to simultaneously observe both locations. In this case it is necessary to adjust the distance between the alignment masks to the field splitting provided by the microscope.

- The precision of the alignment mark suffers during subsequent etching processes. If several mask layers are to be aligned it is recommended that one uses a single pair of alignment marks for each pair of mask layers. With the second layer, a further alignment structure for the next layer is generated at a different position.

- In many cases it is useful to introduce one layer containing all the alignment marks for the following processing steps. These alignment marks can be processed for optimum visibility, e.g., as amplitude masks. Especially if low contrast phase profiles are fabricated, facilitation of the alignment task will outweight the effort necessary for this additional processing step.

Generally, a positive mask structure is aligned to overlap perfectly with a negative structure. Fig. 3.10 shows typical examples of such alignment masks. The mask contains features of different dimensions so that during the alignment it is possible to proceed stepwise from aligning the coarse structures to the finest possible structures. In Chapter 5 we discuss the effects of alignment errors on the performance of multilevel diffractive optical elements.

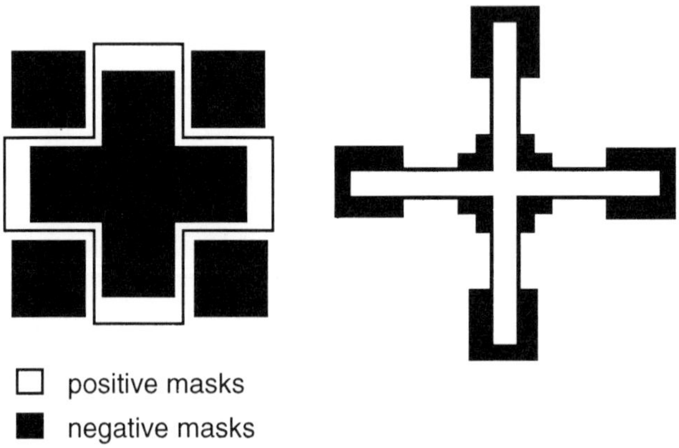

Fig. 3.10: Typical structures used in alignment masks. Positive and negative masks are aligned on top of each other [25].

3.3.5 Through-wafer alignment

For some microoptical applications it is of interest to align two distant layers with optical components or to structure both sides of the same substrate (e.g., in planar optics, Chapter 6) [26]. To this end it is necessary to align two mask planes, with a relatively large distance (e.g., several mm). The depth of focus of the observation microscope is not large enough to observe both masks simultaneousely. We shall discuss this problem using the example of the alignment of two masks on the top and bottom of a thick glass substrate. For transparent substrates this is possible by shadow casting of the alignment mask or by Fresnel diffraction. In the case of the shadow casting technique a relatively coarse symmetrical alignment mask, e.g., a ring mask, is illuminated by a collimated beam [27]. The pattern of the mask is projected into the photosensitive layer on the opposite surface of the substrate. Just as in proximity printing (Section 3.3.1) the mask structures are blurred during propagation through the substrate, due to diffraction (Fig. 3.11a). The symmetry in the alignment mask and the binary characteristics of the photoresist, however, allow one to achieve an alignment precision which exceeds the resolution of the structures transferred through the shadow casting process. Good collimation and perpendicular incidence of the illuminating beam is very important for this method. Experimentally this can be achieved using the beam reflected back from the substrate. The system is aligned if the reflected beam is directed back through the pinhole in the entrance pupil. With this simple shadow casting technique, a relative alignment precision $\delta x_{\text{align}} \leq 1\ \mu\text{m}$ has been achieved for substrates with a thickness of $t = 3$ mm.

A second technique for through-wafer alignment uses Fresnel diffraction. An alignment mask (e.g., a Fresnel zone lens) is designed which generates a well defined image (e.g., a focal spot) on the opposite surface of the wafer. This pattern can be used as alignment mark for this side (Fig. 3.11b).

3.4 Pattern transfer

In the previous chapters we discussed the processing steps necessary to prepare the substrate for the actual structuring step. We have seen how the desired computer-generated pattern is transferred into a mask pattern or directly written into the photosensitive layer on the substrate wafer. In this section we will focus on techniques applicable for the transfer of the pattern into the substrate or the coating on the substrate. Firstly we address "etching" processes, which play an important role in lithography for the pattern transfer. In the second section we focus on "laser micromachining" as an alternative approach to generate a micropattern directly onto a substrate wafer. "Mechanical micromachining", such as diamond turning, the "replication of microrelief structures" and "diffusion" processes are further topics of this section.

Before we focus on the technologies, we briefly want to introduce the two different fundamental procedures for pattern transfer (Fig. 3.12). In the conventional

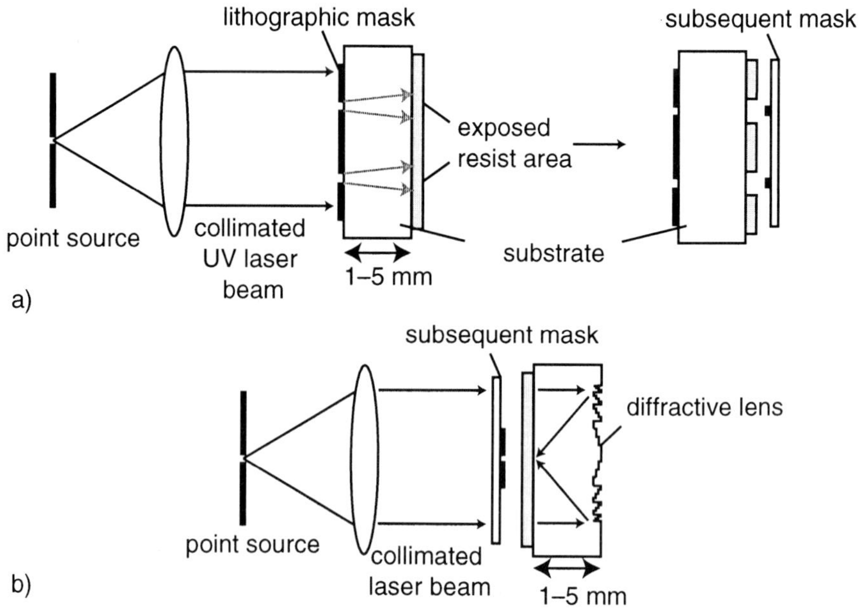

Fig. 3.11: Two methods for through-wafer alignment: a) projection of a symmetrical pattern through the substrate which serves as alignment pattern for the subsequent mask; b) diffractive alignment lenses are used to focus or image a pattern onto the opposite substrate surface.

way to structure the coating on a substrate, a corresponding mask pattern is fabricated, e.g., in photoresist. This mask protects some areas of the coating in the subsequent etching process. After the etching the photoresist layer is removed and the pattern in the coating layer remains on the substrate (Fig. 3.12a).

In the lift-off technique the sequence of the processing steps is changed (Fig. 3.12b). The first step is to spin-coat a photoresist layer onto the substrate. After exposure and development of the photoresist, a coating of the desired material is generated on top of the photoresist layer, e.g., through evaporation of a metallic coating. The final step of the lift-off procedure is to remove the photoresist. During this process those areas of the coating which have been generated on top of the photoresist are also removed. What remains on the substrate is the (metallic) coating in those areas where the photoresist had been removed previousely. The main advantage of the lift-off process is that it allows one to structure layers of materials which are resistant to etching chemicals. Very often the etchants necessary to structure the coating are harmful to the substrate. In the lift-off process it is possible to pattern the coating with chemically less aggressive photoresist developers.

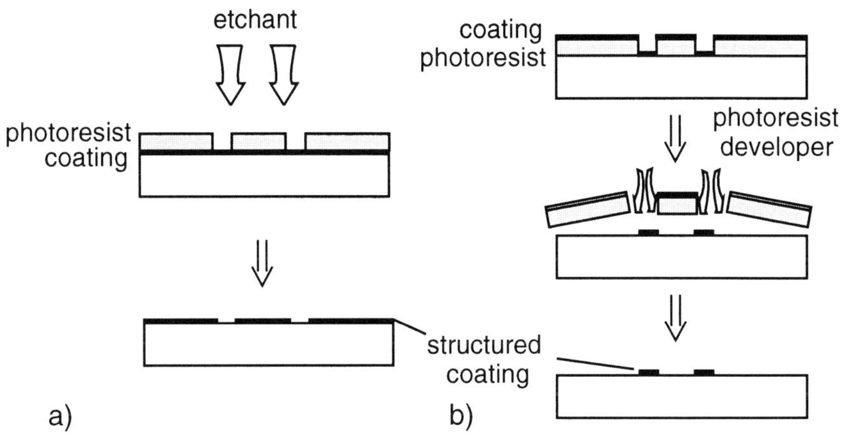

Fig. 3.12: Pattern transfer trough a) conventional and b) lift-off etching.

3.4.1 Etching

We now address possible technologies for the etching of substrates to generate microrelief patterns. Generally we distinguish between isotropic and anisotropic etching processes (Fig. 3.13). For isotropic etching the substrate material is removed at the same speed (i.e., the etch rate) in all directions. This results in an underetching (or undercutting) of the masks. For small mask openings circular symmetrical etching grooves result from isotropic etching. In the case of anisotropic etching the etch rates are different in different directions. The directional dependence may firstly be a result of the etching process itself which can be anisotropic in nature. A typical example for an anisotropic process is reactive ion etching or ion-beam milling.

A second reason for anisotropic etching can be found in an anisotropy of the substrate material. In crystalline substrates (e.g., a crystalline silicon wafer), different crystal planes are more or less resistant to the etching process. Even if all directions are equally exposed to the etchant, the etch rates are strongly dependent on the orientation of the crystal directions. The amount of anisotropy in an etching process can be defined as the ratio between the etch rates in different directions. Both anisotropic and isotropic etching processes are important for the fabrication of microoptical components. Lenses, e.g., have a smooth spherical phase profile which can be fabricated by isotropic etching (wet etching in isotropic substrates). For the fabrication of diffractive optical elements, phase steps with steep edges have to be etched. For this, anisotropic etching processes are used.

Apart from isotropy, a second criterion for the etching process is the selectivity, i.e., the relation between etch rates for different materials. Generally, photoresist

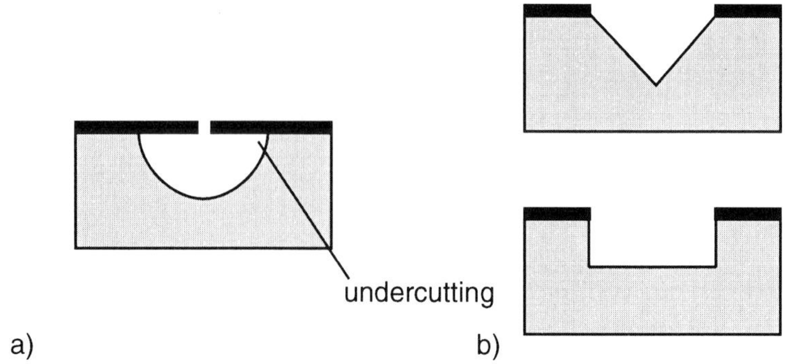

Fig. 3.13: Isotropic (a) and anisotropic (b) etching processes. The bold lines indicate the mask during the etching process.

is used as mask coating for the etch process. The photoresist coating is easy to pattern using alkaline developers. After baking, the photoresist layer exhibits good resistance against acidic etchants. Typical substrate materials such as fused silica or silicon can be etched with acidic etchants such as HF or NHO_3. Typical etch rates range from 1–100 $\frac{\mu m}{min}$. For deep structuring large differences in the rates for the etching of mask and substrate material are necessary ("differential etch").

One of the most important materials for optics is glass. Since glass is an amorphous, isotropic material it does not introduce any anisotropy. Any deviation from the spherical etching groove depends on the process. We will mostly focus on glass etching. The various different etch processes can be classified by the physical state of the active solvent:

- **Wet etching** takes place in a liquid solvent. Since spreading of the liquid solvent has no favourable direction wet etching is an isotropic process. The masked substrate is put into a liquid which contains the active components. In the areas where the substrate is not protected by a mask (e.g., a metal mask) it is exposed to the etching liquid. A chemical reaction between the etching liquid and the substrate material takes place which dissolves the substrate [2]. The dissolved material is removed from the substrate in a diffusion process which can be supported by stirring the liquid. Obviously the choice of solvent strongly depends on the substrate material. For glass the typical solvent is hydrofluoric acid (HF). The corresponding chemical reaction taking place at the glass surface is:

$$\begin{aligned} SiO_2 + 6\,HF + 6\,H_2O &\rightarrow SiO_2 + 6\,H_3O^+ + 6F^- \\ &\rightarrow H_2SiF_6 + 8\,H_2O \end{aligned} \tag{3.4.1}$$

  Wet etching techniques are used for isotropic etching of thin layers (e.g., metal masks etc.). The etching of very fine, deep structures in glass substrates (i.e., structures with high aspect ratio), however, is not possible because the etch rates are equal in different directions. Long etching times necessary for deep structures result in wide etching grooves. The lateral resolution cannot be very high for deep structures. In amorphous substrates such as glass the isotropic nature of the wet etching process can be used for the fabrication of spherical etch grooves which serve as lenses.

- **Wet etching in silicon**

Silicon is the most important material for electronic circuitry, however, it is also used in microoptical applications. The structuring of silicon wafers has been thoroughly investigated. Crystalline silicon wafers are commercially available with any orientation of crystal lattice. Due to the crystalline structure a large variety of different etching patterns can be fabricated through wet etching. The complexity as well as high aspect ratio of the achievable structures caused interest, e.g., for applications in micromechanics. In microoptics silicon structures are used e.g., for the alignment of optical fibers or for beam deflecting mirrors. Without going into detail we discuss some of the structures which can be fabricated in silicon wafers [18].

A typical etchant for anisotropic etching of silicon is a KOH/H_2O solution. For this solution the etch rate in the $\{111\}$ direction of the crystal lattice is about 400 times lower than in the $\{100\}$ lattice direction. Wet etching thus results in a highly selective etch process. A silicon wafer with a $\{100\}$ surface orientation has four $\{111\}$ crystal planes cutting the surface at an angle of 54.74°. If a mask window is aligned properly along these $\{111\}$ planes, during the anisotropic etch process we obtain an etching window limited by walls which are inclined by an angle of 35.36° relative to the surface normal (Fig. 3.14a). If the mask window is of rectangular shape etching results in so-called V-grooves. For short etching times the groove becomes trapezoidal in shape, since the the point of intersection of the two $\{111\}$ planes has not been reached (Fig. 3.14a, right).

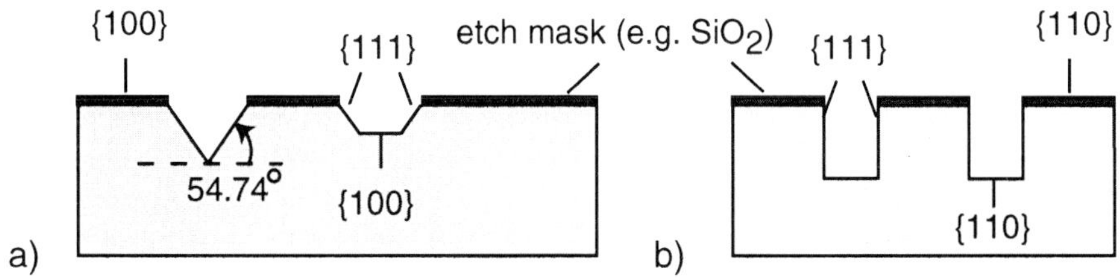

Fig. 3.14: Shape of etching grooves resulting from anisotropic etching in crystalline silicon wafers: a) Si V-grooves etched in $\{100\}$ wafers; b) high aspect ratio grooves etched in $\{110\}$ wafers.

For {110} silicon wafers the situation is even more complex. The {110} surface is cut perpendicularly by two {111} planes, which assume an angle of 109.47° relative to each other. Two other {111} planes are oriented at an angle of 35.26° relative to the surface normal. With suitable orientation of the mask window it is possible to etch long, narrow grooves with perpendicular walls and high aspect ratio (Fig. 3.14b). The aspect ratio of the structure is defined as the ratio between depth and width of the groove. Such high aspect ratio grooves can serve as alignment structures in microoptical systems (Chapter 7).

In addition to the crystal orientation in silicon wafers, the etch-rates are also strongly depending on the doping of the silicon substrate. Through precise control of the dopant concentration, it is possible to introduce well defined etch stops or sacrificial layers which are etched at very high rates. With this technique complex micromechanical structures can be fabricated, including flexible Si-tongues and micro-gears.

- In **dry** or **plasma-enhanced** etching processes the etching of the substrate is performed in a plasma. Typical representatives of this group of technologies are **sputter etching** and **ion-beam etching**, which is sometimes referred to as ion-beam milling. The principle of plasma etching is similar to the depletion of the coating material in a sputtering system (see Section 3.2.2). A plasma is formed in the chamber containing the substrate (sputter etching) or in a separate chamber (ion-beam etching or milling) [2]. The chemically inert plasma ions (e.g., Ar^+) are used to physically destroy the bonds of the substrate materials and thus deplete the material. The purely physical nature of this etching technique is highly anisotropic. Etching occurs mainly in the direction of the accelerated plasma ions. This allows one to fabricate structures with high aspect ratios. The main disadvantage is the low selectivity of the etch rates for different materials. Suitable masks must have high aspect ratios which makes mask fabrication difficult.

- In **reactive ion etching (RIE)** and **reactive ion-beam etching** (sometimes also referred to as **chemically assisted ion-beam etching (CAIBE)**) a combination of physical and chemical processes is utilized for the etching of the substrate material. The etch gases in this case consist of chemically reactive ions, e.g., CHF_3 for glass etching or O_2 for photoresist etching. No etching process occurs at the substrate without the physical impact of the ions on the substrate. The reason for this is twofold. Firstly, the etching gas chemically reacts with the substrate. The products of this reaction, however, bond with the surface so that they have to be removed by physical sputtering. A second reason is the supply of the activation energy necessary for the chemical reaction by the ion-beam. RIE is especially interesting for the etching of diffractive optical elements in glass.

One of the most interesting aspects of reactive ion etching is that the etch rates for different materials can be chosen within a wide range by selecting the gas composition [28]. This is also known under the term "differential etching". The ratio between reactive and inert gas ions determines the anisotropy of the etch process as well as the selectivity of the process to different materials. This is especially interesting when continuous relief structures (e.g. melted photoresist lenses) are transferred into a glass substrate.

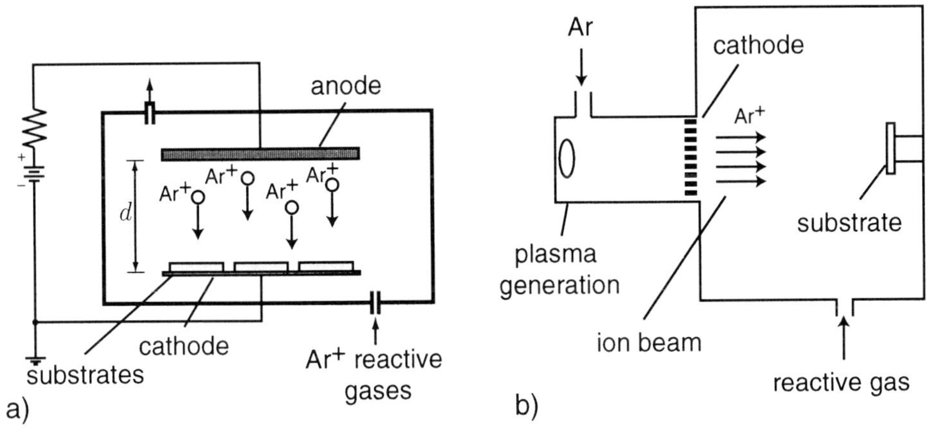

Fig. 3.15: Schematic setup of a) ion and b) ion-beam etching techniques.

mechanism	chemical reaction	physical reaction	reactive or inert chemicals
wet (chem.) etching	yes	no	HF (glass); e.g.: KOH (silicon); H_3PO_4 (Al)
ion (beam) etching (milling), sputtering	no	yes	e.g.: Ar^+
RIE, CAIBE	yes	yes	e.g.: Ar^+ + CHF_3; CF_4, CCl_4

Tab. 3.2: Categorization of the different pattern transfer technologies.

For a summary of the variety of etching techniques which are available for pattern transfer we refer to Table 3.2. Wet etching is based exclusively on chemical dissociation of the substrate by the etchant. The ion etching processes are supported by or completely based upon the physical interaction of a chemically inert plasma gas with the substrate. According to Fig. 3.15 we can distinguish between ion etching and ion-beam etching techniques. In ion-beam etching the plasma ions are generated in a separate reaction chamber and accelerated towards the substrate. Due to the better collimation of the ions in the ion-beam techniques, the anisotropy of the process is better than in ion etching techniques. Both ion etching techniques can be modified to be reactive techniques by addition of reactive chemicals.

3.4.2 Laser micromachining — laser initiated ablation

Throughout recent years, the micromachining of substrates by laser ablation experienced enormous technological progress. Under the title "rapid prototyping" the potential use of the laser for shaping 3D objects has attracted interest in research and for commercial applications. With constant improvement of the precision of the fabricated structures, this technology can be applied to the fabrication of microoptical elements. For laser micromachining applications generally excimer lasers or frequency doubled Nd:YAG lasers are used [29]. For both types of laser the fundamental working principle of laser ablation is the same. Here we concentrate on excimer laser sources. Suitable wavelengths used for micromachining range from 193–351 nm.

The interaction processes which are utilized for structuring a variety of substrate materials with laser radiation are classified into pyrolytic and photolytic processes [30]. In pyrolytic processes the chemical bonds in the target material are decomposed due to thermal heating. In photolytic processes the molecules of the material are excited to occupy higher electronic or vibrational states by directly absorbing the radiation. With these mechanisms it is possible locally to provide the activation energy for chemical reactions. Due to the physical nature of the process, photolytic interaction is highly restricted to the area where the energy is deposited. In pyrolytic processes, on the other hand, diffusion processes cause spreading of the energy beyond the borders of the exposed areas. Depending on the material system as well as on the energy deposition rate it is possible to use laser-induced processes for material deposition as well as removal (i.e., etching or ablation). For example, laser-assisted chemical vapour deposition has been used for the fabrication of microoptical elements. CO_2-laser radiation is used to deposit silicon-nitride and silicon-oxide films on quartz or silicon [31].

Photoablation is initiated by the fast deposition of a large amount of energy into the substrate material, e.g., a polymer. The incident photons cause a dissociation of the substrate molecules. If this dissociation occurs fast enough, the material is removed from the substrate in a "micro-explosion". The ablative reaction is caused by the amount of energy deposited per volume. The time scale for the energy deposition is a critical parameter for the quality of the ablated region. For long energy deposition times, pyrolythic effects occur which generally cause a degradation of the edge quality. Very short low energy pulses from a Ti:sapphire laser have been used to demonstrate high quality machining of metal substrates

[32]. Excimer laser photoablation has been used to structure Teflon cylinders for the fabrication of reflow microlenses (Chapter 4) [33]. Because of the subsequent reflow process surface roughness generated by the ablation process is not a critical issue in this case.

For several reasons polyimide film is an interesting material for the fabrication of microoptical components using excimer laser ablation [34]. Most importantly the absorption characteristic is favourable since the films are transparent in the visible and near infrared and highly absorbing for the excimer wavelengths. At the same time polyimide provides good chemical and physical stability. Especially for the 193 nm wavelength, a very clean photolytic ablation can be observed without any thermal effects. Thermal effects are more frequently observed for longer excimer laser wavelengths (e.g., 308 nm). Direct laser scanning as well as mask projection methods have been demonstrated to yield high quality DOEs [35, 36]. Surface roughness of < 25 nm has been achieved. Submicron minimum feature sizes are possible with a depth resolution of 0.1 μm. Research is also being carried out on laser ablation in glass materials [37]. With further progress in the near future, the photoablative processing can become an interesting approach towards obtaining rapid and inexpensive profiling, even if optical quality of the microstructures is required.

3.4.3 Mechanical micromachining — diamond turning of microoptical components

Although micromachining does not belong to the class of lithographic fabrication processes, its importance to microoptics fabrication is significant. We therefore devote the following short section to this technological approach.

The fabrication of classical macroscopic optical components is traditionally based on mechanical techniques such as grinding, polishing, etc. Most of these techniques are not applicable for the fabrication of microoptical components with dimensions < 1 mm. Nevertheless, being the traditional way of fabrication, a lot of effort has been devoted to the refinement of mechanical processes for microoptics fabrication. In particular, single point diamond turning is able to meet many of the requirements [38, 39]. The process is based on the use of a diamond tool for cutting surface profiles into a substrate, e.g., a glass or silicon wafer. The fabrication of microoptical components becomes possible because of a number of improvements in the technique:

- The sharpness of the diamond tool is optimized in order to be able to cut features in the micrometre range with the smallest possible surface roughness. Very often, tips made of crystalline diamond are used, which show very high precision along the crystal surfaces.
- The movements of the substrate during the diamond turning process is automatically controlled.

- Vibrations of the system are minimized.
- The temperature and other environmental parameters are controlled for optimum turning precision.

In modern diamond turning machines all of these parameters are optimized. The precisely shaped diamond tool can be used to pattern a variety of different materials. For microoptics manufacturing, metal substrates are used which serve as masters for the replication process. The position and movement of the diamond tool is constantly monitored with laser interferometers. The diamond tools nowadays can be fabricated to high perfection. Microoptics fabrication by diamond turning has been demonstrated with minimum feature sizes of $w_{\min} < 5\ \mu$m and rms surface roughnesses of ≈ 40 Å [40, 41, 42].

There are several problems related to micromechanical fabrication. Due to the mechanical alignment of the diamond tool and the movement of the substrate, the precision of the fabricated microrelief is strongly dependent on the geometry of the surface relief profile. Circularly symmetrical structures are fabricated with the highest precision. More complex pneumatic two-axis guidance of the diamond tool also allows other geometrical shapes such as ellipsoids and hyperboloids with good precision. Like direct writing processes, diamond turning is a time sequential fabrication technique. Batch processing is possible in combination with replication techniques (see next section). In summary, diamond turning is an interesting alternative to the lithographic fabrication of microoptical elements, especially if deep surface-relief profiles are required as, e.g., in the case of microprisms (Chapter 4) or blazed diffractive elements (Chapter 5).

3.4.4 Replication of microrelief structures

After discussing the various processing steps necessary for the lithographic 3D profiling of microoptical elements, we now focus on the replication of these elements. Replication techniques are important for the cheap mass production of elements used in microoptics. Mass fabrication, on the other hand, is a critical issue for a commercial breakthrough of microoptical solutions. Many of the replication techniques which are applied nowadays for microoptical components have evolved from techniques developed for the replication of more coarse structures such as, e.g., analog audio discs. A thorough overview of the different technological approaches used in the replication of microoptical components appeared recently in [43].

The first step towards a high quality replication of microoptical components involves the fabrication of a metal preform, also called stamper or shim (Fig. 3.16). Electroforming is a well established technology for the fabrication of these shims. To this end a conductive base layer, e.g., Au or Ag coating of several nm thickness, is generated on the microoptical element. For best results good adhesion of this base layer is indispensable. The electroforming of the shim is performed in a galvanic bath by deposition of Ni or Cu onto the microoptical component [44]. The Ni or Cu sources form the anode of the galvanic bath. The cations released into the

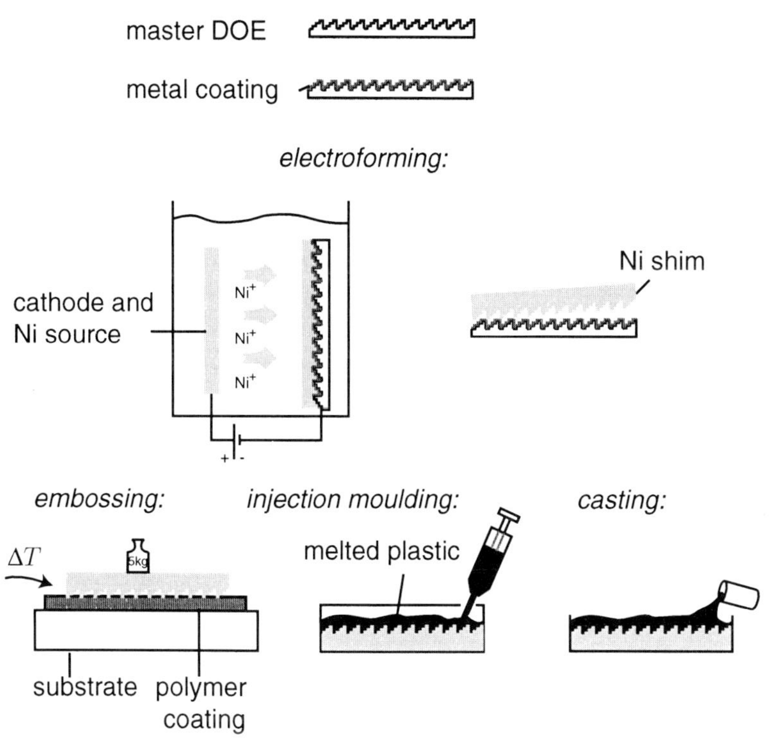

Fig. 3.16: The main processing steps for the replication of microoptical components.

bath perform the current transport and are deposited on the microoptical element, which forms the cathode of the system. With this technique the microoptical pattern is transferred into Ni shims of typically 50–100 μm thickness. These shims are well suited for the subsequent replication. The electroforming procedure has been developed to high standards. The increase in rms surface roughness of the shim compared to the original microoptical component is as low as approx. 1 nm. Subsequent generations of shims can be fabricated with even less additional error.

The replication of the shim can be performed in a variety of different processes. The choice of the right process depends on the desired replication material. Polymer materials are best suited because of the relatively low softening temperature, good transmission in the visible region, as well as easy handling. In the following we will discuss the most important replication technologies (Fig. 3.16):

- During the **embossing** process the Ni preform of the microoptical element is pressed onto a polymer foil. In order to avoid the moulding of surface roughnesses, the foil is placed onto an optically flat surface [45]. While the pressure is sustained, the polymer is heated to temperatures above the softening temperature. This causes the surface profile of the Ni preform to be moulded into the soft polymer substrate. After removal of the pressure and cooling of the polymer, the separation of the preform from the moulded polymer is the most critical part of the process. A Teflon-like coating applied to the shim eases the separation. The achievable lateral resolution of the embossing process depends on the depth of the replicated element. Shallow structures as small as 25 nm have been achieved. Commercial embossing machines such as roller embossers generally achieve submicron resolution with aspect ratios smaller than 1. The replication of components with high aspect ratios with this approach requires precise process control and slower cycle times [46].

- **Injection moulding** is commercially the most advanced replication technique for microoptical elements [47]. It is the standard replication technology for compact discs (CDs). A melted plastic material is injected into a mould containing the Ni preform. The soft material fills all the spaces of the mould and is then cooled down. After removing the mould a high quality replica of the preform is achieved. In other approaches such as **compression moulding** the plastic material is introduced in the solid state, e.g., as powder and melted through heat and pressure applied in the mould. Commercial injection moulding machines currently provide submicron lateral resolution but problems occur for high aspect ratios. The feasibilty for microoptical components with high aspect ratios, however, has been demonstrated in a variety of laboratory experiments.

- **Casting** is the third replication method discussed for microoptical components. In this process the replication polymer material is applied to the Ni preform in the liquid state. After a curing step, e.g., by UV illumination, the shim is released and the replica is available.

- An alternative material for the replication of microoptics is used in the so-called **sol-gel** process. Instead of using a dissolved or melted polymer a special type of soluble glass material is applied. A widely used type of silica base material (so called precursor) is tetraethylorthosilicate ($Si(OC_2H_5)_4$). If added to deionized water the material enters the "sol" phase in a "hydrolysis" reaction:

$$Si(OC_2H_5)_4 + 4H_2O \longrightarrow Si(OH)_4 + 4C_2H_5OH \tag{3.4.2}$$

The sol phase does not fully correspond to a dissolved phase. It rather is a suspension of small particles in a liquid phase. This suspension can be cast in a mould. During the condensation process the solid particles arrange themselves into a continuous 3D network and start to link together randomly.

$$Si(OH)_4 \longrightarrow SiO_2 + 2H_2O \tag{3.4.3}$$

With progressing condensation (called "gelation"), the material becomes a soft solid "gel", which maintains the shape of the mould. The condensation phase is followed by an aging step in which the silicate glass matrix is built up. The aging results in a porous solid glass material. The last processing step is called "densification". By thermal treatment of the stiff gel the material impurities are removed and a dense silicate glass matrix is formed [48].

Apart from the densification process, the whole processing of the sol-gel is performed at room temperature. The major advantage of the sol-gel casting is that it results in a solid glass replica. Sol-gel silicas have been demonstrated to exhibit properties similar to fused silica materials. In particular, the transmission characteristics and low scattering losses compared to polymer materials make them a very interesting material for microoptics replication [49].

Replication of microoptical components is already performed commercially. The main applications are, e.g., reflective holograms for counterfeit protection (e.g., on credit cards). Recently the replication of an integrated microoptical system has been demonstrated by UV casting [50].

3.4.5 Diffusion — ion-exchange processes

So far we have discussed ways to structure the surface of a substrate. An alternative way to fabricate an optical component is to modify its refractive index. This is possible by changing the ion composition of the substrate, e.g., in an ion-exchange process. The ion-exchange technology in glass substrates has been developed for the fabrication of optical waveguides [51, 52]. Meanwhile, it is also used for the fabrication of microlens arrays [53]. In the thermal ion-exchange process a glass substrate, locally protected by a metal mask, is placed, e.g., into a $AgNO_3$ melt [54]. At those locations where the glass is exposed to the melt, a diffusion process takes place. Sodium ions contained in the glass are exchanged by the silver ions from the melt. Depending on the ion composition and the structure of the glass, this ion-exchange process generates index distributions which can serve as good quality microlenses. The properties of the resulting lenses are mainly influenced by the choice of the glass substrates and the exchanging ion pairs. In field-assisted ion-exchange the ion diffusion is supported by an electric field. This accelerates the diffusion significantly and generates steeper phase profiles than a thermal diffusion process. Further details of this fabrication technique will be discussed in the following chapter on refractive microoptics.

3.5 Bonding of planar structures

To conclude this chapter we focus on a group of lithographic techniques which can be considered as mounting technology. In microelectronics as well as microoptics it is necessary to align and combine separate substrates containing different types of components. In electronics this situation occurs, e.g., if components fabricated on different types of substrate wafers need to be linked together. A typical example is the combination of optoelectronic components fabricated in GaAs wafers with CMOS electronics (Chapter 6). In microoptics a similar situation occurs if substrates with different optical or optoelectronic components need to be assembled to make optical systems. For this task hybrid bonding technologies such as "flip-chip bonding" or "thermo-anodic bonding" can be exploited.

3.5.1 Flip-chip bonding

In flip-chip bonding, solder-bump bonding or controlled-collapse chip connection (C4), identical patterns of bond pads are fabricated lithographically on two planar substrates. One of the substrates is flipped and aligned face down with the bond pads precisely on top of the pads on the other substrate. After bringing the substrates in contact, pressure and an optional thermal treatment is applied, to have the bond pads stick together and fix the alignment (Fig. 3.17a). Originally this technique was developed in electronics for the implementation of short distance high bandwidth electronic interconnections [55, 56].

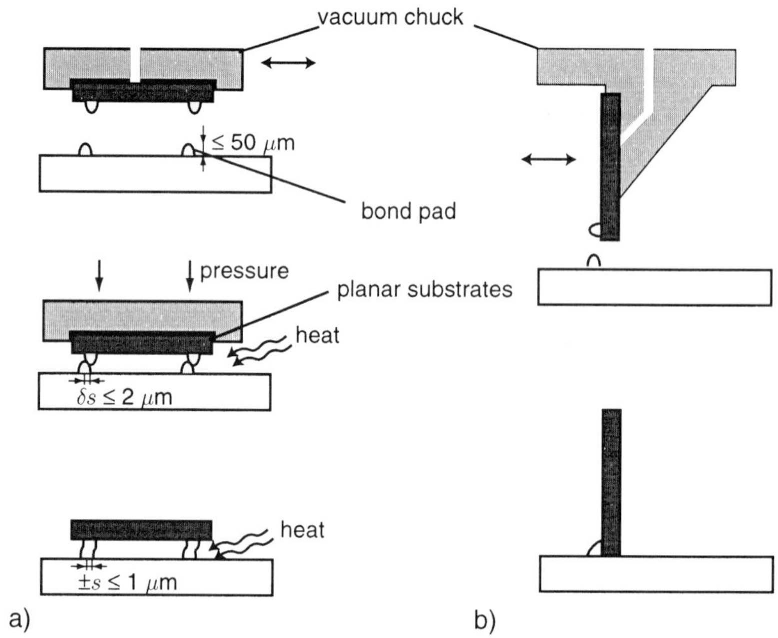

Fig. 3.17: a) Conventional flip-chip bonding and the reflow process; b) vertical flip-chip mounting.

The most critical issue in this process is the choice of suitable solder material and the generation of the solder bumps. The flip-chip interconnections are used for electrical interconnects as well as for the stable assembly. Thus, the solder bump material needs to provide good conductivity as well as physical stability. The thermal stability requirements are especially critical for the flip-chip process. Low melting points are helpful in order to establish the bonds. Reasonable thermal stability is required, on the other hand, during subsequent processing and during running applications. Typically tin/lead (SnPb) solders with different Pb concentrations are used. Alternatively indium (In) is also of interest due to its low melting point and good sticking properties. Recently electrically conducting adhesives have been developed which show interesting properties for flip-chip bonding. Another critical issue during the solder bump generation is the wetting behaviour of the different materials. In order to obtain stable contacts, it is necessary that the bond pads wet the surfaces. Often this has to be achieved using additional contact coatings on the substrates [57].

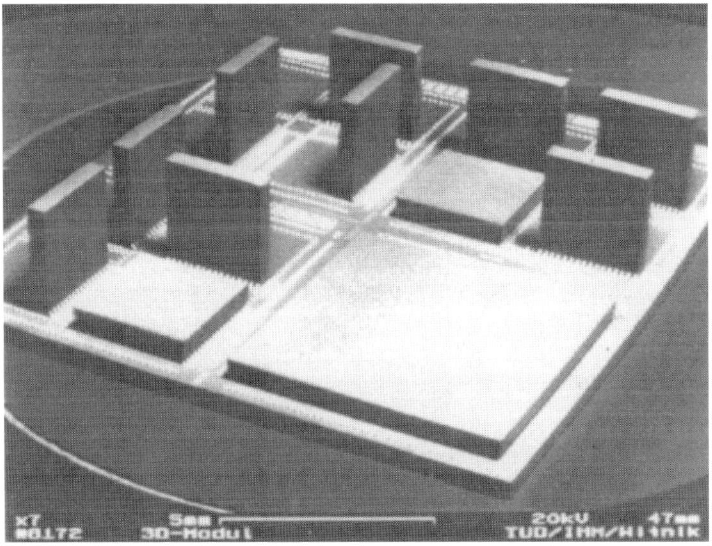

Fig. 3.18: Demonstration of planar and vertical flip-chip bonding as an approach to mounting 3D microsystems (photograph courtesy of Prof. E. Meusel Universität Dresden).

Flip-chip bonding can be applied in free-space microoptics as a hybrid surface mounting technology which provides high lateral alignment precision and mechanical stability. It has been shown that with flip-chip bonding the two planar substrates can be aligned with a precision of 1–2 μm relative to each other. Best results are possible if the so-called "reflow process" is applied. After the first bonding, the chuck which held the top substrate is removed. If the lower substrate is heated so that the bond pads are melting, surface tension will cause an additional realignment of the substrates. Experiments where an electronic chip containing surface emitting lasers has been bonded to a microoptical system demonstrate that flip-chip bonding can meet the requirements of microoptics on alignment precision and stability [58]. An interesting extension of the flip-chip technique to the bonding of vertically mounted chips has been demonstrated recently. This offers the potential for integration of microsystems (Fig. 3.17b). Vertical flip-chip mounting may become an assembling approach which is also interesting for microoptical systems, e.g., for stacked microoptics (Chapter 6) [59].

3.5.2 Thermo-anodic bonding

Thermo-anodic bonding is a planar bonding technology to link glass and silicon wafers to each other [60]. The planar surfaces of the substrates are aligned and pressed together in a vacuum chamber (Fig. 3.19). Thermal heating to temperatures of about 200–500°C and the application of an electric field (200–1000 V) accross the planar contact area initiate an ion-diffusion process which forms a very stable electrochemical bonding. While the alkali ions within the glass are attracted to the cathode, the oxygen ions carrying negative charges are significantly less mobile. At the interface they cause a strong electrostatic field which causes the bonding forces [61].

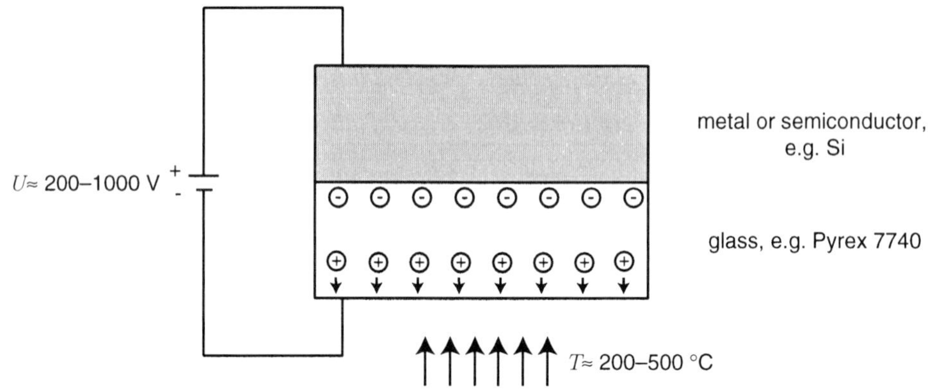

Fig. 3.19: Schematic of the thermo-anodic bonding process.

Thermo-anodic bonding is a surface bonding technique. In contrast to wire-bonding techniques, the whole surface areas of the substrates are bonded. This results in severe requirements on the planarity and cleanness of the surfaces. Very few dust particles can destroy the contact between the substrates which affects the quality of the bonding. On the other hand, the large contact areas lead to very stable bonds.

The material combinations which can be bonded with this technology have to meet additional requirements. It is necessary that the thermal expansion coefficients of the bonded substrates are of the same magnitude. Otherwise strong mechanical stress occurs during the thermal treatment which would destroy the bonds. Thermo-anodic bonding is only applicable to selected wafer combinations. One of the most widely used combinations of materials is with, e.g., Pyrex glass substrates bonded to silicon wafers. For optoelectronic applications thermo-anodic bonding has been discussed frequently for the bonding of optical and electronical substrates. However, a thorough investigation of the influence of the ion diffusion process on the quality of the optical substrate is still to be performed.

With this quick look at possible bonding techniques we want to conclude this chapter on the fabrication technology. In the following chapters we shall focus more specifically on microoptical components, by discussing the fundamentals of refractive and diffractive optical components.

3.6 Glossary

- **anisotropic etching**
  Etching process with etch rates varying in different directions. The origin of the anisotropy can be found in the etching process (i.e., physical vs. chemical etching) or the structure of the substrate material (e.g., crystalline or amorphous).

- **aspect ratio of microstructures**
  The ratio between width and depth of the microstructure. Structures with high aspect ratio require high anisotropy in the etching process.

- **CAIBE**
  acronym for "chemically assisted ion-beam etching"; an approach for lithographic pattern transfer, where the substrate material is depleted by a combination of the physical impact from an ion-beam, generated in a separate reaction chamber (ion-**beam** etching), and the chemical reaction of reactive gases contained in the plasma. The amount of anisotropy in the etching process can be manipulated by a control of the inert plasma ions versus the reactive gases.

- **direct-writing**
  technological approach used in "scanning lithography". Instead of using lithographic masks for exposure of the substrate, laser or e-beams are used as writing "pens" in a scanning mode.

- **lithography**
  planar processing technology for structuring substrate wafers in a series of processing steps.

- **lift-off**
  technique to fabricate a structured coating on a substrate. The coating is applied on top of a structured photoresist layer. Upon removing the photoresist layer in these areas, the coating layer is also removed. The remaining coating layer shows the photographic negative of the original photoresist pattern.

- **minimum feature size**
  smallest lateral extension of individual feature elements, which can be generated in a lithographic process. Generally the minimum feature size also depends on the aspect ratio of the pattern.

- **proximity printing**
  in mask lithography: Exposure geometry where the lithographic mask is aligned at a very small distance from the substrate. The goal is to avoid damage to either substrate or mask, which occur during contact printing. In order to reduce the effect of diffraction, the distance between substrate and mask ("proximity distance") has to be reduced to a few micrometres.

3.7 List of new symbols

A_m:	complex amplitude of the m-th diffraction order; Fourier coefficient
Al:	aluminum
Ar^+:	Argon ions carrying positive charges, Ar-cations
α:	diffraction angle
α, β:	parameters describing the electron scattering in the proximity effect
c:	speed of light in vacuum: $c = 2.998 \cdot 10^8$ ms^{-1}
C:	proportionality constant
CaF_2:	calcium fluoride
eV:	electron volts, unit of energy
E[units]:	particle energy in units
ΔE:	latent heat of vaporization
epsf:	electron point spread function, describing shape and extension of the focus of an e-beam
η:	ratio between forward and backward scattering in the proximity effect
HF:	hydrofluoric acid
HNO_3:	nitric acid
$f(x)$:	complex amplitude of spatial function, e.g., a diffraction grating
Δf:	frequency detuning
g_p:	proximity gap
h:	Planck's constant: $h = 6.626176 \cdot 10^{-34}$ Joule $\times$ second
I:	electric current
k:	Boltzmann's constant: $k = 1.381 \cdot 10^{-23} \frac{\text{Joule}}{\text{Kelvin}}$
K:	material constant
m:	integer valued index usually denoting the diffraction order
M:	molar mass
MgF_2:	magnesium fluoride
N:	integer value, denoting the number of phase levels in a diffractive component
NA:	numerical aperture of the optical system
p:	period of a diffraction grating
$p(T)$:	vapour pressure dependent on the temperature
$t,\ t_r$:	thickness of a coating (e.g., photoresist)
R:	electric resistance
R:	material deposition rate
rpm:	rotations per minute
SiO:	silicon monoxide
T:	temperature
$V_g,\ V_l$:	molar volume of a material in the gas and liquid phase
W:	width of the bar of a grating
$\frac{W}{p}$:	"duty cycle", i.e., the ratio between the width of a grating bar and the period of the grating
$\Delta x, \Delta y, \Delta z$:	lateral shift or lateral extension
$\delta x, \delta y, gz$:	lateral resolution
z_{focus}:	depth of focus of an optical system

3.8 Exercises

1. **Photoresist**
   What is a photoresist?

2. **Resolution limits in lithography**
   What limits the resolution in optical and e-beam lithography?

3. **Lift-off processing**
   What are the most important advantages of the lift-off technique for generating structured coatings on a substrate? What problems do you see for this process?

4. **Reactive ion etching**
   What are the differences between reactive ion eching and chemically assisted ion-beam etching (or milling)? Why are these the most important techniques for lithographic pattern transfer?

References

[1] U. Krackhardt, "Phasenquantisierung und Herstellungsfehler von periodischen computererzeugten dünnen Phasenhologrammen", Ph.D. thesis, University Erlangen-Nürnberg (1993).

[2] G. E. Anner, "Planar processing primer", Van Nostrand Reinhold, New York (1990).

[3] T. H. P. Chang, "Proximity effect in electron-beam lithography", *J. Vac. Sci. Technol.* **12** (1975), 1271–1275.

[4] G. Owen, "Methods for proximity effect correction in electron lithography", *J. Vac. Sci. Technol. B* **8** (1990), 1889.

[5] M. Ekberg, F. Nikolajeff, M. Larsson and S. Hård, "Proximity-compensated blazed transmission grating manufacture with direct-writing, electron-beam lithography", *Appl. Opt.* **33** (1994), 103–107.

[6] W. Däschner, M. Larrson and S. H. Lee, "Fabrication of monolithic diffractive optical elements by the use of e-beam direct write on an analog resist and a single chemically assisted ion-beam-etching step", *Appl. Opt.* **34** (1995), 2534–2539.

[7] E. B. Kley and W. Dorl, "Einsatz der Elektronenstrahllithography zur Herstellung mikrooptischer Bauelelemente", VDI-Berichte (1992).

[8] G. Farin, "Curves and surfaces for computer aided geometric design: a practical guide", Academic Press, San Diego (1997).

[9] L. B. Lesem, P. M. Hirsch and J. Jordan, "The kinoform: a new wavefront reconstruction device", *IBM J. Res. Dev.* **13** (1969), 150–155.

[10] H. Anderson, M. Ekberg, S. Hård, S. Jacobsson, M. Larsson and T. Nilsson, "Single photomask, multilevel kinoform in quartz and photoresist: manufacture and evaluation", *Appl. Opt.* **29** (1990), 4259–4267.

[11] T. J. Suleski and D. C. O'Shea, "Gray-scale masks for diffractive optics fabrication: I. Commercial slide imagers", *Appl. Opt.* **34** (1995), 7507–7516.

[12] V. E. Shrauger, L. L. Erwin, J. L. Ahn and C. Warde, "Computer-generated multiple-phase-level holograms by use of colour-printer techniques", *Appl. Opt.* **33** (1994), 5318–5327.

[13] K. Reimer, U. Hofmann, M. Jürss, W. Pilz, H. J. Quenzer and B. Wagner, "Fabrication of microrelief surfaces using a one-step lithographic process", *Proc. SPIE* **3226**, Microelectronic structures and MEMS for optical processing III, Austin, Tx, USA (1997), 2–10.

[14] D. C. O'Shea and W. S. Rockward, "Gray-scale masks for diffractive optics fabrication: II. Spatially filtered halftone screens", *Appl. Opt.* **34** (1995), 7518–7526.

[15] W. Däschner, P. Long, M. Larsson and S. H. Lee, "Fabrication of diffractive optical elements using a single optical exposure with a gray level mask", *J. Vac. Sci. Technol. B* **13** (1995), 2729–2731.

[16] W. Däschner, P. Long, R. Stein, C. Wu and S. H. Lee, "Cost-effective mass fabrication of multilevel diffractive optical elements by use of a single optical exposure with a gray-scale mask on high-energy beam-sensitive glass", *Appl. Opt.* **36** (1997), 4675–4680.

[17] H.-L. Huber, "Tiefenlithographie und Abformtechnik", in Mikromechanik, A. Heuberger (ed.), Springer Verlag, Berlin (1991), 236–265.

[18] S. Büttgenbach, "Mikromechanik", Teubner Verlag, Stuttgart (1991).

[19] L. Geppert, "A second coming of contact lithography", *IEEE spectrum*, 9 (1998), 18.

[20] K. Suss, "Operators Manual, MA4 and MA6 Mask Aligners" (1995).

[21] B. E. A. Saleh and M. C. Teich, "Fundamentals of Photonics", John Wiley & Sons, New York (1991).

[22] H. Naumann and G. Schröder, "Bauelemente der Optik", Carl Hanser Verlag, Munich (1992).

[23] W. M. Moreau, "Semiconductor lithography, principles, practices and materials", Plenum Press, New York (1988).

[24] K.-H. Brenner, M. Kufner, S. Kufner, J. Moisel, A. Müller, S. Sinzinger and M. Testorf, "Application of three-dimensional micro-optical components formed by lithography, electroforming, and plastic molding", *Appl. Opt.* **32** (1993), 6464–6469.

[25] M. Feldman, "Alignment techniques in optical and x-ray lithography", in VLSI Electronics microstructure science 16, N. G. Einspruch (ed.), Academic Press, Orlando (1987), 229–264.

[26] M. W. Farn, J. S. Kane and W. Delaney, "Dual-sided lithography: a method for evaluating alignment accuracy", *Appl. Opt.* **31** (1992), 7295–7300.

[27] J. Jahns and W. Däschner, "Precise alignment through thick wafers using an optical copying technique", *Opt. Lett.* **17** (1992), 390–392.

[28] M. B. Stern and T. R. Jay, "Dry etching of coherent refractive microlens arrays", *Opt. Eng.* **33** (1994), 3547–3551.

[29] F. v. Alvensleben, M. Gonschior, H. Kappel and P. Heekenjann, "Laser micromachining", *Opt. Phot. News* (1995), 23–27.

[30] H.-C. Petzold, "Laserinduzierte Prozesse", in Mikromechanik, A. Heuberger (ed.), Springer Verlag, Berlin (1991), 265–329.

[31] A. Sugimura, Y. Fukuda and M. Hanabusa, "Selective area deposition of siliconnitride and silicon-oxide by laser chemical vapour deposition and fabrication of microlenses", *J. Appl. Phys.* **62** (1987), 3222–3227.

[32] C. Momma, B. N. Chichkov, S. Nolte, F. v. Alvensleben, A. Tünnermann, H. Welling and B. Wellegehausen, "Short-pulse laser ablation of solid targets", *Opt. Comm.* **129** (1996), 134–142.

[33] S. Mihailov and S. Lazare, "Fabrication of refractive microlens arrays by excimer laser ablation of amorphous Teflon", *Appl. Opt.* **32** (1993), 6211–6218.

[34] R. Braun, R. Nowak, P. Hess, H. Oetzmann and C. Schmidt, "Photoablation of polyimide with IR and UV laser radiation", *Appl. Surf. Sc.* **43** (1989), 352–357.

[35] X. Wang, J. R. Leger and R. H. Rediker, "Rapid fabrication of diffractive optical elements by use of image-based excimer laser ablation", *Appl. Opt.* **36** (1997), 4660–4665.

[36] G. P. Behrmann and M. T. Duignan, "Excimer laser micromachining for rapid prototyping of diffractive optical elements", *Appl. Opt.* **36** (1997), 4666–4674.

[37] R. Nowak, S. Metev, G. Sepold and K. Großkopf, "Excimer laser processing in BK7 and BGG31 glasses", *Glastech. Ber.* **66** (1993), 227–233.

[38] T. T. Saito, "Diamond turning of optics: the past, the present, and the exciting future", *Opt. Eng.* **17** (1978), 570–573.

[39] R. J. Benjamin, "Diamond turning at a large optical manufacturer", *Opt. Eng.* **17** (1978), 574–577.

[40] P. P. Clark and C. Londoño, "Production of kinoforms by single point diamond turning", *Optics News* **12** (1989), 39–40.

[41] G. Blough and M. Morris, "Hybrid Lenses offer high performance at low cost", *Laser Focus World* (1995), 67–74.

[42] G. C. Blough, M. Rossi, S. K. Mack and R. L. Michaels, "Single point diamond turning and replication of visible and near-infrared diffractive optical elements", *Appl. Opt.* **36** (1997), 4648–4654.

[43] M. Gale, "Replication", in Microoptics, H.-P. Herzig (ed.), Taylor & Francis, London (1997), 152–177.

[44] C. Budzinski and H. J. Tiziani, "Radiation resistant diffractive optics generated by micro electroforming", *Laser und Optoelektronik* **27** (1995), 54–61.

[45] J. Jahns, K.-H. Brenner, W. Däschner, C. Doubrava and T. Merklein, "Replication of diffractive microoptical elements using a PMMA moulding technique", *Optik* **89** (1991), 98–100.

[46] S. Y. Chou, P. R. Krauss and P. J. Renstrom, "Imprint of sub-25 nm vias and trenches in polymers", *Appl. Phys. Lett.* **67** (1995), 3114–3116.

[47] I. Rubin, "Injection moulding", John Wiley & Sons, New York (1972).

[48] L. C. Klein (ed.), "SOL-GEL Optics: Processing and Applications", Kluwer Academic Press, Boston (1994).

[49] J.-L. Nogues, "Pure silica diffraction gratings and microlenses made by the sol-gel process", *OSA Techn. Digest* , **9**, Diffractive optics: design, fabrication, and application (1992), 15–17.

[50] M. T. Gale, M. Rossi, L. Stauffer, M. Scheidt and J. R. Rogers, "Integrated micro-optical systems fabricated by replication technology", *OSA, Techn. digest* , **10**, Diffractive optics and micro-optics, Kona Surf, Hawaii (1998), 183–185.

[51] T. Findakly, "Glass waveguides by ion-exchange: a review", *Opt. Eng.* **24** (1985), 244–250.

[52] R. V. Ramaswamy and R. Srivastava, "Ion-exchanged glass waveguides", *IEEE J. Lightwave techn.* **LT-6** (1988), 984–1002.

[53] K. Iga, Y. Kokubun and M. Oikawa, "Fundamentals of microoptics", Academic Press, Tokyo (1984).

[54] J. Bähr, K.-H. Brenner, S. Sinzinger, T. Spick and M. Testorf, "Index-distributed planar microlenses for three-dimensional microoptics fabricated by silver-sodium ion-exchange in BGG35 substrates", *Appl. Opt.* **33** (1994), 5919–5924.

[55] L. S. Goldman and P. A. Totta, "Area array solder connections for VLSI", *Solid State Technol.* (1983), 91–97.

[56] V. C. Marcotte, N. G. Koopman and P. A. Totta, "Review of flip chip bonding", *ASM international,* Microelectronic packaging technology: materials and processes, Philadelphia, PA, USA (1989), 73–81.

[57] M. Pecht, A. Dasgupta, J. W. Evans and J. Y. Evans (eds), "Quality conformance and qualification of microelectronic packages and interconnects", John Wiley & Sons, New York (1994).

[58] J. Jahns, R. A. Morgan, H. N. Nguyen, J. A. Walker, S. J. Walker and Y. M. Wong, "Hybrid integration of surface-emitting microlaser chip and planar optics substrate for interconnection applications", *IEEE Photon. Technol. Lett.* **4** (1992), 1369–1372.

[59] B. Waidhas, M. Boettcher and E. Meusel, "3D packaging technologies for microsystems", *VDE Verlag* , Micro systems technologies 96, Potsdam (1996), 349–356.

[60] G. Wallis and D. I. Pomerantz, "Field assisted glass-metal sealing", *J. Appl. Phys.* **40** (1969), 3946–3849.

[61] W. Menz and P. Bley, "Mikrosystemtechnik für Ingenieure", VCH Verlagsgesellschaft, Weinheim (1993).

4 Refractive microoptics

In conventional macrooptics, the use of refractive elements (lenses, prisms, beam-splitter cubes) dominates, while diffractive elements (gratings) are only used for specific applications (mostly in spectroscopy). In microoptics, diffraction plays a much stronger role. This is mostly due to the large functionality of computer-designed diffractive elements which can be used to fabricate unusual individual elements (for example, beam splitters with high splitting ratios). However, refractive microoptics does play an important role, which may increase as the variety of refractive hardware becomes larger. Until recently, the hardware pool of refractive microoptics consisted essentially of lenslets and lenslet arrays. This is beginning to change as newer fabrication techniques can be applied to make deflecting elements such as microprisms and micromirrors. From a systems point of view, refractive optics offers several features that may be of importance: a significantly reduced wavelength sensitivity compared to diffractive optics (necessary for broadband applications), the possibility of realizing large numerical apertures, and, in general, a high light efficiency.

A large variety of fabrication techniques have been applied to the fabrication of refractive optical elements (ROEs). It is the goal of this chapter to give a comprehensive overview of the different types of elements, their fabrication and the theory. Despite the plethora of techniques used for fabrication, there are only a few underlying principles. Fabrication of refractive microlenses is often based on some (analog) physical process, such as mass transport due to surface tension and diffusion. Based on these two processes, two classes of refractive microlenses can be identified: those with a surface profile and those with a refractive index gradient within the material. Due to the analog nature of most fabrication techniques used for ROEs, their fabrication is often more difficult compared to the fabrication of diffractive optics, since one has to control the process parameters with very high precision in order to achieve good reproducibility and uniformity. Furthermore, process and geometrical parameters are usually confined to certain ranges, thus leading to restrictions in design and geometry. Recently, direct writing techniques in combination with analog grey-scale lithography have been applied to make ROEs. These techniques have allowed one to make refractive microoptical components with non-symmetrical profiles (such as microprisms) and more flexibility in shape and geometry.

Throughout most of this chapter we are going to discuss a variety of techniques for the fabrication of "surface profile microlenses" and "GRIN-microlenses". Unusual fabrication techniques for "microprisms and micromirrors" are described in Section 4.3.

4.1 Surface profile microlenses

Surface profile ROEs are the miniaturized versions of the classical refractive optical components. This group of ROEs is further subdivided according to the fabrication technique. Since the basic functionality is equivalent for all surface profile elements we will discuss their performance using the example of the "melted photoresist lenses". For the other types of surface profile ROEs we will only point out specific differences in the functionality which stem from the fabrication processes.

4.1.1 Melted photoresist lenses — reflow lenses

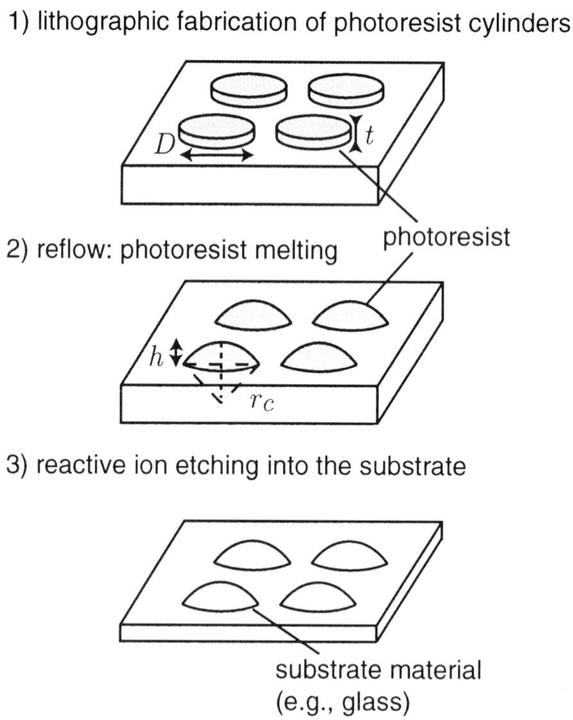

Fig. 4.1: The reflow fabrication process for refractive lenses.

One way to fabricate refractive microlenses is by melting cylindrical "islands" of photoresist or similar polymer materials (Fig. 4.1) [1, 2, 3]. To this end a layer of photoresist is patterned lithographically or interferometrically [4] to form small cylinders on a substrate. This structure is heated to temperatures above the glass temperature of the photoresist ($T \approx 160°$C). At these temperatures mass transport mechanisms can take place. Due to surface tension the shape of the photoresist cylinders changes to minimize the surface energy. In a good approximation the photoresist islands assume spherical shape. The focal length f of the resulting microlenses is determined by the radius of curvature r_c of the resulting spherical profile.

$$f = \frac{r_c}{n-1} \tag{4.1.1}$$

Here, n is the refractive index of the lens material. r_c can be calculated from the volume of the photoresist droplet. Before melting, the photoresist volume is calculated from the volume of the cylinder:

$$V_{\mathrm{cyl}} = \pi \left(\frac{D}{2}\right)^2 \cdot t \tag{4.1.2}$$

t is the thickness of the photoresist layer and D the diameter of the patterned photoresist islands. The volume of the spherical photoresist cap after melting is calculated to be:

$$V_{\mathrm{cap}} = \pi\, h^2\, \frac{(3r_c - h)}{3} \tag{4.1.3}$$

h is the "sag" of the lens (i.e., the height of the droplet). h is related to the radius of curvature and the diameter of the lens by basic geometrical considerations (Fig. 4.1):

$$r_c = \frac{h^2 + \frac{D^2}{4}}{2h} \tag{4.1.4}$$

We assume that the photoresist volume does not change during fabrication. In this case from Eq. 4.1.2–4.1.4 the relation between the thickness of the photoresist cylinders and the resulting lens sag is calculated as [5]:

$$t = \frac{h}{6}\left(3 + 4\,\frac{h^2}{D^2}\right) \tag{4.1.5}$$

Using Eqs 4.1.1 and 4.1.5 it is possible to determine the necessary thickness of the photoresist layer if a specific focal length f is desired. Fig. 4.2 shows the dependence of the focal length f on the diameter D of the cylindrical photoresist islands and the thickness t of the photoresist coating, respectively. From these diagrams we see that the thickness of the photoresist needs to be controlled precisely in order to achieve the desired focal lengths. Especially for lens diameters $D > 10\ \mu$m it is important to achieve large coating thicknesses ($t > 5\ \mu$m), in order to reach an area where the $f(t)$ curves are approximately constant. Only in this case can small

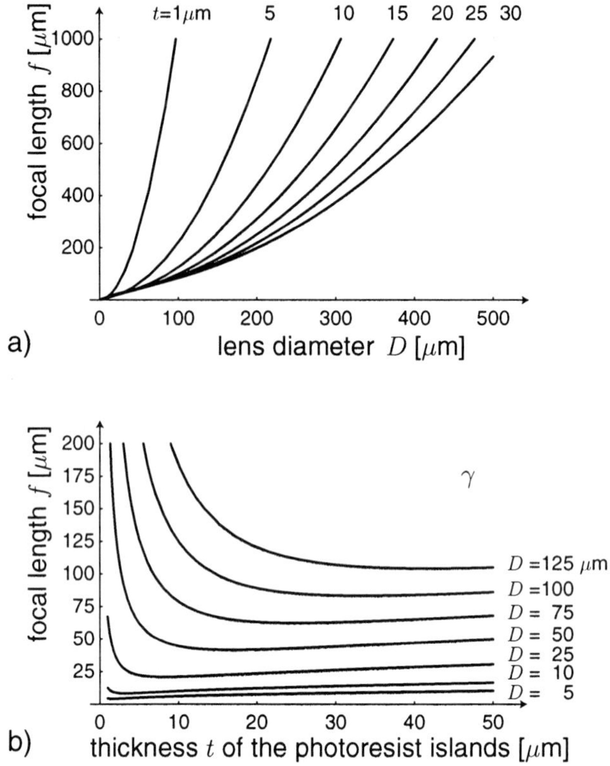

Fig. 4.2: Microlenses fabricated by melting of photoresist. Simulation of: a) the focal length f depending on the diameter D of the photoresist cylinders (thickness t as a parameter); b) the focal length f as a function of the thickness t of the initial coating (diameter D as a parameter)

variations of the coating thickness be tolerated during fabrication.

With the help of the theory presented in Eq. 4.1.1–4.1.5 it is possible to estimate the relationship between the different process parameters. However, it is important to note that it is based on some simplifying assumptions. For a more thorough mathematical description of the process it is necessary to take several additional parameters into account. During the melting process, e.g., the photoresist volume changes due to evaporation of the solvent. Additional influence on the lens shape can be expected from the temperature and the substrate material. The photoresist cylinder changes its shape during melting in order to reduce its energy. Due to the surface tension at the resist-air interface the quasi-spherical shape develops. For a precise calculation, the interface between the resist and the substrate needs to be taken into account, e.g., in an energy balance [6]. The energy at the interface between two materials can be calculated from:

$$E = \tau A \tag{4.1.6}$$

Here A is the area of contact between the two materials and τ is the "surface tension" of the material. Eq. 4.1.6 is valid for all types of interfaces. For the photoresist cylinders on the substrate, the surface energy consists of the energy of the interfaces between photoresist-air (surface tension: τ_{pa}) and photoresist-substrate (τ_{ps}) :

$$E_{\text{cyl}} = \tau_{\text{pa}} \left(\pi \frac{D_{\text{cyl}}^2}{4} + \pi D t \right) + \tau_{\text{ps}} \pi \frac{D_{\text{cyl}}^2}{4} \tag{4.1.7}$$

During melting of the resist, the mass transport takes place in order to find an energetically preferable shape. The surface energy of the photoresist cap formed during this melting process is calculated by means of:

$$E_{\text{cap}} = \tau_{\text{pa}} \pi (h^2 + \frac{D_{\text{cap}}^2}{2}) + \tau_{\text{ps}} \pi \frac{D_{\text{cap}}^2}{4} \tag{4.1.8}$$

Here the possibility of a change of the contact area between the substrate and the photoresist is taken into account by assigning two different diameters D_{cyl} and D_{cap} to the different shapes. From Eqs 4.1.7 and 4.1.8 we can calculate the difference in the surface energies (ΔE_{S}) of the two shapes of the photoresist. For the resulting shape of the microlens, the reduced surface energy is balanced by an increase in the gravitational energy ΔE_{G} of the photoresist cap. The gravitational energy (G_{cyl}) in the photoresist cylinder is:

$$G_{\text{cyl}} = g \varrho \pi \int_0^t \frac{D_{\text{cyl}}^2}{4} z dz = g \varrho \pi \frac{D_{\text{cyl}}^2}{4} \frac{t^2}{2} \tag{4.1.9}$$

Here ϱ denotes the density of the photoresist and g is the acceleration due to gravity. Due to the swelling of the photoresist cap the gravitational energy (G_{cap}) is increased to:

$$\begin{aligned} G_{\text{cap}} &= g \varrho \int_0^h V(z) dz \\ &= g \varrho \int_0^h \pi z \left[\frac{3D^2}{4} + z^2 + 3 \left(r_c^2 - (r_c - (h - z))^2 \right) \right] z dz \\ &= g \varrho \pi (\frac{h^3 r_c}{6} + \frac{h^2 D^2}{16}) \end{aligned} \tag{4.1.10}$$

The energy balance taking into account the surface energies as well as the gravitational energies reads:

$$\Delta E_{\text{S}} + \Delta E_{\text{G}} = (E_{\text{cyl}} - E_{\text{cap}}) - (G_{\text{cyl}} - G_{\text{cap}}) = 0 \tag{4.1.11}$$

This theoretical approach helps us to understand the influence of the physical boundary conditions on the shape of the resulting photoresist microlenses. It is useful when investigating the influence of different substrates or processing temperatures. The shape of the microlenses depends on the ratio between the surface tension coefficients of the polymer-air and polymer-substrate interfaces. The surface energy of the substrate-photoresist interface varies for different materials. In particular, surface roughness, which increases the surface area, has an important influence on the energy balance. The processing temperature can also be expected to be a critical parameter, since it directly influences the surface tension. From the energy balance we find that the lens parameters can be varied if, e.g., the photoresist is melted with the substrate flipped upside down. In this case the graviational energy enters in Eq. 4.1.11 with the opposite sign. Larger lens sags and shorter radii of curvature can be expected.

The reflow process is an interesting approach for the fabrication of large arrays of microlenses. During melting, the ideal spherical shape of the photoresist cap can only be formed for specific ratios between lens diameter and photoresist thickness. Especially if the resist thickness is too low, can significant deviations from the spherical shape occur and hamper the optical performance. Therefore, the generation of high quality photoresist coatings is the most challenging aspect of the fabrication process. In order to achieve good quality lenses with diameters (D = 100–1000 μm) typically used in microoptical systems, thick photoresist coatings (t > 10 μm) are necessary. Special types of photoresist (e.g., Hoechst AZ 4562) with high viscosity have been developed for thick coatings, allowing layer thicknesses of 10–30 μm. Using a sequence of multiple coating steps it is possible to increase the thickness further. With photoresist coatings as thick as 60μ m, lens diameters of up to 2 mm have been achieved [7]. Due to the large volume of the photoresist cylinders additional problems arise during the melting process. The reflow process yields best results if the whole resist volume reaches the glass temperature at approximately the same time. During the melting of large photoresist cylinders it may happen that the outer parts of the cylinder reach the solid phase (when the solvent is completely evaporated) before the center areas reach the melting temperature. This is a fundamental problem which might be controlled by a suitable heating setup.

Surface tension generally leads to relatively short focal lengths for the resulting microlenses (i.e., high numerical apertures). This parameter can be influenced by the use of different substrates or resist materials. Coating of the substrate with different base layers also allows one to influence the surface tension between substrate and photoresist. Preshaping is another way to influence the parameters of the microlenses. To this end, the photoresist cylinder is structured like a multiple phase element (Fig. 4.4). If the resulting discrete phase profile is heated above the glass temperature of the resist the discrete phase steps disappear in the reflow process. With this approach the fabrication of microlenses with longer focal lengths (i.e., smaller numerical apertures) and aspheric profiles is possible. In order to be able to extend the parameter range of the microlenses by preshaping, it is especially important to control the melting temperature and the processing time. For temperatures too high or long processing times the influence of the preshaping is

Fig. 4.3: REM photograph of an array of melted photoresist microlenses (diameter: 100 μm).

lost. Instead of heating the whole preshaped component, it is also possible to heat the surface of the preshape locally. Local heating of the substrate surface can be provided for example through laser or e-beam radiation.

The reflow process is also suitable for the fabrication of nonspherical elements such as cylindrical lenses. To this end, the shape of the photoresist islands before the melting is changed, e.g., into an elliptical shape. Table 4.1 shows a list of microlenses fabricated with this technology.

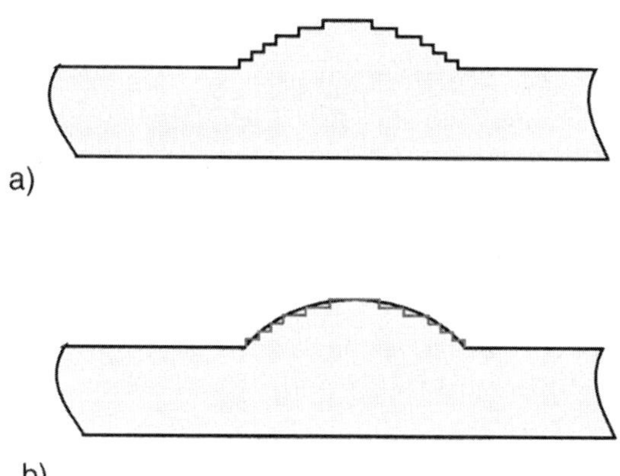

Fig. 4.4: The reflow process with a preshaped profile.

technique	*diameter*	*focal length*	*resist thickness (t)*	*reference*
base layer	30 μm , stop at 15 μm	36 μm	12 μm	Popovic et al. (1988)[2]
no base layer	65 μm	110 μm	5 μm	Daly al. (1990) [5]
	125 μm	180 μm	8 μm	
	280 μm	570 μm	20 μm	
	750 μm	630 μm	40–50 μm	
no base layer	100 μm	100 μm	14.8 μm	Daly at al. (1991) [8]
		200 μm	5.3 μm	
		400 μm	2.6 μm	
	50 μm	50 μm	6 μm	Hutley et al. (1992) [9]
	100 μm	200 μm	8 μm	
	150 μm	250 μm	12 μm	
	200 μm	300 μm	15 μm	
	250 μm	400 μm	20 μm	
RIE in SiO_2	100 μm	280 μm		Mersereau et al. (1992) [10]
	10–400 μm	f/# $\approx$ 1.5–6		Mersereau et al. (1996) [11]
base layer	272 μm	666 μm	11 μm	Haselbeck et al. (1993) [12]
no base layer	200 μm	520 μm	9.1 μm	Jay et al. (1994) [13]
		496 μm	9.6 μm	
		440 μm	10.9 μm	
		400 μm	11.3 μm	
preshaped		900 μm	9.4 μm	
base layer	150 μm	500 μm		Nussbaum et al. (1997) [7]
	250 μm	350 μm		
	2–2000 μm	$<$ 9000 μm	$<$ 60 μm	
RIE in GaAs	100 μm	60 μm		Strzelecka et al.(1997) [14]
	150 μm	125 μm		
	200 μm	225 μm		
	200 μm	117 μm		
	200 μm	151 μm		

Tab. 4.1: Parameters of reflow microlenses reported by various research groups.

Due to the limited stability and the absorption characteristics, photoresist is not an ideal material for microoptical elements. Therefore, efforts have been made to transfer the continuous surface profile of photoresist reflow lenses into more stable substrates. The transfer of the lens pattern into fused silica, silicon, gallium arsenide (GaAs) or indium phosphide (InP) substrates has been demonstrated. Reactive ion etching (RIE) is a suitable technology for this purpose [14, 15, 16]. For transferring of the analog lens profile into the substrate, the differential etching potential of RIE is especially important (Chapter 3). For a one-to-one replication of the photoresist pattern in the substrate, equal etch rates for the photoresist and the substrate material are required. At the same time the etching process needs to be performed with high anisotropy to avoid lateral spreading of the structure. In the RIE process the etch rates are determined through the composition of the reactive etching gas [15]. In modern RIE systems high precision control units allow one to determine the gas composition precisely. The possibility of choosing different etch rates for photoresist and substrate and to vary them during the etching process allows one to optimize the lens quality further through the transfer process [11].

Plasma-enhanced dry etching processes such as ion-beam milling or chemically assisted ion-beam etching (CAIBE) have also been investigated to transfer the analog photoresist profiles into substrates. A possible advantage lies in the free choice of the ion acceleration energy. This additional free parameter allows one to influence the resulting profile. Through an increase of the anisotropic etching parameter, i.e., the physical depletion effect of the ion-beam, the photoresist profile can be etched deeper into the substrate than with the RIE technique. At the same time the selectivity of the etching process to different materials is controlled by the plasma composition.

In the previous discussion of the fabrication process for reflow microlenses we found that the melted photoresist cap has a natural tendency to form quasi spherical shapes. This is caused by the surface tension of the materials involved. For high quality refractive microlenses, surface structures very close to spherical profiles are required. Surface tension is, therefore, the underlying physical principle of many of the analog fabrication processes for microlenses. In the following sections we discuss several other fabrication processes.

4.1.2 Microlens fabrication by mass transport mechanisms in semiconductors

For many applications, for example, in laser or interconnection technology, it is desirable to put microoptical elements directly onto a device made of a semiconductor material. Mass transport is one of the techniques that have been used to fabricate microlenses in semiconductors. This process is similar to the thermal reflow technique for photoresist. In a first processing step, a binary or multilevel preshape is fabricated in the semiconductor material (e.g., InP or GaP). Ion-beam etching is a suitable technology for this process. After preshaping, the substrate is heated to temperatures as high as $T > 1000°\mathrm{C}$. Since the physical and chemical nature of the semiconductor material is different from polymer material, the effect of the heating is quite different. At the processing temperatures the photoresist is melting without decomposing chemically. The melted resist can readily assume the energetically preferred spherical shape. The GaP or InP wafer, however, is chemically decomposed upon heating before reaching the melting temperature ($T = 1460°\mathrm{C}$ for GaP). The phosphorous evaporates, leaving metallic Ga or In behind, which have much lower vapour pressures. If, however, the heating is performed in a phosphorous atmosphere, the phosphorous evaporation is slowed down sufficiently to allow mass transport to take place. The evaporating phosphorous leaves behind Ga or In atoms at the substrate surfaces. Due to the surface pattern, the atoms are not uniformly distributed over the surface. Surface energy effects support the decomposition process so that the semiconductor atom concentration is higher in convex corners of the preshape than in concave corners. The concentration gradient drives diffusion of the Ga or In atoms to the areas of lower concentration, i.e., the concave corners. There, a recombination with phosphorous from the atmosphere takes place, resulting in material growth. Just as with photoresist reflow, this process supports the formation of energetically preferred smooth spherical surface profiles [17, 18]. The mass transport technique has been successfully used for the fabrication of refractive microlenses in InP and GaP substrates. These semiconductor compound materials have large refractive indices ($n \approx 3.2$) and are transparent for wavelengths $\lambda > 920$ nm and $\lambda > 550$ nm, respectively. GaP is an especially interesting substrate, e.g., for the fabrication of microoptical components which can be used in combination with GaAs laser diodes at $\lambda = 850$ nm. Due to the large refractive index, large numerical aperture lenses can be fabricated with moderate lens sags.

In a modified experimental approach the preshaped GaP substrate as well as some elementary phosphorus is sealed in a quartz ampoule before the annealing process. With this "sealed ampoule" technique it is possible to perform the mass transport at higher temperatures and phosphorus vapour pressures. The fabrication of larger numerical aperture lenses is possible with this technique [19]. Recently the sealed ampoule approach has also been investigated for the fabrication of microprisms in GaP. Some examples of microlenses fabricated with this approach can be found in Tab. 4.2.

technique	*diameter*	*focal length*	NA	*reference*
mass-transport	130 μm	200 μm	0.325	Liau et al. (1989) [17]
sealed ampoules	300 μm	210 μm	0.71	Swenson et al. (1995) [19]
mass-transport	140 μm	152 μm	0.46	Tsang at al. (1996) [20]
sealed amploule	200 μm	300 μm	0.33	Ballen et al. (1998) [21]

Tab. 4.2: Parameters of microlenses fabricated by mass-transport in GaP.

4.1.3 Microlenses formed by volume change of a substrate material

The formation of microlenses can also be initiated by a local volume change of the substrate material. This causes local swelling of the substrate. Surface tension is responsible for the swellings, which assume quasi spherical shapes and can act as microlenses. Volume change can be induced in numerous materials by various lithographic exposure and developing processes.

One of the first applications of this principle to microlens fabrication was reported by Borrelli et al. [22]. They used photosensitive glass material (Fotoform^{TM}

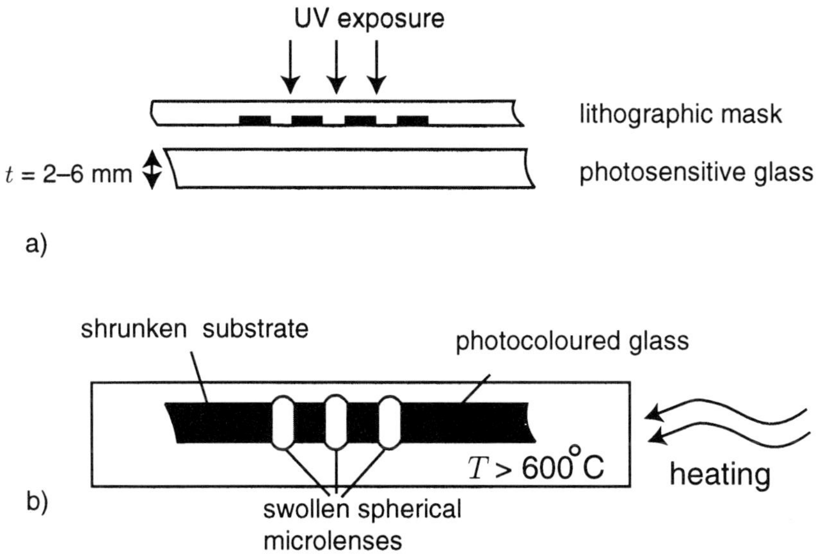

Fig. 4.5: Photothermal fabrication of microlenses. a) Lithographic UV exposure of a photosensitive glass; b) Thermal treatment to induce crystallization and shrinking of the exposed areas. Non-exposed areas swell to form spherical caps.

by Corning, Inc.), which changes colour under intense UV illumination (e.g., with a 1000 W mercury-xenon lamp) and thermal treatment ("photocolouration"). This change of the absorption behaviour is induced by metall colloids (e.g., of Ag, Au), which are formed during the process. In a tempering step after UV exposure the glass substrate is heated to temperatures close to the melting point ($T \geq 600°\text{C}$). The metal colloids act as seeds for regional crystal growth. This is accompanied by an increase of the substrate density in the exposed regions. The stress caused by this volume shrinkage in the exposed glass areas causes compression of the non-exposed areas which are surrounded by exposed glass. The soft glass material escapes at the substrate surfaces and causes a swelling of spherical shape. Again, surface tension is the physical reason responsible for the shape of the glass surface. An analysis similar to the one outlined for the melted photoresist lenses can be made. The volume V_{cap} of the spherical glass caps is determined by the volume (ΔV) by which the unexposed glass cylinder with radius r is reduced. ΔV can be calculated from the densities ϱ_0 and ϱ of the exposed and the unexposed glass substrates.

$$\Delta V = r^2 \pi t \left(1 - \frac{\varrho}{\varrho_0}\right) \tag{4.1.12}$$

If the intensity of the UV exposure is high enough to expose the glass over the whole thickness, spherical microlenses are formed on both sides of the substrate. ΔV is equal to the volume of the two spherical caps with diameter $2r$ and height

h above the glass substrate:

$$\Delta V = 2\pi h \frac{3r^2 + h^2}{6} \tag{4.1.13}$$

Eqs 4.1.12 and 4.1.13 allow one to relate the lens sag h to the change of the material densities. Borrelli et al. used a variational approach to relate the pressure built up in the non-crystallized glass to the radius of curvature of the microlenses. In this way the ratio between the thickness of the glass substrate t and the lens sag h can be calculated by:

$$\frac{h}{t} = \frac{2}{3}\left(1 - \frac{\varrho}{\varrho_0}\right) \tag{4.1.14}$$

Typical lens diameters fabricated with this approach range from 400–800 μm with numerical apertures of about 0.11–0.19. Due to the photocolouration, the UV exposed glass areas automatically serve as stops for the lens apertures. One of the most interesting aspects of the photothermal fabrication of microlenses is the possibility for fabricating thick bi-convex stacked lenses at opposite surfaces of the substrate. Arrays of such lens doublets have been demonstrated for erect imaging, e.g., in copying machines. Bi-convex stacked lenses with diameters of $400--800$ μm have been fabricated with numerical apertures of 0.28 to 0.36 [23].

Photothermal fabrication of microlenses can also be modified by a combination with an ion-exchange technique. To this end, the UV exposed and crystallized glass is exposed to a potassium melt. The lithium ions in the glass partially exchange with the potassium ions from the melt. Due to the crystallization of the UV exposed glass, the mobility of the ions is much lower so that hardly any exchange takes place in these areas. In the other substrate areas a significant portion of the lithium ions in the glass is exchanged by potassium ions from the melt. The larger size of the potassium ions causes a volume swelling of the glass substrate. Surface tension is responsible for the formation of spherical caps. A detailed analysis of a similar ion-exchange process used for the fabrication of gradient index optics will be given later on in this chapter. The swelling induced by ion diffusion into a lithographically structured substrate is a principle which is also applied to the fabrication of microlenses in polymer substrates.

4.1.4 Lithographically initiated volume growth in PMMA for microlens fabrication

PMMA (polymethyl methacrylate) is an organic material which consists of chains of MMA (methyl methacrylate). The length of these chains determines the physical properties of the material, such as stability, solubility etc. Since the chemical bonds within the chains and between the various polymer chains are highly sensitive to various sorts of radiation, PMMA is an interesting material for lithographic processing. Without giving a detailed discussion of the chemical structure we want to mention the most significant effects of radiation on the structure of the polymer.

High energy radiation, such as UV or synchrotron radiation, x-ray, electron or proton beams, incident on a polymer substrate, causes polymer chain scission. This results in a reduction of the molecular weight of the polymer material in the exposed areas. Broken chemical bonds can also be the origin for new crosslinks to neighbouring chains which strengthen the overall polymer matrix. This corresponds to an increase in the molecular weight of the polymer matrix. The probabilities for these two chemical reactions determine the effect of the radiation on the polymer material. Many physical properties (e.g., solubility and stability) of polymer materials are directly related to the molecular weight of the material. In PMMA materials the overall molecular weight of the polymer is reduced by the exposure. The stability of the illuminated polymer areas is reduced. This is utilized, for example, in microlithography, where PMMA-based polymers are used as positive resists. The areas exposed to UV light can be removed easily in a suitable developer. The mask pattern is transferred into a PMMA structure. This process is generally used for the fabrication of binary structures which are necessary, e.g., in diffractive optics (Chapter 5) [24].

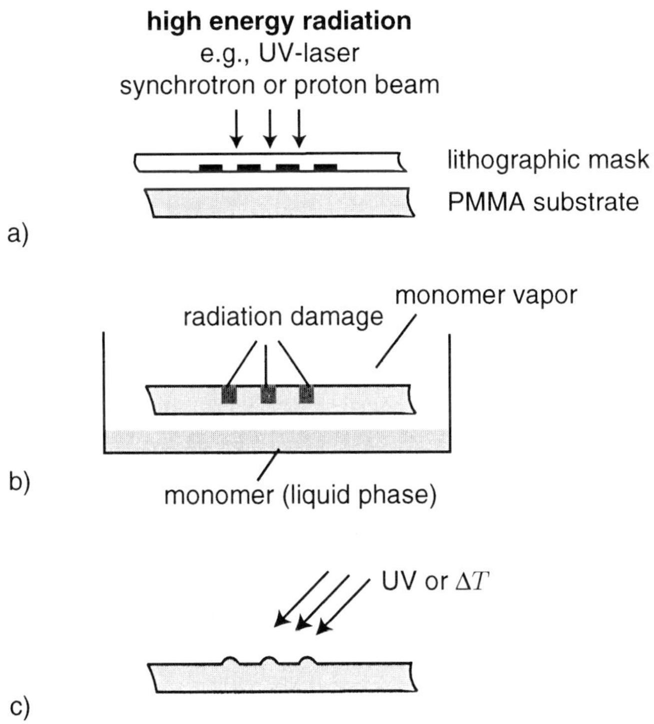

Fig. 4.6: Microlens fabrication by volume swelling in PMMA: a) illumination; b) diffusion; c) curing.

The reduced physical stability in the exposed areas of the PMMA substrate can be exploited for microlens fabrication [25, 26]. If exposed to a monomer vapour the PMMA material swells due to monomer molecules diffusing into the substrate. The amount of swelling directly depends on the stability of the PMMA substrate. Low molecular weight corresponds to few bonds between the polymer chains confining the material. These confinement forces balance the osmotic pressure which drives the diffusion. Consequently, in the illuminated areas of the PMMA substrate, the volume increase due to the diffusion process is significantly higher than in the non-exposed areas with high molecular weight. The swelling of the PMMA substrate in the monomer (e.g., styrene) vapour can be directly calculated from the local molecular weight M of the PMMA [27]:

$$\frac{V_0 + \Delta V}{V_0} \propto M^{\frac{3}{5}} \tag{4.1.15}$$

The volume growth leads to the formation of quasi-spherical microlenses at the substrate surfaces. Surface tension is responsible for the shape of the swollen volume. After monomer diffusion, the swollen microlens surface is cured in order to prevent the evaporation of the monomer molecules outside the reaction chamber. The PMMA structure is activated chemically to include the monomer molecules into the matrix. This process is called polymerization and is initiated either thermally or through flash exposure with a UV light source. Suitable chemical initiators are added to the monomer vapour. It has been shown that with this technique good long term stability of the microlenses can be achieved.

Microlens formation using this technique can be initiated by exposure to any radiation, provided the energy is large enough to cause breaks in the polymer chains of the PMMA substrate. Proton beams and UV laser radiation have been demonstrated to yield high quality microlenses [26, 28]. The main difference between these types of radiation is found in the interaction mechanisms with the PMMA substrate. For UV laser radiation the energy deposited in the PMMA material has its maximum at the front surface and experiences an exponential decay with increasing penetration depth. The molecular weight of the PMMA has its minimum at the front surface of the substrate. This is different for interaction with a particle beam such as proton radiation. In this case the energy deposition remains approximately constant over the penetration depth. Most of the energy, however, is deposited in the area very close to the maximum penetration depth. No chain breaks occur beyond this maximum penetration depth. For a thorough mathematical analysis of the relationship between the lens parameters and the energy of the illuminating radiation, these different deposition mechanisms need to be considered [27].

The parameters of lenses fabricated in PMMA by monomer diffusion after local exposure with high energy radiation are listed in Table 4.3.

Technology	diameter	focal length	reference
Proton beam	300 μm	255 μm	Kufner et al. [26]
	300 μm	438 μm	
	500 μm	620 μm	
	50–1000 μm	NA = 0.2–1	
UV excimer laser	200 μm	1000 μm	Lazare et al. [28]

Tab. 4.3: Parameters of refractive microlenses fabricated by monomer diffusion in PMMA after illumination with different sources of radiation.

4.1.5 Dispensed or droplet microlenses

The droplet method is yet another way to exploit surface tension for the fabrication of microlenses. A small amount of a transparent liquid material (e.g., photoresist, UV curing adhesive) is released onto a suitable substrate (e.g., fused silica glass). Due to surface tension, droplets of spherical shape are formed by the liquid. After a curing step the remaining spherical caps can serve as microlenses [29]. The important parameters in this technology are the surface tension coefficients of substrate and liquid as well as the amount of deposited liquid. The former can be controlled through temperature, base layers and the viscosity of the photoresist. The latter requires precise control of the deposition process.

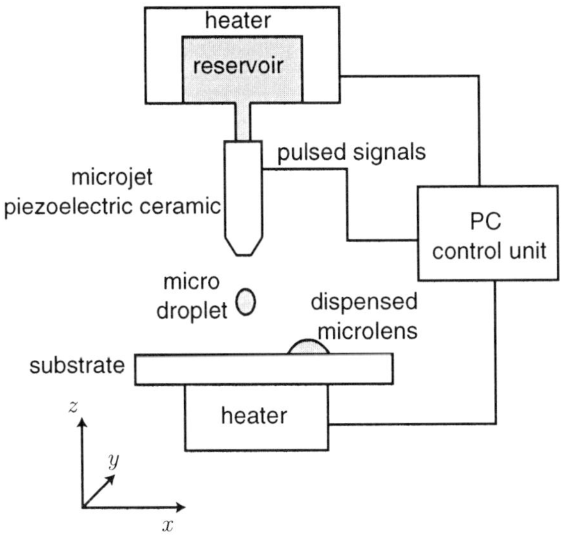

Fig. 4.7: Automated fabrication of droplet lenses.

Ink jet technology which was originally developed for printers has been applied to this fabrication concept [30]. The microjet system consists of a piezoelectric ceramic with a microchannel (Fig. 4.7). This microchannel connects a reservoir of the polymer liquid with a nozzle directing the liquid towards the substrate. Pulsed signals cause contractions of the piezoelectric ceramic so that polymer droplets are emitted through the nozzle. These droplets hit the substrate to form spherical caps acting as plano-convex microlenses. The substrate is fixed to a translation stage to control the position of the droplet. Temperature control of the reservoir and substrate allows one to control the viscosity of the dispensed liquid and its cooling on the substrate. Since these parameters directly influence the surface tension coefficients, they yield control of the resulting lens shape. The liquid volume dispensed for each microlens is controlled by the number of droplets applied per microlens [31]. Depending on the polymer material, curing of the microlenses can be performed by heating or by UV illumination. With this microjet technology the fabrication of polymer microlenses with diameters D in the range 70 to 150 μm and focal lengths f = 50–150 μm has been reported [30]. The uniformity of the microlens parameters is within about 12%. A large variety of materials (dispensed polymers as well as substrates) can be used in this fabrication process. The parameter range for the dispensed microlenses is, therefore, likely to be extended significantly. An extension of the multidroplet technique to arbitrary non-spherical shapes is possible with online control of the curing process [32].

In a slightly different technological approach, various types of liquid UV curing optical adhesives (e.g., Norland NOA 63, 65, 68) with different viscosities are dispensed via a pressurized syringe. The parameters of the lenses fabricated this way cover diameters of D = 100–3000 μm with f-numbers $f/\#$ = 0.9–30 [33].

An extension of the droplet technique to the fabrication of microdoublet lenses is possible by using preshaped substrates [29]. The liquid droplets are dispensed onto isotropically etched microgrooves. The liquid fills the groove and at the same time forms spherically shaped caps on the surface. This structure corresponds to a microdoublet lens consisting of a concave lenslet and a biconvex microlens formed by the droplet. With a suitable choice of materials this allows one to fabricate achromatic microdoublets [6].

The dispension of a liquid polymer offers an interesting fabrication technology for cheap and large-scale mass production of microlens arrays. In order to improve the stability of the lenses, the transfer into different substrate materials through RIE techniques can be considered.

4.1.6 Direct writing techniques for refractive microoptics

Due to the small dimensions, classical fabrication processes, e.g., grinding and polishing are not suitable for the fabricaton of microlenses. Nevertheless, lithographic profiling techniques have been developed to perform a similar task on a microscopic scale. Two different approaches can be taken to perform the profiling. On the one hand, physical, chemical or mechanical techniques can be used to ablate a planar layer of a suitable material. On the other hand, a surface profile can be generated through structured material deposition onto the substrate.

The fabrication techniques for lithographic masks, such as laser or e-beam writing, are used for profiling in a direct writing mode. The challenge of writing refractive optical elements lies in the continuous surface profile of these elements. The process parameters, such as exposure and development of the photoresist layer, need to be controlled with high precision. An additional requirement for the fabrication of refractive microlenses is related to the maximum profiling depth. As we have discussed previously, the height of a spherical photoresist cap (i.e., the lens sag h) determines the focal length f (see Eqs 4.1.1 and 4.1.4). From Fig. 4.8 we see that for reasonable lens parameters structuring depths $h > 5$ μm are necessary. For the simulation in Fig. 4.8 we assumed a refractive index of the lens material of $n \approx 1.6$.

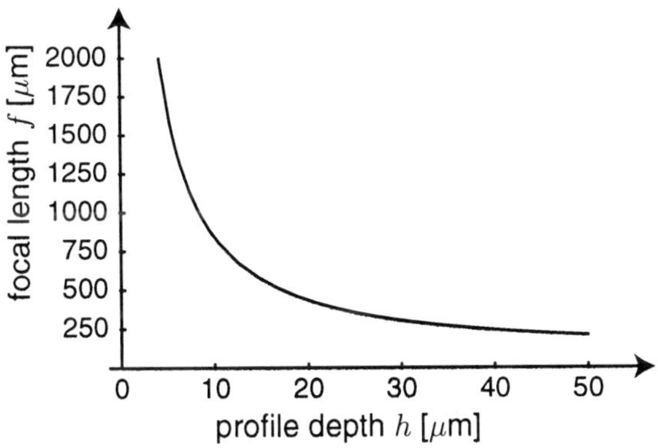

Fig. 4.8: Simulation of the focal length f of a microlens (diameter 200 μm) with spherical profile, depending on the profiling depth of the fabrication process.

The setup for direct writing by laser or e-beam is equivalent to the laser or e-beam lithographic scanners described in the previous chapter (Fig. 3.3). Special care has to be taken to the positioning requirements for continuous profile writing. Typically, a lateral alignment precision of $\delta x < 100$ nm is necessary for high quality elements. For mask writing the laser beam is often modulated in a binary way, i.e., switched between the "on" and "off" state. For direct writing of refractive components a quasi continuous intensity modulation of the laser beam is necessary. This is generally performed using an acousto-optic modulator. Such a device typically allows 256 different intensity values.

One of the most critical issues in direct writing techniques is related to the photoresist processing. Depending on the type and concentration of the developer many of the commercially available photoresists can be processed to yield an approximately linear characteristic. We call this an analog response curve of the photoresist ("analog lithography"). After a precise characterisation, the relation between exposure intensity and etching depth is used to calculate the laser intensity distribution for a necessary phase profile. Exposure is performed in the scanning mode. The sensitivity of the analog processing technique can be seen, for example, from the fact that even a dependence of the achieved etching depth on the delay time between exposure and developing needs to be taken into account. Nevertheless, through precise control of the processing parameters a phase resolution of $\delta\varphi \approx 10$ nm can be achieved [34].

The e-beam scanners for lithographic mask generation can be modified in an analogous way to allow the writing of continuous profiles. Again, the photoresist processing has to be adjusted for non-binary structures. The locally variable exposure of the photoresist can be regulated through the electron current in the beam ("variable dose writing"). Recently "variable energy writing" has been investigated as an alternative writing scheme. The penetration depth of the electrons is directly related to the acceleration energy. It is, therefore, possible to control the structuring depth of the process by a variation of the acceleration energy. Typical energies range from 0.5 to 20 keV. In this case, the profiling structure is mostly independent of the photoresist developing process. Multiple scanning techniques with a constant dose are also feasible due to the high repositioning precison of the e-beam scanning systems. The increase of the writing time, however, is a significant problem for this approach to dose regulation. Alternatively a variable dose deposition can be achieved through variation of the scanning speed. Slow scanning causes longer exposure times per pixel.

One of the major problems of direct e-beam scanning stems from the proximity effect. As described previously, due to scattering of the electrons within the resist layer the minimum feature size $\delta w_{\min}$ of the e-beam writing system is significantly larger than the minimum focus extension $\delta w_{\text{focus}} > 2$ nm of the electron beam. In particular in thick photoresist layers which are necessary for refractive microlenses this proximity effect causes an enlargement of the minimum feature sizes to $\delta w_{\min} > 1$ μm. For continuous relief elements this is a rather poor lateral resolution.

In summary, the direct laser or e-beam writing techniques for the fabrication of purely refractive microlenses are currently limited to low numerical apertures (see Table 4.4). They can provide an interesting complement to the reflow process which is restricted to large numerical apertures. The main application of this technological approach at the current stage of development, however, is the generation of blazed diffractive optical elements. Especially for the e-beam writing technique further progress in the control of the proximity effect, e.g., with the "variable energy writing mode", can be expected in the near future.

Technology	diameter	focal length	reference
direct laser writing	500 μm	22500 μm	Gale et al. [35]
direct e-beam writing	352 μm	16100 μm	Kaschlik et al. [36]
laser ablation	15 μm 500 μm	$\approx$ 20 μm $\approx$ 7000 μm	Zimmer et al. [37]
optical grey-scale lithography	50 μm	$\approx$ 140 μm	Däschner et al. [38]
+ CAIBE	50 μm	$\approx$ 55 μm	Däschner et al. [38]

Tab. 4.4: Typical parameters of refractive microlenses fabricated by analog lithography using direct-writing techniques or grey-scale masks, or by laser machining.

Various other ablative techniques, such as ion-beam milling, laser ablation [37] etc. are being investigated for the fabrication of continuous relief profiles. None of these techniques, however, seems capable of generating continuous profiles with sufficiently large phase depths. Mechanical micromachining is an alternative to the lithographic methods discussed so far. For example, with the diamond turning technique, deep continuous structures are feasible. The shape in this case is limited by the shape of the turning tool. Thus, the technological challenge lies in the fabrication of this tool (Chapter 3).

4.1.7 Grey-scale lithography for ROE fabrication

Another approach for the fabrication of continuous relief microoptical elements through analog lithography is possible using grey-scale masks (Chapter 3). The exposure through the grey-scale mask results in a continuous intensity distribution in the photoresist coating. An optimized development process generates a surface profile in the photoresist. If the pattern in the mask can be encoded with sufficient dynamic range it is possible to fabricate refractive elements. The dynamic range of the grey-scale masks determines the achievable profiling depth. For grey-scale masks fabricated in HEBS glass the fabrication of refractive microlens arrays with a lens sag of 4 μm and a diameter of 50 μm has been demonstrated. During the subsequent etching transfer of the photoresist structure into the substrate using CAIBE, the sag could be increased to 24 μm allowing high numerical apertures of NA$\approx$ 0.5 (Table 4.4). For increasing lens diameters the achievable numerical aperture decreases since the maximum lens sag remains constant. This example shows that grey-scale masks written by laser lithography in combination with reduction steppers can be used for ROE fabrication. Due to the limited dynamic range, however, a more interesting application for grey-scale lithography is the fabrication of blazed diffractive optical components (Chapter 5).

4.2 Gradient-index (GRIN) optics

Gradient-index (GRIN) elements are a different type of refractive optical elements. Rather than being refracted at a surface profile, in GRIN components the light is deflected by a refractive index gradient within the substrate. In this section we will focus on such index gradients and their application in microoptics. In the introductory part we discuss light propagation in gradient-index media. Then we focus on two ways to fabricate GRIN microlenses. The possibility of using short pieces of optical GRIN fibers for imaging applications is an immediate result of the analysis of light propagation in GRIN fibers. This type of miniaturized lens is called GRIN rod or SELFOC™ lens. A further section is devoted to "planar GRIN microlenses", well known under the trademark "PML™". They are better suited for free-space microoptics, since they can be fabricated by means of lithography.

In order to illustrate the light propagation in a gradient-index material we can imagine that it consists of a large number of transitions between materials with slightly different refractive indices. At each of these boundaries the light beam is refracted and gradually changes direction (Fig. 4.9). The light propagation within a GRIN media is calculated by simply applying Snell's law of refraction. The propagation angle $\theta(x)$ of the ray changes at each propagating step to $\theta(x + \Delta x)$. This deflection is described by the law of refraction:

$$n(x)\cos[\theta(x)] = n(x + \Delta x)\cos[\theta(x + \Delta x)] \tag{4.2.1}$$

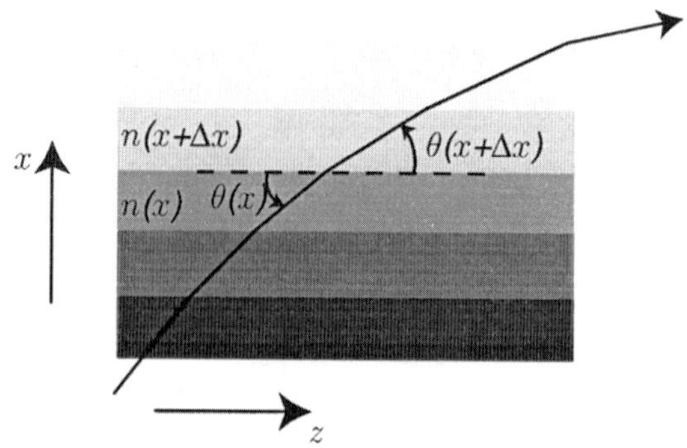

Fig. 4.9: Light propagation in a GRIN material approximated by a large number of layers with constant refractive index.

Expanding the functions $n(x + \Delta x)$ and $\cos[\theta(x + \Delta x)]$ we obtain:

$$n(x)\cos[\theta(x)] = \left[n(x) + \frac{dn}{dx}\Delta x\right]\left[\cos[\theta(x)] - \frac{d\theta}{dx}\Delta x \sin[\theta(x)]\right] \tag{4.2.2}$$

Using the paraxial approximation, we can approximate:

$$\theta(x) \approx \frac{dx}{dz} \quad \text{and} \quad \tan(\theta) \approx \theta \tag{4.2.3}$$

In the limit of very small propagation steps ($\Delta x \rightarrow 0$) Eq. 4.2.2 is reduced to the paraxial ray equation in GRIN media [39]:

$$\frac{dn(x)}{dx} = n\tan(\theta)\frac{d\theta}{dx} = n\frac{d^2x}{dz^2} \tag{4.2.4}$$

This equation can also be derived using Fermat's principle and the variational calculus [39].

From Eq. 4.2.4 we see immediately that the shape of the refractive index profile $n(x)$ directly influences the optical properties of the GRIN component. GRIN profiles are especially interesting for the fabrication and optimization of microlenses.

4.2.1 GRIN rod lenses

In this section we are going to discuss GRIN rod lenses. According to what was said in Chapter 1, they belong to the class of miniature optics rather than microoptics. Nonetheless, it is justified to include them in our discussion. One reason is their importance for many applications, for example, copiers or fax machines. Second, the analysis of GRIN rods helps one to understand the principles involved in lithographically fabricated GRIN-type microlenses.

The optical properties of index gradients have been exploited for many years in gradient-index fibers. In GRIN fibers the refractive index is high at the center and gradually decreases towards the cladding. Consequently, an axial mode, propagating at small angles relative to the optical axis, most of the time experiences high refractive indices. Higher order modes propagate at larger angles relative to the axis. Throughout significant portions of their path through the fiber, they propagate closer to the fiber cladding and experience lower refractive indices. The refractive index profile generates differences in the average group velocities for the different modes. The differences in propagation speed compensate for phase differences due to different propagation distances. With suitable index profiles the phase differences can be completely eliminated. GRIN fibers are ideally free from modal dispersion. The index distribution in GRIN fibers is described by [40]:

$$n^2(x) = n_0^2(1 - \alpha^2 x^2) \tag{4.2.5}$$

with α and n_0 being parameters of the distribution. Usually α is very small ($\alpha^2 x^2 \ll 1$) so that $n(x)$ can be approximated by the parabola:

$$n(x) \approx n_0(1 - \frac{1}{2}\alpha^2 x^2) \tag{4.2.6}$$

At the same time, n_0 is sufficiently large so that the fractional change of the index is very small. Introducing this index distribution into Eq. 4.2.4 we obtain the following differential equation for the light propagation in GRIN media:

$$\frac{d^2x}{dz^2} = -\alpha^2 x \tag{4.2.7}$$

The solutions of this propagation equation are the harmonic functions described by:

$$x(z) = x_0 \cos(\alpha z) + \frac{\theta_0}{\alpha} \sin(\alpha z) \tag{4.2.8}$$

$x(z)$ describes the oscillating path of the light around the axis (Fig. 4.10). All the harmonic solutions in Eq. 4.2.8 have the same period. At distances $z_L = 2\pi/\alpha$ all rays return to the same axial position. These locations can be interpreted as focal points. This describes the compensation of the modal dispersion in GRIN fibers. The compensation of the delays for the various propagating modes corresponds to the performance of a lens in free-space optics. Consequently, pieces of gradient-index optical fibers can be used as lenses. These miniaturized lenses are called GRIN rod lenses or SELFOC^{TM} lenses.

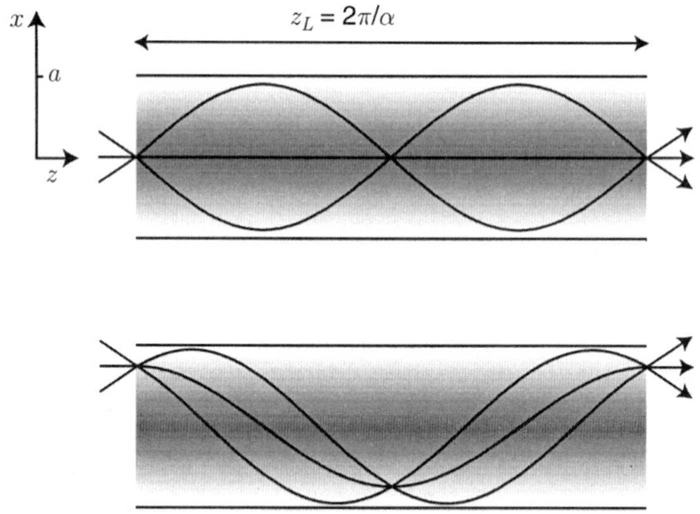

Fig. 4.10: Ray paths in a GRIN rod lens for an on-axis (top) and an off-axis point (bottom).

Fig. 4.10 shows the ray paths through a GRIN rod lens for an on-axis and an off-axis point source. After a propagation distance $z = z_L/2$ an inverted image is formed which is inverted again in the next $z_L/2$ propagation step to form an upright image. At intermediate positions (e.g., $z = 1/4z_L$ and $z = 3/4z_L$) all rays are propagating parallel to each other, corresponding to a collimated beam. The light propagation in the fiber is strongly dependent on the refractive index profile. The focal length of a GRIN rod lens additionally depends on the length z_{R} of the fiber rod [41]:

$$f_{\mathrm{rod}} = \frac{1}{n_0 \alpha \sin(\alpha z_{\mathrm{R}})} \tag{4.2.9}$$

Typically the rod length is chosen as $z_{\mathrm{R}} = z_L/4$ so that:

$$f_{\mathrm{rod}} = \frac{1}{n_0 \alpha} \tag{4.2.10}$$

The numerical aperture NA of the GRIN rod lens can be calculated analogous to the NA of the GRIN fiber as:

$$\mathrm{NA} = 2\pi n_0 / z_L = n_0 a \alpha \tag{4.2.11}$$

Here a is the radius of the lens (Fig. 4.10).

The key to the optimization of the lens performance lies in the index distribution inside the GRIN material. Wave-optical as well as ray-optical approaches have been used to discuss the effect of various index distributions on the imaging performance of the lenses [41]. In the paraxial approximation the ideal index distribution is described by Eq. 4.2.6. For imaging with GRIN rod lenses with this profile no coma and astigmatism occurs. However, field curvature is responsible for a longitudinal shift of the ideal image point for off-axis points. Another index distribution provides optimum imaging for non-paraxial meridional rays (i.e., rays lying in one plane with the optical axis). This optimized distribution is described by the hyperbolic secant ($\mathrm{sech}(x)$) (Eq. 4.2.12):

$$n^2(x) = n_0^2 \mathrm{sech}^2(\alpha x) = n_0^2 \left[1 - (\alpha r)^2 + \frac{2}{3}(\alpha r)^4 - \frac{17}{45}(\alpha r)^6 + \dots\right] \tag{4.2.12}$$

Slightly different index distributions which differ in higher order terms from this profile have been discussed for skew ray imaging.

In general, the aberration characteristics of GRIN fibers are superior to conventional homogeneous refractive lenses. Tomlinson, e.g., showed that a GRIN rod lens with parabolic index distribution theoretically generates a significantly smaller on-axis focus spot than a pair of homogeneous refractive singlets [41]. This can be understood by considering the fact that GRIN rods are thick lenses. The light is interacting with the lens material over a long distance. According to our description of light propagation inside a GRIN material, refraction takes place at a large number of material interfaces. In the case of conventional lenses the interaction with the lens media only takes place at the surface of the lens. Experience shows that the design freedom for the optimization of the image quality increases with the number of refracting surfaces, e.g., in a multielement objective lens.

The major drawback for GRIN rod lenses lies in the difficult fabrication. GRIN fibers can be fabricated with a variety of different processes, e.g., ion-exchange techniques [42]. Standard techniques generally yield approximations of the parabolic or the hyperbolic secant distributions. However, the precise fabrication of prespecified profiles is a complex technological problem. Variations in the index profile caused by the fabrication process are the reason why commercially available SELFOC™ lenses exhibit stronger aberrations than expected theoretically [43]. The fabrication of arrays of GRIN rods is not possible in batch processing. 1D arrays can be fabricated by subsequent alignment of single elements. 2D arrays are even more difficult and expensive to fabricate.

Because of their good imaging properties GRIN rods have many applications especially for fiber optical devices in optical communication systems. For these applications the spot size needs to be optimized in oder to reduce coupling losses [44]. Another commercially successful application is found in copying or fax machines or optical scanners. There, 1D arrays of GRIN rods are used for the formation of images of widely extended objects.

4.2.2 Planar GRIN lenses

In order to avoid problems related to the fabrication of 2D arrays of GRIN rod lenses, Oikawa et al. fabricated radially symmetrical GRIN profiles through ion-exchange in planar substrates [45, 46]. These radial index distributions can be used as microlenses and are called planar (GRIN) microlenses or PML™. Due to the planar fabrication process, 2D arrays of microlenses can be fabricated without additional technological effort.

The fabrication process

The fabrication of planar (GRIN) microlenses is possible by an ion-exchange process. To this end, the substrate (generally a glass substrate with well controlled properties and ion composition) is coated with a mask providing local protection from the subsequent ion-exchange process. After etching the mask structures (e.g., small circular openings are used for microlens fabrication), the substrate is immersed in a melt containing ions suitable to replace some of the ions from the substrate (Fig. 4.11a). The ion-exchange takes place at temperatures of several hundred degrees in a diffusion-like process. At those temperatures the ion mobility is large enough so that some of the ions (generally the alkali ions, such as Na^+) from the glass are replaced by ions from the melt (e.g., Ag^+ or Tl^+). Due to different polarisation characteristics of the ions, this exchange influences the dielectric properties of the glass matrix and effectively changes the refractive index of the

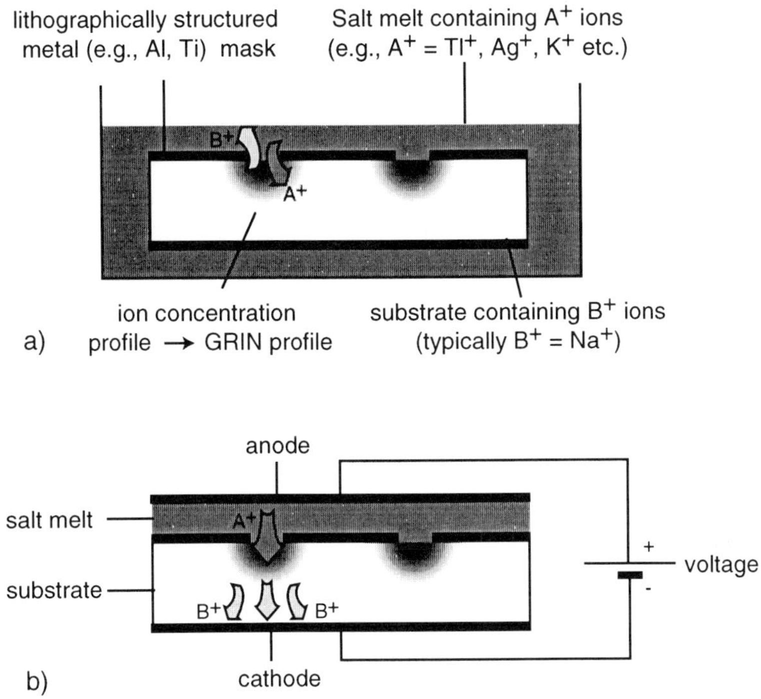

Fig. 4.11: Fabrication of planar GRIN microlenses with a) the thermal ion-exchange process and b) the electromigration (electric field-assisted ion-exchange) process (b).

substrate. For a given material and ion combination the amount of index change is proportional to the percentage of exchanged ions. Since the ion-exchange is an isotropic process, it does not stop at the edges of the mask. The ratio of exchanged ions will rather drop gradually from a maximum at the surface of the substrate areas with no mask layer to zero at the maximum penetration depth of the ions inside the substrate (typically of the order of 100 μm) or at similar distances in a lateral direction under the mask. A radially symmetrical ion concentration profile is generated which causes a refractive index distribution of corresponding shape. The refractive index profile is characteristic of the substrate and the exchanging ion pairs (see following section). If the exchange process is sufficiently nonlinear, the radial index distribution acts as a microlens. Some modifications to the thermal ion-exchange process described have been used in order to optimize the resulting index profiles. In the field-assisted ion-exchange, also called electromigration, the ion mobility is increased by an electric field across the interface between the substrate and the salt melt (Fig. 4.11b). Due to this electric field the movement of the ions from the melt into the substrate is accelerated and the penetration depth of the ions is increased significantly.

The ion pairs used in the exchange process and the type of substrate glass are the most important parameters in the ion-exchange process. A variety of different ion pairs have been investigated for the fabrication of planar GRIN microlenses. The most promising results were achieved with sodium-thallium and sodium-silver ion-exchange. Both ion combinations yield comparable maximum index changes of about $\Delta n = 0.1$ at the substrate surface in commercially available glass substrates [42, 47]. Extremely large index changes (e.g., $\Delta n > 0.2$) have been achieved in glasses with high sodium content (≈ 25 vol%) [48, 49]. These substrates tend to suffer from high internal tension, which builds up because of the different diameters of the exchanging ions. Most of the microlenses, therefore, have been fabricated with lower index changes achieved in glasses with lower sodium content.

Modelling of the refractive index profile

During the thermal ion-exchange process a flow of ions is caused by the concentration gradient between the melt and the glass substrate. One of the requirements for this ion flow to take place is a sufficiently large ion mobility, which, e.g., is achieved at high temperatures $T = 300 - 500°C$. The ion concentration $C(\boldsymbol{r}, t)$ is a function of the location $\boldsymbol{r}$ and the diffusion time t. The variation of $C(\boldsymbol{r}, t)$ with time is determined by the gradient of the ion current so that we obtain a general expression of the diffusion equation which reads:

$$\frac{\partial C(\boldsymbol{r}, t)}{\partial t} = -\nabla \boldsymbol{J} = \nabla[D(\boldsymbol{r}, t)\nabla C(\boldsymbol{r}, t)] \tag{4.2.13}$$

Here $\boldsymbol{J}$ denotes the ion current and $D(\boldsymbol{r}, t)$ the diffusion coefficient, which is highly temperature dependent. For an ion-exchange process which consists of two different ion propagations $D(\boldsymbol{r}, t)$ is composed of the two diffusion constants and is called the "interdiffusion coefficient" [42].

The general diffusion equation (Eq. 4.2.13) can be solved analytically only for the case of the linear diffusion for which the interdiffusion coefficient is constant in time and location ($D(\boldsymbol{r}, t) = \mathrm{D_0} = \mathrm{const.}$). For this linear diffusion the concentration profile resulting from diffusion through a spherical mask (diameter d) can be described by [47, 50]:

$$C(\mathbf{r}, t) = \mathrm{C_0}\,\frac{d}{2r}\,\mathrm{erfc}\left(\frac{r - d/2}{2\sqrt{\mathrm{D_0}t}}\right) \tag{4.2.14}$$

$\mathrm{C_0}$ is the saturation concentration of the ions in the substrate and erfc is the error function.

However, the assumption of a constant diffusion coefficient is valid only for few material combinations. Most of the time the interdiffusion coefficient varies strongly with diffusion time and location. In a phenomenological model an exponential relation between the diffusion coefficient and the concentration is assumed:

$$D(C) = e^{k\frac{C}{C_0}} \tag{4.2.15}$$

Since the nonlinearity of the process is described by the parameter k this is often referred to as the k-model [51]. For nonlinear diffusion, the diffusion equation has to be solved numerically. It is possible to determine the k-parameters for the specific processes by a comparison of numerical results with experimental data for the achieved index profiles. Some of the k-values found in the literature for different ion-exchange systems are listed in Table 4.5.

	k-value	reference
Na^+/Ag^+ in BGG31	k=0	Bähr et al. [52]
Na^+/Ag^+ in BGG35	k=2.36	Testorf (1994)[50]
Na^+/Ag^+ in Hoya substrates	k=2.97	Testorf (1994) [50]
Na^+/Tl^+	k=4.5	Oikawa et al. (1984) [53]

Tab. 4.5: The values of the k-parameters, describing the nonlinearity of the ion-exchange process for different material systems.

The characteristics of the ion-exchange process in various material systems is strongly influenced by the application of an electric field across the substrate-melt interface. The E-field $\boldsymbol{E}$ supports an ion current $\boldsymbol{J}_{\mathrm{el}}$ additional to the one which stems from the concentration gradient [54, 55].

$$\boldsymbol{J}_{\mathrm{el}} = \mu \boldsymbol{E} C \tag{4.2.16}$$

Here μ denotes the ion mobility.

This additional ion current has to be considered in the general diffusion equation (Eq. 4.2.13):

$$\frac{\partial C(\boldsymbol{r},t)}{\partial t} = -\nabla(\boldsymbol{J} + \boldsymbol{J}_{\mathrm{el}}) = \nabla[D(\boldsymbol{r},t)\nabla C(\boldsymbol{r},t) - \mu \boldsymbol{E} C] \tag{4.2.17}$$

A precise mathematical description of the process needs to take into account the different mobilities of the ions. They cause the formation of intrinsic electric fields which have a significant influence on the index profile formed during the exchange process.

Even though this is an approximation, Eq. 4.2.17 helps us to understand the most important aspects of the field-assisted exchange process. First of all it is important to notice that the ion current triggered by the electric field is no longer isotropic but rather directed parallel to the field lines. Because of the low thermal diffusion coefficient D, the influence of the electrically initiated ion migration dominates even for low electric fields. In a first approximation the solution of the one-dimensional case yields a step profile described by a Fermi or a Heavyside distribution.

During the fabrication process the gradient of the refractive index distribution is determined by the relation between isotropic and anisotropic contributions to the ion-exchange process. With field-assisted ion-exchange very high nonlinearities in the index profile can be achieved for specific orientations and intensities of the electric field. For the optimization of the index distribution several possibilities are available. By controlling the intensity of the electric field it is possible to achieve the desired concentration distribution. Another approach is the generation of highly nonlinear step-like index distributions which are optimized in a subsequent isotropic thermal post diffusion process to yield the desired profile [56].

Influence of the diffusion parameters on the lens properties

The refractive index change in a GRIN substrate is proportional to the concentration of exchanged ions. The radius of curvature (r_c) of the light path through the GRIN media is described by Eq. 4.2.18:

$$|r_c| = |\nabla \log n(\boldsymbol{r})| \tag{4.2.18}$$

With an index profile generated by linear diffusion (Eq. 4.2.14) no suitable lens-like focusing behaviour can be achieved. This is illustrated in Fig. 4.12 where the deflection of two light rays is shown for a GRIN profile generated by linear diffusion, i.e., with a concentration-independent diffusion coefficient (Fig. 4.12a) and by nonlinear ion-exchange (Fig. 4.12b) [47].

In order to simulate the light propagation through the lenses, Singer et al. approximated the radial refractive index distribution $n(r)$ by a Fermi-distribution [57]:

$$n(r) = \mathrm{n_0} + \frac{\delta n}{1 + \exp(\frac{r - \mathrm{r_0}}{a})} \tag{4.2.19}$$

The decay parameter a is related to the linearity of the diffusion process. For a limited parameter range the Fermi-distribution yields a good approximation to the distributions resulting from the k-model. Fig. 4.13 shows the shape of the index distributions for different decay parameters a in Eq. 4.2.19.

The Fermi model can be used as the basis for simulations of the optical performance of GRIN lenses. Singer et al. [57] found optimum performance of the GRIN

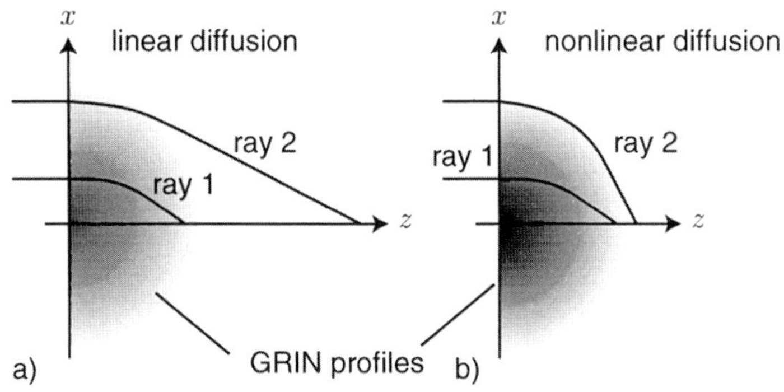

Fig. 4.12: Illustration of the deflection of two light rays by a GRIN profile generated by a) linear and b) nonlinear ion-exchange in glass.

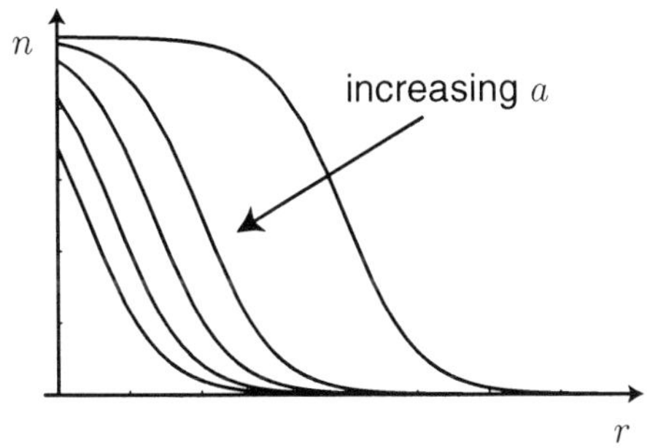

Fig. 4.13: Radial index distributions in GRIN lenses described by the Fermi-distribution for different decay parameters a. The decay parameter approximately corresponds to the linearity of the diffusion process.

lenses for a ratio of 1:4.3 for the Fermi decay length to the radius of the lenses.

The goal for microlens fabrication is to generate index profiles which provide optimized lens performance. If approximated as thin elements the numerical aperture achieved in a planar GRIN lens is determined by the maximum achievable index change and a diffusion parameter which determines the shape of the index profile [58]. Thermal ion-exchange in glasses with suitable nonlinear behaviour is one way to achieve good microlenses. However, the number of design parameters is low in this case. Significantly more design freedom is available for the field-assisted exchange process with a subsequent post diffusion process.

Microlens parameters achieved for planar GRIN microlenses

Planar GRIN microlenses fabricated by field-assisted sodium-thallium ion exchange have been commercially available from Nippon Sheet Glass (NSG) Inc. for more than 10 years. The specifications of the lenses offered are listed in Table 4.6. There is still a significant amount of research being carried out on optimizing the performance of these lenses for specific applications. One way to increase the numerical aperture of planar GRIN lenses is by stacking two lens elements on top of each other [48]. Since the surfaces of the GRIN lenses are generally plane, the alignment and fixing of two substrates relative to each other can be performed easily. Another way to fabricate GRIN doublets can be achieved by ion-exchange from both sides of the substrate. To this end two identical mask openings need to be aligned on the opposite surfaces of the substrate. During the thermal ion-exchange the ions from the melt enter the substrate through the mask openings at both surfaces. Thus, a GRIN doublet is formed.

Microlenses with large numerical apertures can be obtained by ion-exchange technique if a secondary physical effect is utilized. The ions which are exchanged during the fabrication process generally have different diameters. For example, if larger thallium or silver ions are used to replace the sodium ions in the glass, the volume will be increased. This causes surface swelling of the order of 1 μm which contributes to the focusing effect of the element. By combining the surface effect and the gradient-index profile, high numerical apertures have been achieved [59].

In recent years the silver-sodium exchange process has also been optimized to yield good microlenses. This has the advantage that the process is more cost-effective, since the materials involved are less toxic than those involved in sodium-thallium exchange. Additional efforts have been directed towards the optimization of index profiles through mask design. Applications may be found for astigmatic lenses formed by ion-exchange through elliptical mask openings. They may be applied for laser diode beam shaping [60].

One of the special features of GRIN microlenses is that the surfaces of the elements are planar. This has two consequences for applications in microoptics. On the one hand, GRIN elements can be integrated with additional surface relief components, which are fabricated lithographically on the plane surface above the index gradient. Index distributions have been applied for the optimization of refractive surface profile lenses. On the other hand, the fact that the index gradient is completely buried in the substrate is an obstacle for the replication of GRIN elements. Since this route to cheap mass fabrication is not feasible, a lot of effort has to be concentrated on cost-reduction for GRIN fabrication. Electric field-assisted ion-exchange is one important step in this direction. Apart from the possibility of optimizing the index gradient, it accelerates the exchange process by several orders of magnitude, and thus helps to reduce the production costs significantly.

GRIN components can be fabricated in a variety of different substrates, such as

	diameters	NA	focal length	reference
Na^+/Tl^+ field-assisted	10–1000 μm	0.02–0.25	20–4000 μm	Oikawa et al. (1994) [47]
Na^+/Tl^+ surface swelling	50–400 μm	0.4–0.6	55–500 μm	Oikawa et al. (1994) [47]
Na^+/Ag^+ in BGG35 (thermal)	42–138 μm	0.036–0.05	560–1380 μm	Bähr et al. (1994) [58]
Na^+/Ag^+ thermal, astigmatic	8–70 μm	0.015–0.09	280–8230 μm	Sinzinger et al. (1995) [60]
Na^+/Ag^+ field-assisted, post diffusion	< 1000 μm	< 0.2		Bähr et al. (1996) [56]

Tab. 4.6: Parameters of planar GRIN lenses.

polymer or liquid crystal materials. In this section we have restricted ourselves to GRIN profiles in glass substrates, which are currently the most highly developed for microoptical applications. The relationship between the basic parameters, however, is similar in other materials.

4.3 Microprisms and micromirrors

This section is devoted to microprisms and micromirrors. After an analysis of the challenges of microprism fabrication, we will focus on some techniques which have been discussed for this purpose. "Deep lithography", "analog lithography for microprism fabrication", "micromachining", "mass transport techniques" and "anisotropic etching" are some of the possible approaches.

As we have seen previously, a large variety of different fabrication techniques for refractive microlenses exists. These provide alternatives to mechanical methods such as polishing or grinding, which are inapplicable for batch fabrication of microoptical components. The situation is different for the fabrication of microprisms and micromirrors. Only very few fabrication techniques exist for microprism fabrication, with most of them based on mechanical methods. In order to understand the reasons for this situation, we want to compare the requirements on fabrication processes for microlenses and microprisms. For microlens fabrication it is necessary to generate continuous spherical or quasi-spherical surface or index profiles (Fig. 4.14a). Such radially symmetrical shapes are very well supported by a variety of physical effects, e.g., surface tension (utilized in many of the fabrication techniques mentioned for surface profile lenses), or isotropic processes such as diffusion or etching in amorphous materials. As soon as the symmetry of the profile is to be broken, as, e.g., in the case of aspherical lenses optimized for specific imaging situations, the fabrication effort is increased significantly.

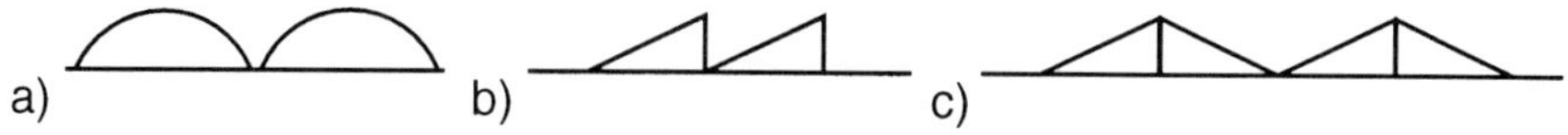

Fig. 4.14: Illustration of the different surface profiles required for arrays of a) microlens, b) microprism and c) microbiprisms.

For prism fabrication the requirements are very different. A microprism is characterised by a linear surface or phase slope which is ended by an ideally very sharp edge (Fig. 4.14b). The necessary overall phase depth has to be at least several wavelengths. For a microprism with a lateral dimension of 50 μm and a moderate wedge angle of $\epsilon = 5°$, e.g., a height $h \approx 4.5$ μm is required. These considerations illustrate the fabrication problems. Isotropic processes suitable for the fabrication of the linear slope cannot be utilized for microprisms since they do not allow the generation of sharp edges. The necessary anisotropy of the fabrication process is determined by the acceptable minimum steepness of the prism edge. For fabrication with an anisotropic process, the prism slope can be quantized in space and phase. However, the requirements on the fabrication process are very demanding. In particular, the necessary phase depth is difficult to achieve with most of the lithographic techniques discussed in Chapter 3.

4.3.1 Lithography for the fabrication of microprisms

Deep synchrotron or proton lithography

Lithographic processes with high energy beams such as synchrotron or proton radiation can be used for the generation of deep surface structures with high aspect ratios [61, 62]. The individual processing steps are very similar to conventional lithography. The illumination process of a polymer substrate is performed with a collimated synchrotron or proton beam. In order to achieve good contrast ratios for this kind of illumination, special thick mask structures have to be used. Such masks can be fabricated, e.g., in an iterative LIGA process (Chapter 3) [63]. Alternatively a thick pinhole mask can be used [27]. The different structures are fabricated by scanning the substrate behind the homogeneously illuminated pinhole. The high energy beam is used in this case as a precise tool for micromachining. The size and precision of the pinhole determines the minimum feature size of the process.

Due to the high penetration depth of the synchrotron or proton radiation, structuring depths of several hundred micrometres can easily be achieved. The quality of the surfaces fabricated with this approach depends on their orientation. The quality of the side walls is determined by the precision of the mask edges and the lateral scattering of the illuminating beam [61, 64]. Good surface quality can be achieved with both synchrotron as well as proton beam irradiation and precise mask fabrication processes. High quality of the surfaces oriented perpendicularly to the illumination beam is more difficult to achieve. Due to the depth distribution of the energy deposition in the substrate, these surfaces show relatively large roughness. Different prism slopes can be achieved with different illuminating angles.

Analog lithography for microprism fabrication

Grey-scale lithography as well as direct writing techniques in combination with analog processing of photopolymers have been suggested for microprism fabrication. The basics of these approaches are discussed in Chapter 3 and in the section on microlens fabrication. In combination with a photoinduced volume growth in a polymer material, grey-scale lithography has been suggested for microprism fabrication [65].

All of the different approaches suffer from similar technological limitations. The trade-off between the profiling depths and the sharpness of the etches currently is the biggest challenge for microprism fabrication. In order to avoid sharp edges in microprism arrays often the use of biprims is considered which are significantly easier to fabricate (Fig. 4.14c).

A promising approach for generating good quality microprisms is the combination of analog lithography with a subsequent etching process. Analog lithography is used for the generation of a shallow microprism. Due to the low phase-depth this is possible with good quality, e.g., using HEBS grey-scale masks. In order to increase the prism angle, an anisotropic selective etching process, e.g., CAIBE or RIE can be applied [38, 66]. Additional gain in the deflection angle of such prisms results if high refractive index substrate materials (e.g., GaAs) are used. The finite edge sharpness still limits the packaging density of such microprisms. For specific applications the optical system can be designed for optimum use of this type of prism arrays. They are interesting, for example, for planar-optically integrated systems, where an array of light beams is deflected. The microprism arrays are arranged at a location where the individual beams are not densely packed. Thus, it is possible to avoid losses at the prism edges and only to use the high quality central parts of the phase slope of the prisms [67].

Reflow and mass-transport techniques

Reflow and mass-transport techniques are also being investigated in order to overcome the problems related to the fabrication of microprisms. Binary or multilevel

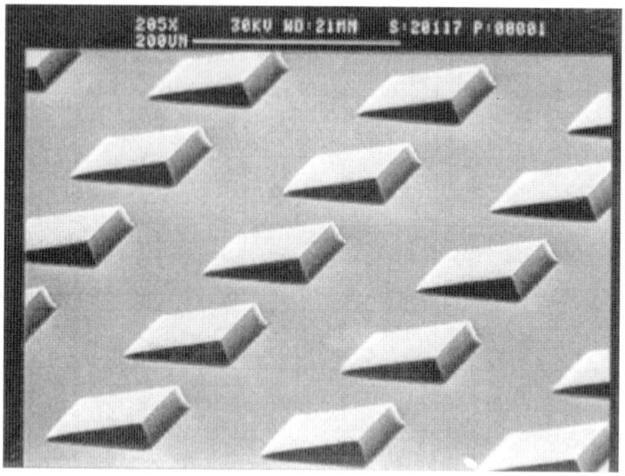

Fig. 4.15: REM micrograph of an array of microprisms (diameter: 100 μm) fabricated by analog lithography using HEBS glass grey-scale masks (courtesy of Dr. E. B. Kley, F. Thoma, Friedrich-Schiller Universität, Jena).

preshapes are generated by lithographic techniques. Depending on the substrate material, a mass transport or reflow process can be initiated by melting or heating of the element (Fig. 4.16) [21]. Since melting is an isotropic process, it will cause some loss of edge sharpness. This effect is reduced if the binary structure is only heated locally in the areas where smoothing of the surface is desired. Laser-, ion- or e-beam polishing techniques are well suited for this application [68]. Similar techniques are applied commercially for microoptics fabrication [69].

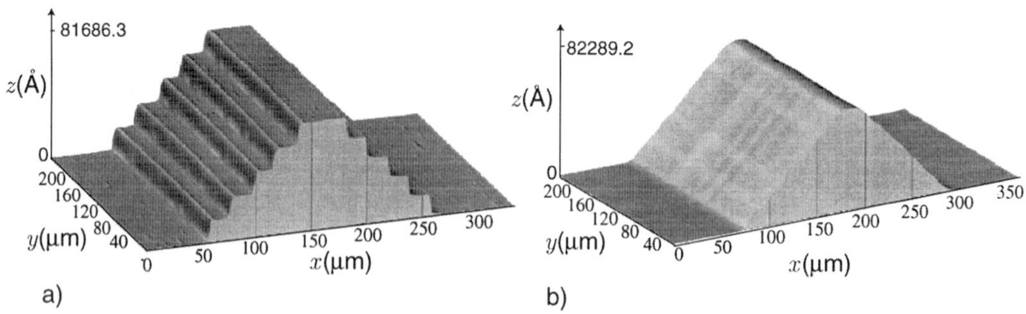

Fig. 4.16: 2D profilometer measurements of microprisms fabricated by mass transport in GaP: a) multilevel preform; b) biprism after mass transport (pictures courtesy of Prof. J. Leger, University of Minnesota).

4.3.2 Micromachining of microprisms using single point diamond turning or embossing

Diamond turning machines developed to high fabrication precision also allow the micromechanical fabrication of prisms with lateral dimensions of $\Delta x \leq 1000\ \mu m$. Profiling depths of several tens of micrometres can be easily achieved. A large variety of prism sizes and angles can be fabricated. Diamond turning with subsequent replication techniques is the most widespread technique for the micromachining of prism gratings.

The availability of sufficiently precise mechanical tools allows one to apply embossing techniques in combination with step and repeat processing for the fabrication of microprism arrays. Suitable embossing tools are crystalline materials providing very good surface quality or polished steel or diamond tools [70].

4.3.3 Anisotropic etching of mirror structures in crystalline materials

Beam deflecting optical elements for microoptic applications can also be fabricated using anisotropic etching processes in crystalline substrates. The most popular material for this purpose is silicon, because silicon processing is well known from electronics. When immersed into an etching liquid such as KOH, the silicon wafer becomes etched anisotropically (Chapter 3). The resulting surfaces with different angles relative to the substrate surface can be used as deflecting mirrors [71]. A large number of angles can be achieved with different wafer orientations, etchants or by doping or electric field-assisted etching [72, 73]. After fabrication, anisotropically etched silicon mirrors can be replicated in a variety of optical materials or bonded to them.

In this chapter we have learned about the problems and possibilities for the fabrication of refractive optical elements. An abundance of methods is available for this purpose. Nevertheless, there are still parameter ranges which cannot be reached by any fabrication technique. Further optimization of the techniques is necessary especially for microprism fabrication. Another aspect is the fact that the different methods require different substrate materials. Monolithic integration of different elements is often impossible. As we will see in the following chapter, this situation is different for diffractive optical elements. The whole variety of diffractive elements can be fabricated with the same technology. This offers new possibilities for systems integration (Chapter 7).

technology	surface quality	profiling depth; depending on:	edge sharpness; depending on:	reference
x-ray or proton lithography	side walls: good bottom: poor–medium	large, < 1000 μm	very good	Brenner et al. [61] Kufner et al. [62]
analog optical lithography	good	< 5–10 μm	medium–low; profiling depth	Rohrbach et al. [65] Gale [34] Jahns et al. [67]
analog lithography + CAIBE	good	< 50 μm	medium	for microlenses: Däschner et al. [38]
anisotropic Si etching	very good	<100 μm; feature size	very good	Kendall et al. [73] Moisel et al. [71]
preforming and mass transport	good	possibly large; preform	good	Ballen et al. [21]
diamond turning	good	possibly large; > 5 μm	good; diamond tool	for kinoforms: Blough et al. [74]

Tab. 4.7: Fabrication technologies for microprisms, strengths and weaknesses.

4.4 Glossary

- **ion-exchange technique**
technique for fabrication of GRIN microlenses, e.g., in glass substrates. To this end some of the ions contained in the glass matrix are replaced by different ions. This changes the material properties such as the refractive index. Generally the process takes place at high temperatures in a salt melt. In field-assisted ion-exchange the ion migration is supported by an electric field.

- **planar GRIN microlenses (PML^{TM})**
gradient-index microlenses fabricated by ion-exchange in planar glass substrates. PML is a trademark of Nippon Sheet Glass Inc., Japan.

- **reflow microlenses**
microlenses fabricated by melting cylindrical islands of photoresist.

- **sag** of a microlens
maximum height of the lens cap (see Fig. 4.1).

- **sealed ampoule technique**
mass transport technique for the fabrication of smooth surface profiles in compound semiconductor materials, e.g., for microlenses or microprisms. The process takes place in a small ampoule which forms the reaction chamber. During heating, e.g., a phosphoric atmosphere is created inside the ampoule so that the mass transport can take place.

- **SELFOC^{TM}** lenses
also called GRIN rod lenses; short pieces of GRIN optical fibers in which the radial GRIN profile is used for imaging purposes. SELFOC is a trademark of Nippon Sheet Glass Inc., Japan.

4.5 List of new symbols

A:	surface area
Ag^+:	silver ions
a:	decay parameter of the Fermi-distribution
α:	parameter of the index distribution in GRIN fibers
$C(\mathbf{r},t)$:	ion concentration
C_0	initial ion concentration
$D(\mathbf{r},t)$:	diffusion coefficient, for ion-exchange called interdiffusion coefficient
D_0	initial diffusion coefficient
E, E_{cyl}, E_{cap}:	energies due to the different states of the photoresist
$\boldsymbol{E}$:	electric field vector
ΔE_S, ΔE_G:	change of surface energy and gravitational energy, respectively
erfc:	error function
ϵ:	wedge angle of a microprism
f:	focal length
g:	acceleration due to gravity: $g = 9.81\frac{m}{s^2}$
h:	height of cylindrical photoresist cap; sag of the lenslet
k:	parameter describing the nonlinearity of the diffusion process
$\boldsymbol{J}$, $\boldsymbol{J}_0$:	ion flow during the ion-exchange process
μ:	ion mobility
Na^+:	sodium ions
$n, n(x)$:	refractive index
∇:	Laplace operator defined as: $\nabla = \left(\frac{\partial}{\partial x}, \frac{\partial}{\partial y}, \frac{\partial}{\partial z}\right)$
Δn:	change of refractive index
r_c:	radius of curvature of spherical photoresist cap; radius of curvature of the light path through GRIN media
$\boldsymbol{r}_c$:	radius of curvature of the light path
ϱ, ϱ_0:	material density
Tl^+:	thallium ions
t:	thickness of photoresist coating; time parameter for the diffusion process
τ, τ_{pa}, τ_{ps}:	surface tension coefficients
$\theta(x)$, $\theta(x+\Delta x)$:	propagation angle of a light ray
V_{cyl}, V_{cap}:	volume of photoresist cylinder or photoresist cap
ΔV:	volume change

4.6 Exercises

1. **Technology for microlens fabrication**

   For setting up a microoptical system we need refractive lenses of 200 μm diameter (focal length: $f = 650$ μm) and larger lenses with diameters of 3000 μm and a focal length of 5000 μm.
   a) Calculate the sags of these lenses, resulting if they were fabricated in fused silica glass substrates ($n \approx 1.46$).
   b) What technology would you choose for the fabrication of those lenses? Discuss your decision.

2. **Deflection by microprisms**

   We want to deflect an optical beam (diameter 200 μm) by an angle of $\alpha = 10°$. Calculate the necessary depth of microprisms made of fused silica glass (n $\approx$ 1.46) and GaAs ($n \approx 3.5$) for this purpose.

3. **Thermal reflow and mass transport mechanisms**

   Describe the similarities and differences between the thermal reflow process in photoresist and the mass transport process in compound semiconductor materials for microoptics fabrication.

References

[1] 0. Wada, "Ion-beam etching of InP and its applications to the fabrication of high radiance InGaAsP/InP light emitting diodes", *J. Electrochem. Soc.* **131** (1984), 2373–2380.

[2] Z. D. Popovic, R. A. Sprague and G. A. N. Connell, "Technique for the monolithic fabrication of microlens arrays", *Appl. Opt.* **27** (1988), 1281–1284.

[3] A. Y. Feldblum, C. R. Nijander, W. P. Townsend and C. M. Mayer-Costa, "Performance and measurement of refractive microlens arrays", *Proc. SPIE* **1544**, Miniature and micro-optics: fabrication and system applications (1991), 200.

[4] M. C. Hutley, "Optical techniques for the generation of microlens arrays", *J. Mod. Opt.* **37** (1990), 253–265.

[5] D. Daly, R. F. Stevens, M. C. Hutley and N. Davies, "The manufacture of microlenses by melting photoresist", *Meas. Sci. Techn.* **1** (1990), 759–766.

[6] A. W. Lohmann, "A lenslet array based on droplet technology", *NEC research institute, technical memorandum* 91-030-2-7002-1 (1991).

[7] P. Nussbaum, R. Völkel, H. P. Herzig, M. Eisner and S. Haselbeck, "Design, fabrication and testing of microlens arrays for sensors and microsystems", *Pure Appl. Opt.* **6** (1997), 617–636.

[8] D. Daly, R. F. Stevens and M. C. Hutley, "The manufacture of microlenses by melting photoresist", *IOP short meetings series*, **30**, Microlens arrays, Teddington, UK (1991), 23–34.

[9] M. C. Hutley, P. Savander and M. Schrader, "The use of microlenses for making spatially variant optical interconnections", *Pure Appl. Opt.* **1** (1992), 337–346.

[10] K. Mersereau, C. R. Nijander, A. Y. Feldblum and W. P. Townsend, "Fabrication and measurement of fused silica microlens arrays", *Proc. SPIE* **1751**, Conference on Miniature and Micro-Optics: Fabrication and Systems Applications II, San Diego (1992), 229–233.

[11] K. Mersereau, C. R. Nijander, W. P. Townsend and A. Y. Feldblum, "Aberration-corrected etched microlens array", *OSA Techn. digest* **5**, Diffractive optics and microoptics, Boston, USA (1996).

[12] S. Haselbeck, H. Schreiber, J. Schwider and N. Streibl, "Microlenses fabricated by melting a photoresist on a base layer", *Opt. Eng.* **32** (1993), 1322–1324.

[13] T. R. Jay and M. B. Stern, "Preshaping photoresist for refractive microlens fabrication", *Opt. Eng.* **33** (1994), 3552–3555.

[14] E. M. Strzelecka, G. D. Robinson, L. A. Coldren and E. L. Hu, "Fabrication of microlenses in semiconductors by mask shape transfer in reactive ion etching", *Microelec. Eng.* **35**, (1997) 385–388.

[15] M. B. Stern and T. R. Jay, "Dry etching of coherent refractive microlens arrays", *Opt. Eng.* **33** (1994), 3547–3551.

[16] M. Eisner and J. Schwider, "Transferring resist lenses into silicon by RIE", *Opt. Eng.* **10** (1996), 2979–2982.

[17] Z. L. Liau, V. Diadiuk, J. N. Walpole and D. E. Mull, "Gallium phosphide microlenses by mass transport", *Appl. Phys. Lett.* **55** (1989), 97–99.

[18] Z. L. Liau and H. J. Zeiger, "Surface-energy-induced mass-transport phenomenon in annealing of etched compound semiconductor structures: theoretical model and experimental confirmation", *J. Appl. Phys.* **67** (1990), 2434–2440.

[19] J. S. Swenson Jr., R. A. Fields and M. H. Abraham, "Enhanced mass-transport smoothing of f/0.7 GaP microlenses by the use of sealed ampoules", *Appl. Phys. Lett.* **66** (1995), 1304–1306.

[20] D. Z. Tsang, Z. T. Liau and J. N. Walpole, "Compact micro-optical diode laser assembly using mass-transported large-numerical aperture refractive microlenses", *OSA Techn. Digest* **5**, Diffractive Optics and Microoptics, Boston, MA, USA (1996), 35–38.

[21] T. A. Ballen and J. R. Leger, "Mass-transport gallium phosphide refractive micro-lenses for diode array reconfiguration", *OSA Techn. Digest* **10**, Diffractive Optics and Microoptics, Kona, Hawaii, USA (1998).

[22] N. F. Borrelli, D. L. Morse, R. H. Bellmann and W. L. Morgan, "Photolytic technique for producing microlenses in photosensitive glass", *Appl. Opt.* **24** (1985), 2520.

[23] N. F. Borelli, R. H. Bellmann, J. A. Durbin and W. Lama, "Imaging and radiometric properties of microlens arrays", *Appl. Opt.* **30** (1991), 3633–3642.

[24] J. O. Choi, J. A. Moore, J. C. Corelli, J. P. Silverman and H. Bakhru, "Degradation of poly(methylmethacrylate) by deepultraviolet, x-ray, electron beam, and proton beam irradiations", *J. Vac. Sc. Technol. B* **6** (1988), 2286–2289.

[25] M. Frank, M. Kufner, S. Kufner and M. Testorf, "Microlenses in polymethyl methacrylate with high relative aperture", *Appl. Opt.* **30** (1991), 2666–2667.

[26] M. Kufner, S. Kufner, M. Frank, J. Moisel and M. Testorf, "Microlenses in PMMA with high relative aperture: a parameter study", *Pure Appl. Opt.* **2** (1993), 9–19.

[27] M. Kufner and S. Kufner, "Micro-optics and lithography", VUB Press, Brussels, Belgium (1997).

[28] S. Lazare, J. Lopez, J.-M. Turlet, M. Kufner, S. Kufner and P. Chavel, "Microlenses fabricated by ultraviolet excimer laser irradiation of poly(methyl metacrylate) followed by styrene diffusion", *Appl. Opt.* **35** (1996), 4471–4475.

[29] A. W. Lohmann and S. Sinzinger, "Arrays of doublet lenses", *Optik* **93** (1992), 84–86.

[30] D. L. Macfarlane, V. Narayan, J. A. Tatum, W. R. Cox, T. Chen and D. J. Hayes, "Microjet fabrication of microlens arrays", *IEEE Photon. technol. lett.* **6** (1994), 1112–1114.

[31] V. Guyenot, "Herstellung von Mikrolinsen durch Klebstofftropfen-Applikation", *Jahresbericht Fraunhofer Institut für angewandte Optik und Feinmechanik (1996)*, 44–46.

[32] W. R. Cox, T. Chen, C. Guan, D. J. Hayes, R. F. Hoenigman, B. T. Teipen and D. L. MacFarlane, "Micro-jet printing of refractive microlenses", *OSA Techn. Digest* , **10**, Diffractive Optics and Microoptics, Kona, Hawaii, USA (1998), 276–278.

[33] B. P. Keyworth, D. J. Corazza, J. N. McMullin and L. Mabbott, "Single-step fabrication of refractive microlens arrays", *Appl. Opt.* **36** (1997), 2198–2202.

[34] M. Gale, "Direct writing of continuous-relief micro-optics", in Microoptics, H.-P. Herzig, Taylor & Francis, London (1997), 87–126.

[35] M. Gale, M. Rossi, J. Pedersen and H. Schütz, "Fabrication of continuous-relief micro-optical elements by direct laser writing in photoresists", *Opt. Eng.* **33** (1994), 3556–3566.

[36] K. Kaschlik and E. B. Kley, "Masken für die Mikrotechnik mit dem Variable-Shaped-Beam Elektronenschreiber ZBA 23", VDI-Berichte (1992).

[37] K. Zimmer, D. Hirsch and F. Bigl, "Excimer laser machining for the fabrication of analogous microstructures", *Appl. Surf. Sci.* **96–98** (1996), 425–429.

[38] W. Däschner, P. Long, R. Stein, W. Chuck and S. H. Lee, "General aspheric refractive micro-optics fabricated by optical lithography using a high energy beam sensitive glass gray-level mask", *J. Vac. Sci. Technol. B* **14** (1996), 135–138.

[39] B. E. A. Saleh and M. C. Teich, "Fundamentals of Photonics", John Wiley & Sons, New York (1991).

[40] E. G. Rawson, D. R. Herriot and J. McKenna, "Analysis of refractive index distributions in cylindrical, graded-index glass rods (GRIN rods) used as image relays", *Appl. Opt.* **9** (1970), 753–759.

[41] W. J. Tomlinson, "Aberrations of GRIN rod lenses in multimode optical fiber devices", *Appl. Opt.* **19** (1980), 1117–1126.

[42] T. Findakly, "Glass waveguides by ion-exchange: a review", *Opt. Eng.* **24** (1985), 244–250.

[43] T. W. Cline and R. B. Jander, "Wave-front aberration measurements on GRIN rod lenses", *Appl. Opt.* **21** (1982), 1035–1041.

[44] W. J. Tomlinson, "Application of GRIN rod lenses in optical fiber communication systems", *Appl. Opt.* **19** (1980), 1127–1138.

[45] M. Oikawa, K. Iga and T. Sanada, "Distributed-index planar microlens prepared from ion-exchange technique", *Jpn. J. Appl. Phys.* **20** (1981), L296–L298.

[46] M. Oikawa and K. Iga, "Distributed-index planar microlens", *Appl. Opt.* **21** (1982), 1052–1056.

[47] M. Oikawa and K. Hamanaka, "Physics of planar microlenses", in Optical computing hardware, J. Jahns and S. H. Lee (eds), Academic Press, San Diego (1994), 113–136.

[48] S. Misawa, M. Oikawa and K. Iga, "Maximum and effective numerical apertures of a planar microlens", *Appl. Opt.* **23** (1984), 1784–1786.

[49] J. L. Coutaz and P. C. Jaussand, "High index gradient in glass by ion-exchange", *Appl. Opt.* **21** (1982), 1063–1065.

[50] M. Testorf, "Analyse des Ionenaustausches in Glas im Hinblick auf eine Synthese refraktiver optischer Mikroelemente", PhD Thesis, University of Erlangen (January 1994).

[51] J. Crank, "The mathematics of diffusion", Oxford University Press (1974).

[52] J. Bähr, K.-H. Brenner, J. Moisel, W. Singer, S. Sinzinger, A. Spick and M. Testorf, "Diffusion elements in glass: comparison of the diffusion response in different substrates", *Proc. SPIE* **1806**, Topical meeting on Optical Computing, Minsk, Belarus (1992), 234–242.

[53] M. Oikawa, K. Iga, M. Morinaga, T. Usui and T. Chiba, "Distributed-index formation process in a planar microlens", *Appl. Opt.* **23** (1984), 1787–1789.

[54] M. M. Abou-el-leil and A. R. Cooper, "Analysis of the field assisted binary ion-exchange", *J. Am. Ceram. Soc.* **62** (1979), 390–395.

[55] A. J. Cantor, M. M. Abou-el-leil and R. H. Hobbs, "Theory of 2D ion-exchange in glass: optimization of microlens arrays", *Appl. Opt.* **30** (1991), 2704–2713.

[56] J. Bähr and K.-H. Brenner, "Realization and optimization of planar microlenses by Ag-Na ion-exchange techniques", *Appl. Opt.* **35** (1996), 5102–5107.

[57] W. Singer, M. Testorf and K.-H. Brenner, "Gradient-index microlenses: numerical simulation of different spherical index profiles with the wave propagation method", *Appl. Opt.* **34** (1995), 2165–2171.

[58] J. Bähr, K.-H. Brenner, S. Sinzinger, T. Spick and M. Testorf, "Index-distributed planar microlenses for three-dimensional microoptics fabricated by silver-sodium ion-exchange in BGG35 substrates", *Appl. Opt.* **33** (1994), 5919–5924.

[59] M. Oikawa, H. Imanishi and T. Kishimoto, "High NA planar microlens for LD array", *Proc. SPIE* **1751**, Miniature and microoptics: fabrication and systems application (1992), 246–254.

[60] S. Sinzinger, K.-H. Brenner, J. Moisel, A. Spick and M. Testorf, "Astigmatic gradient-index elements for laser-diode collimation and beam shaping", *Appl. Opt.* **34** (1995), 6626–6631.

[61] K.-H. Brenner, M. Kufner, S. Kufner, J. Moisel, A. Müller, S. Sinzinger, M. Testorf, J. Göttert and J. Mohr, "Application of three-dimensional micro-optical components formed by lithography, electroforming, and plastic moulding", *Appl. Opt.* **32** (1993), 6464–6469.

[62] S. Kufner, M. Kufner, M. Frank, A. Müller and K.-H. Brenner, "3D integration of refractive micro-optical components by deep proton irradiation", *Pure Appl. Opt.* **2** (1993), 111–124.

[63] J. Göttert, J. Mohr, A. Müller and C. Müller, "Fabrication of microoptical components and systems by the LIGA technique", *Entropie* **192/193** (1995), 20–26.

[64] M. Kufner, S. Kufner, P. Chavel and M. Frank, "Monolithic integration of microlens arrays with fiber holder arrays in poly(methyl methacrylate) with fiber self-centering", *Opt. Lett.* **20** (1995), 276–278.

[65] A. Rohrbach and K.-H. Brenner, "Surface-relief phase structures generated by light-initiated polymerisation", *Appl Opt.* **34** (1995), 4747–4754.

[66] C. Gimkiewicz, D. Hagedorn, J. Jahns, F. Thoma and E. B. Kley, "Fabrication of microprisms for planar-optical interconnections using analog gray-scale lithography with high energy beam sensitive glass", *OSA techn. digest* , **10**, Diffractive optics and microoptics, Kona, Hawaii, USA (1998).

[67] J. Jahns, C. Gimkiewicz and S. Sinzinger, "Light efficient parallel interconnect using integrated planar free-space optics and vertical cavity surface emitting laser diodes", *Proc. SPIE* , Vertical cavity surface emitting lasers, San Jose, CA, USA (1998).

[68] R. A. Hoffman, W. J. Lange, J. G. Gowan and C. J. Miglionico, "Ion polishing as a surface preparation for dielectric coating of diamond turned optics", *Opt. Eng.* **17** (1978), 578–585.

[69] For example, AMS Mikrooptik GmbH, Saarbrücken, Germany and LIMO, Mikrooptik, Dortmund, Germany.

[70] C. Passon, J. Moisel, N. McArdle, W. Eckert, K.-H. Brenner and M. Kuijk, P. Heremans, "Integration of refractive micro-optical elements with differential-pair optical-thyristor arrays", *Appl. Opt.* **35** (1996), 1205–1211.

[71] J. Moisel, C. Passon, J. Bähr and K.-H. Brenner, "Homogeneous concept for the coupling of active and passive single mode devices by utilizing planar gradient-index lenses and silicon V-grooves", *Appl. Opt.* **36** (1997), 4736–4743.

[72] H. Seidel and R. Voss, "Anisotropic silicon etching techniques", *VDE Verlag*, Microsystems Technologies 91, Berlin, Germany (1991), 291–301.

[73] D. L. Kendall, W. P. Eaton, R. Manginell and T. G. Digges Jr., "Micromirror arrays using KOH:H_2O micromachining of silicon for lens templates, geodesic lenses, and other applications", *Opt. Eng.* **33** (1994), 3578–3587.

[74] G. C. Blough, M. Rossi, S. K. Mack and R. L. Michaels, "Single point diamond turning and replication of visible and near-infrared diffractive optical elements", *Appl. Opt.* **36** (1997), 4648–4654.

5 Diffractive microoptics

In Chapter 4 we focused on refractive microoptics. Our main concern was to analyse the potential of a variety of different lithographic approaches to the miniaturization of conventional refractive optical elements (e.g., lenses and prisms). This chapter is devoted to diffractive optics. Rather than being refracted at continuous surface profiles, in diffractive optical elements (DOEs) light is diffracted at the periodic microstructure of the element. It is the periodicity and spatial structure rather than the surface profile which determines the optical performance of the DOE. This leads to a number of interesting features of diffractive optics which are the topic of this chapter. In Section 1 we learn what is meant by diffractive optics. We understand diffractive optics as a concept for "trading spatial resolution for reduced phase thickness". The spatial periodicity of DOEs results from phase quantization as a consequence of the periodicity of the light wave. In Section 2 we focus on the "fabrication" of DOEs. We will understand the importance of DOEs for lithographically fabricated microoptics. Sections 3 and 4 are devoted to the "modelling" and the "design" of diffractive elements. Here our main concern will be with scalar diffraction theory.

5.1 Trading spatial resolution for reduced phase thickness

5.1.1 Blazing and phase quantization

Diffractive optics can be viewed as an approach to the fabrication of optical components, which is optimized for the application of lithographic techniques. As described in the previous chapters, in refractive optical elements (ROEs) the light is manipulated by analog phase elements of considerable thickness (in relation to the optical wavelength). We have seen that even for microoptical elements with diameters of only tens to hundreds of micrometres, phase structures with thicknesses $> 10\ \mu m$ are necessary for the implementation of refractive microlenses or microprisms. For optimized components (e.g., aspherical lenses or complex phase profiles of multiple beam splitters) in particular, the fabrication is impossible with mechanical profiling techniques (due to the small lateral extension) and very challenging with microlithography (due to the large phase depth of the component). The solution to this fabrication problem lies in the periodic nature of the light wave $U(x)$. If a light wave is delayed by one wavelength (corresponding to a phase lag of $\varphi = 2\pi$), no difference to the original wave can be found (Eq. 5.1.1).

$$U(x,\varphi) = A_0(x)e^{i\varphi} = A_0(x)e^{i\varphi+2\pi} = U(x,\varphi + 2\pi) \tag{5.1.1}$$

Retardation occurs, for example, when the wave passes through a dielectric material (e.g., glass or photoresist). The insensitivity of the light wave to phase jumps of $N \cdot 2\pi$ (N: integer) allows one to reduce the thickness of an optical element without changing its effect on a monochromatic wave. In transmission the maximum thickness of the corresponding optical component can be reduced to $t_{\max} = \lambda/(n-1)$, where n denotes the refractive index of the substrate (Fig. 5.1) [1, 2]. For a monochromatic wave with $\lambda = 633$ nm, a DOE made of fused silica glass (SiO_2 with n=1.457) has a thickness of $t_{\max} \leq 1.4$ μm. Such profiling depths are readily fabricated with lithographic techniques. Reduction of the thickness of optical components by sampling the phase values at multiples of 2π is referred to as blazing [3, 4, 5]. The blazing of continuous phase functions results in laterally periodic elements.

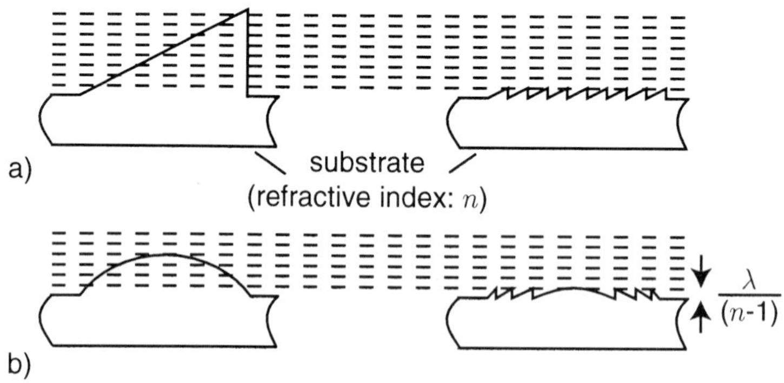

Fig. 5.1: The blazing of a) a prism and b) a lens, resulting in a reduced thickness of the elements.

The reduction of the phase depth through blazing is the first step towards an easier lithographic fabrication of the optical components. For good performance of the blazed elements it is necessary to fabricate the phase jumps with sharp edges. This requires a fabrication process with high lateral resolution. From the point of view of element fabrication, in blazed components we are trading lateral resolution for a reduction of the necessary phase depth. This is interesting for microoptics, since microlithography generally provides high lateral resolution, whereas it is difficult to achieve phase profiling depths > 5–10 μm.

Due to mapping of the phase into the range $[0, 2\pi]$, the originally continuous phase function is subdivided into features of small lateral extension which are periodically replicated. Each of the periods generates a beamlet. The individual beamlets interfere to form the output light distribution.

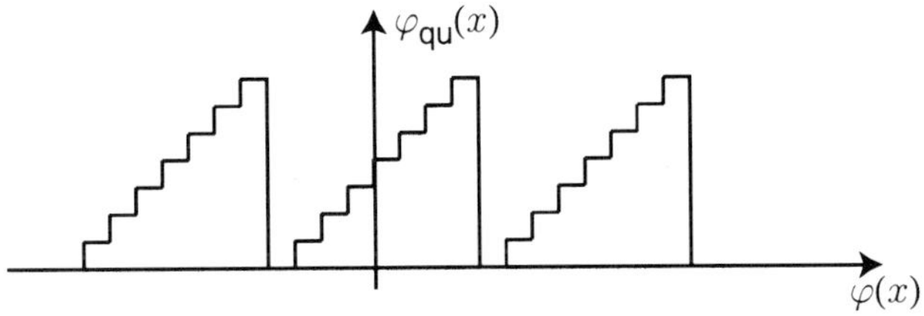

Fig. 5.2: Mapping of the continuous phase profile onto the quantized phase.

As we have seen in Chapter 3, the fabrication of arbitrarily shaped continuous phase profiles is difficult and expensive even for features with low thickness. The next step to ease the fabrication of the microoptical component is, therefore, the quantization of the phase profile [6, 7, 8]. The continuous phase values in the interval $[0; 2\pi]$ are reduced to an integer number of discrete values. This quantization corresponds to a nonlinearity which is imposed on the phase profile (Fig. 5.2) [9]. Mathematically this mapping of the non-quantized phase $\varphi(x)$ into a quantized phase $\varphi_{\mathrm{qu}}(x)$ can be described by:

$$e^{i\varphi_{\mathrm{qu}}(x)} - \sum_{q=-\infty}^{q=+\infty} e^{i\,q\frac{2\pi}{N}} \operatorname{rect}\left[\frac{\varphi(x) - q\frac{2\pi}{N}}{\frac{2\pi}{N}}\right] \tag{5.1.2}$$

The phase steps are uniformly distributed over the interval $[0, 2\pi]$, so that for N quantization steps a step height of $\Delta\varphi = 2\pi/N$ results. A periodic continuation of the mapping relation allows us to expand Eq. 5.1.2 in a Fourier series of functions $\exp[2\pi\, i\, n/(2\pi)\ \varphi(x)]$ [10]:

$$e^{i\varphi_{\mathrm{qu}}(x)} = \sum_{n=-\infty}^{+\infty} A_n e^{2\pi\, i\, n\frac{\varphi(x)}{2\pi}} \tag{5.1.3}$$

with:

$$A_n = \frac{1}{2\pi}\int_0^{2\pi} e^{i\,\varphi_{qu}(x)}\, e^{2\pi i\, n\frac{\varphi(x)}{2\pi}}\, d\varphi(x, y) \tag{5.1.4}$$

Inserting the expansion of the rect function into Eq. 5.1.2 yields:

$$e^{i\varphi_{\mathrm{qu}}(x)} = \sum_n \operatorname{sinc}\left(\frac{n}{N}\right) e^{in\varphi(x)} \left[\frac{1}{N}\sum_{k=0}^{N-1} e^{2\pi i\frac{k}{N}}(1-n)\right] \tag{5.1.5}$$

where $\operatorname{sinc}(x) = \operatorname{sinc}(\pi x)/(\pi x)$.

Eq. 5.1.5 is further simplified by recognizing that the sum over k only yields nonzero results in diffraction orders n, where $n-1=0,\pm N,\pm 2N$ etc. Introducing the new index $m=(n-1)/N$ we obtain:

$$\begin{aligned} e^{i\varphi_{\mathrm{qu}}(x)} &= \sum_m \operatorname{sinc}\left(m+\frac{1}{N}\right) e^{i(Nm+1)\varphi(x)} \\ &= \operatorname{sinc}\left(\frac{1}{N}\right)\sum_m \left[\frac{(-1)^m}{(Nm+1)}\right] e^{i(Nm+1)\varphi(x)} \end{aligned} \tag{5.1.6}$$

If $\varphi(x)$ represents the phase of a Fourier element $g(x)$, we can see the influence of phase quantization by calculating the Fourier transform $\hat{\mathcal{G}}(\nu_x)$ of the quantized phase function $\hat{g}(x)$:

$$\hat{g}(x)=e^{i\varphi_{\mathrm{qu}}(x)} \quad \longrightarrow \quad \hat{\mathcal{G}}(\nu_x)=\operatorname{sinc}\left(\frac{1}{N}\right)\sum_m \left[\frac{(-1)^m}{(Nm+1)}\right]\hat{G}_m(\nu_x) \tag{5.1.7}$$

The signal $\hat{\mathcal{G}}(\nu_x)$, reconstructed from the quantized Fourier element, is a superposition of a number of "ghost" signals $\hat{G}_m(\nu_x)$, with:

$$\hat{G}_m(\nu_x)=\int e^{i(Nm+1)\varphi(x)}e^{2\pi i m\nu_x x}dx \tag{5.1.8}$$

From Eq. 5.1.8 we see that for $m=0$ the quantized phase element always yields the undistorted signal $\tilde{g}(\nu_x)$ [10]. For $m\neq 0$, ghost images are reconstructed which are centered in the diffraction orders $n=Nm+1$. If they are not clearly separated from the signal order, these ghost images contribute to background noise in the signal plane [11]. The amplitudes of the ghost images are weighted according to Eq. 5.1.7 with the factor $\operatorname{sinc}(1/N)\cdot[\frac{(-1)^m}{(Nm+1)}]$. Neglecting reflection losses, we define the diffraction efficiency of the element as the ratio between the light intensity $I_{n=1;m=0}$ in the desired signal orders and the incident light I_{inc}:

$$\eta=\frac{I_1}{I_{\mathrm{inc}}}=\operatorname{sinc}^2\left(\frac{1}{N}\right) \tag{5.1.9}$$

For a high number of quantization steps ($N\rightarrow\infty$), η approaches unity so that all of the light is then contributing to the signal [9, 10]. This basic rule of phase quantization can be applied to any quantized diffractive element.

5.1.2 Alternative quantization schemes for microlenses

We now apply the blazing procedure described above to the circularly symmetrical phase function of a microlens. The phase is first quantized to a phase depth of 2π and then approximated with discrete steps of equal phase depth. The result is a number of concentric rings called Fresnel zones. The phase slope of the microlens becomes increasingly steep towards the edges (Fig. 5.3a). Thus, the 2π quantization results in Fresnel zones which decrease with increasing radial distance from the centre of the lens. The limit for the fabrication of diffractive lenses is reached if the period in the outermost zones becomes too small to be fabricated with the available technology. For lithographically fabricated multilevel elements it is possible to extend the lens diameter by reducing the number of phase quantization steps per period. In this case, however, the efficiency of the outermost zones is reduced. The areas at the edges of the lens contribute less than the center areas to the lens performance. This affects the imaging behaviour of the lens as compared to homogeneously quantized lenses [12].

A similar effort to overcome the fabrication limit of diffractive lenses is the concept of so-called superzone lenses [13]. In order to avoid the decrease of the period size in the outer zones of the lens, the structuring depth is increased towards the edges. As illustrated in Fig. 5.3b this yields larger periods. This concept is beneficial for techniques which allow the fabrication of deep continuous relief profiles such as diamond turning. In the case of multilevel phase elements the trade-off becomes obvious since the superzone structure corresponds to a reduction in the number of quantization steps per 2π period. This again leads to lower diffraction efficiency in the outer zones.

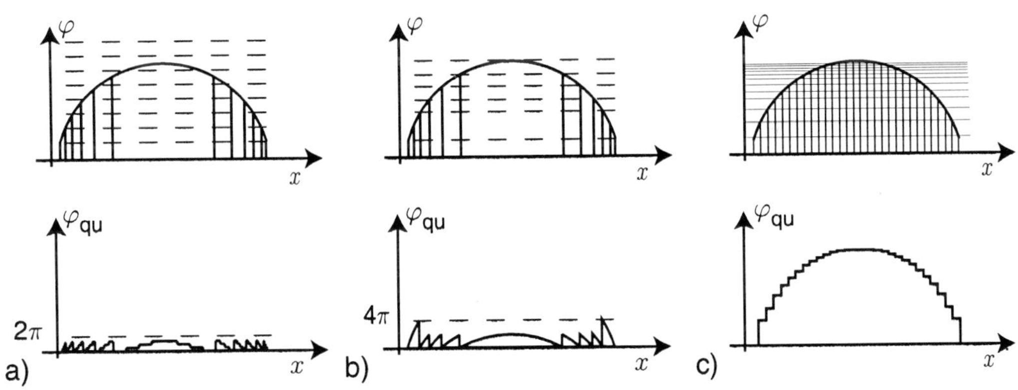

Fig. 5.3: Different quantization schemes for diffractive lenses: a) conventional quantization based on the Fresnel zones; b) superzone concept to increase the aperture of diffractive lenses; c) constant pixel size quantization, e.g., for implementation in spatially quantized liquid crystal phase modulators.

In other concepts for the implementation of diffractive lenses a quantization procedure is applied, which keeps the pixel size constant over the whole element (Fig. 5.3c). In this case, phase quantization automatically results from the spatial quantization. Because of the analogy to phased array microwave antennas, this approach to quantization is sometimes referred to as phased array-like binary optics [14]. Due to the lower phase slope, the phase steps are much lower in the central area than in the outer periods of the lens. We again observe an increased efficiency in the central areas. If the phase is not mapped into the region $[0,2\pi]$, such lenses do not show any periodicity. Because of this lack of periodicity, some of the features typical for DOEs, e.g., dispersion, do not occur with these lenses. They are not typical representatives of DOEs. Nevertheless, due to the spatial quantization, some diffractive effects such as stray light in higher orders can be observed. This quantization concept is especially interesting for the implementation of diffractive lenses in phase modulators such as ferroelectric liquid crystals [15, 16, 17]. Such modulator devices are usually laterally quantized to a relatively coarse square pixel grid and possibly provide large phase modulation depth. Fig. 5.3c shows the extreme case. Most of the spatial light modulator based lenses in addition to the relatively coarse spatial quantization also show a 2π quantization, to reduce the required phase modulation depth of the modulator. The quantization of diffractive lenses based on constant pixel dimensions has also been demonstrated for optimization of the lens performance [18].

5.1.3 Examples of diffractive optical components

So far during our discussion of diffractive optics we have learnt about diffractive implementation of microlenses and prism gratings. The potential of diffractive optics, however, lies in the fact that any phase profile can be fabricated as a DOE, even if refractive implementation is impossible due to constraints in the fabrication process. On the other hand, DOEs for any functionality are implemented as optimized gratings. The fabrication is possible with the same technological approach, independent of the functionality. This makes diffractive optics a versatile concept for microoptics. In the following chapters of this book we will discuss several applications of diffractive microoptics, such as multiple beam splitting (Chapter 7), beam shaping and steering (Chapter 8). In Fig. 5.4 we summarize some of the frequently used types of DOEs, while Fig. 5.5 shows REM photographs of multiple phase-level diffractive optical elements.

5.2 Fabrication of diffractive optics

With the concept of phase quantization described in the previous section it is possible to encode an arbitrary phase profile in a diffractive element [19]. Most of the microlithographic techniques for pattern transfer can be used for the fabrication of these kinoform DOEs. Microlithographic techniques are optimized for the fabrication of step profiles and are, therefore, well suited for kinoform elements. Phase steps may be generated by anisotropic lithographic etching or thin layer deposition [20, 21, 22, 23]. Depending on the substrate material and the requirements on the DOE it is possible to choose from the large variety of lithographic methods introduced in Chapter 3.

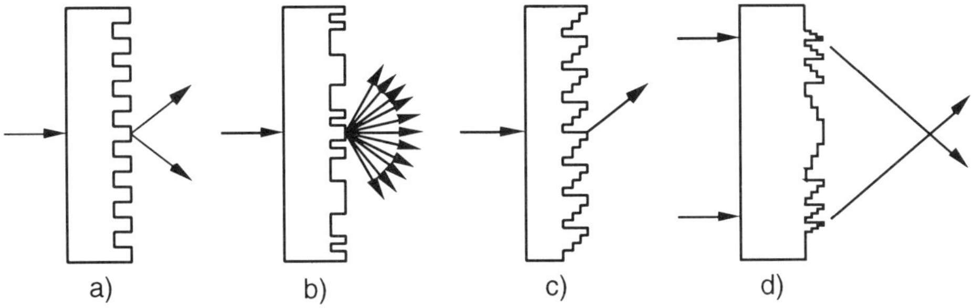

Fig. 5.4: Examples of DOEs: a) 1×2 beamsplitter; b) 1×N beamsplitter (e.g., Dammann grating); c) beam deflector; d) diffractive lens.

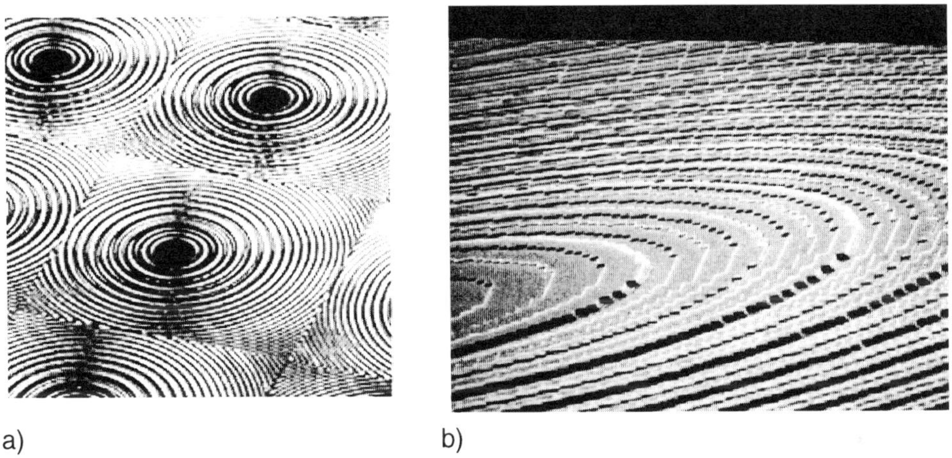

Fig. 5.5: REM pictures of diffractive optical elements: a) microlens array with 4 phase levels; b) section of an 8 phase level DOE combining the functionality of a beam splitter and a diffractive microlens.

5.2.1 Multimask processing for kinoform DOEs

A sequence of binary processing steps can be used to generate multilevel profiles. Since each phase step has a specific height, an intuitive approach is to process each step separately. To this end we need one lithographic mask for each phase step in the DOE (Fig. 5.6a). This linear mask sequence requires $N-1$ processing steps for the fabrication of a DOE with N different phase levels [24]. It allows one to fabricate multiple phase level diffractive elements with arbitrary irregular step heights.

As in the previously described quantization scheme, for DOE fabrication generally steps of equal height are used. In this case the number of processing steps can be reduced to $\log_2 N$ for an element with N phase levels (Fig. 5.6b). A logarithmic mask sequence can then be applied, where each etching step generates 2 phase levels [24]. For a transmission grating we can calculate the necessary etching depths t_N for the N-th mask layer by means of:

$$t_N = \frac{\lambda}{2^N(n-1)} \tag{5.2.1}$$

Obviousely, for gratings with phase depths larger than $\Delta\varphi = 2\pi$ and for gratings used in reflection other etching depths result (see Exercises). The logarithmic mask sequence significantly reduces production cost by saving processing time and lithographic masks. Mask patterns with higher frequencies are etched to lower depths, which is advantageous for high resolution lithography. By a suitable decomposition of the etching depths, phase gratings with varying step heights can also be fabricated in a logarithmic mask sequence [25]. For practical purposes it is recommended that one should process the high frequency shallow phase steps first and progress to deeper and coarser features in subsequent lithographic steps. This order helps to reduce fabrication errors. For the fabrication of very fine features in particular, the uniformity of the photoresist coating is very critical. Good resist coatings require good flatness of the substrate surface. After previous profiling steps, however, the substrate surface is no longer flat. The application of very homogeneous coatings becomes increasingly difficult with the size and number of existing profile features.

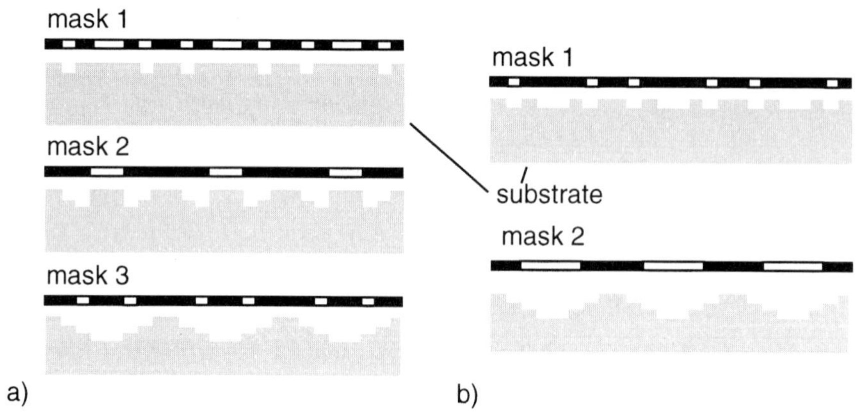

Fig. 5.6: Fabrication of a 4 level phase element using a) three lithographic steps in a linear mask sequence and b) two steps in a logarithmic mask sequence.

Due to the use of a sequence of binary fabrication steps this branch of optics is often referred to as "binary optics"[26]. We find this term sometimes misleading, since it might cause confusion if used in the context of truly binary elements, i.e., elements implemented with only two different phase levels. In the following, we will, therefore, use the terms multilevel phase elements or kinoform elements as opposed to two level phase elements or binary phase elements.

5.2.2 Fabrication errors for kinoform elements

The precision of the lithographic processes applied to DOE fabrication is limited, so that the surface profile generated in reality deviates from an ideal profile. In this section we shall categorize the most important fabrication errors, investigate their effect on the kinoform profile and describe a method to analyse their impact on the performance of the DOE. As described above, the fabrication process for multilevel phase elements generally consists of a sequence of binary fabrication steps each of which requires an alignment and an etching step. Both processing steps are possible with some finite precision and thus introduce fabrication errors. Fig. 5.7 illustrates the effect of the most important errors for the fabrication of a 4 phase level linear grating using a logarithmic mask sequence [27, 28, 24].

- **Alignment errors** occur if the masks for subsequent lithographic structuring processes are slightly shifted relative to existing structures. Due to this lateral misalignment, additional structures are generated in the phase element. Fig. 5.7b shows the deviations occurring in a four phase level kinoform grating for a small lateral misalignment (δs) of the second mask layer. By analysing the difference between the profile generated in reality and the ideal element we find the phase error caused by the misalignment.
- **Etching errors** occur if the phase step introduced during individual etching processes deviates from an ideal step profile with precisely defined etching depth and vertical edges.
  - **Errors in the phase depth** are caused by the finite depth resolution of the etching process. Instead of an ideal rectangular phase profile with depth t a rectangular profile of the correct width but with the erroneous depth $t + \delta t$ is generated. This leads to a phase error ($\delta\varphi$) in the DOE illustrated in Fig. 5.7c.
  - **Over- or underetching** (Fig. 5.7d and e) causes a change of the lateral extensions of the mask structures upon transfer onto the glass substrate. The ideal widths w of the structures in the DOEs are increased or decreased by δw.
  - **Partial isotropy of the etching process** generates slopes of finite steepness at the edges of the rectangular phase steps. As illustrated in Fig. 5.7f this again introduces a deviation from an ideal phase front, which is periodic with the mask structure of the respective layer.

As we can see from Fig. 5.7 the different types of fabrication errors introduce deviations $\delta\varphi$ in the phase profile which reflect the periodicity of the mask structure used in the respective process. Mathematically the real phase profile can be described by multiplication of the ideal phase distribution with the phase error:

$$e^{i\varphi_{\text{real}}(x,y)} = e^{i(\varphi_{\text{ideal}}(x,y)-\varphi_{\text{error}}(x,y))} = e^{i\varphi_{\text{ideal}}(x,y)}e^{-i\varphi_{\text{error}}(x,y)} \qquad (5.2.2)$$

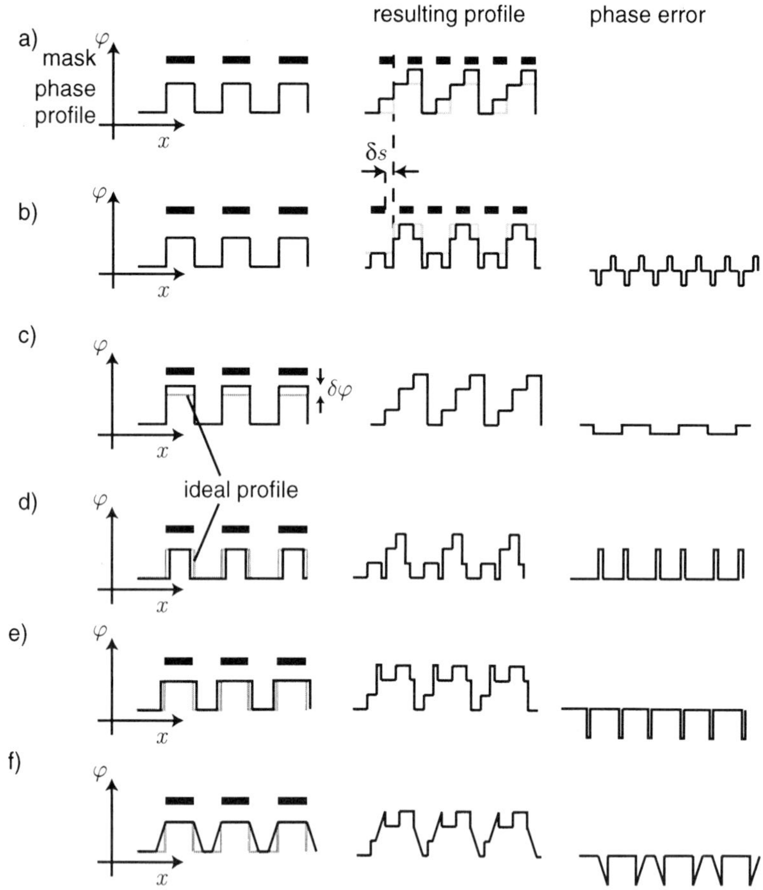

Fig. 5.7: The effect of the various possible errors during the multimask fabrication process of a 4-level phase grating: a) ideal process; b) lateral misalignment of the two masks; c) error in the etching depth; d) over-etching; e) under-etching; f) partial isotropy during etching. In c), d), e), f) we assume that the errors only occur in mask layer 1.

In a scalar approximation (see following section) the effect of the phase errors on the optical performance can be calculated by a convolution of the ideal signal reconstructed from $e^{i\varphi_{\text{ideal}}}$ and the error in the output plane generated by $e^{i\varphi_{\text{error}}}$. For example, we want to calculate the diffraction pattern of the phase function introduced by an error $\delta\varphi$ in the etching depth of the N-th mask with period $p/2^N$. $e^{-i\varphi_{\text{error}}(x)}$ can be described by:

$$
\begin{aligned}
e^{i\varphi_{\text{error}}(x)} &= \sum_m \left[\text{rect}\left(\frac{x2^{N+1}}{p}\right)e^{i\delta\varphi} + \text{rect}\left(\frac{x - p/2^{N+1}}{p/2^{N+1}}\right)\right] \\
&= \sum_n A_n e^{2\pi i n \frac{x2^N}{p}}
\end{aligned}
\tag{5.2.3}
$$

The amplitudes in the diffraction orders of this phase profile can be calculated by the Fourier transform:

$$\begin{aligned} A_n &= \frac{2^N}{p} \int_{-\frac{p}{2^{N+2}}}^{\frac{p}{2^{N+2}}} e^{i\delta\varphi} e^{-2\pi i n x \frac{2^N}{p}} dx + \frac{2^N}{p} \int_{\frac{p}{2^{N+2}}}^{\frac{3p}{2^{N+2}}} e^{-2\pi i n x \frac{2^N}{p}} dx \\ &= \frac{1}{2} e^{i\delta\varphi} \mathrm{sinc}\left(\frac{n}{2}\right) + \frac{1}{2} e^{i\pi n} \mathrm{sinc}\left(\frac{n}{2}\right) = \frac{1}{2} \mathrm{sinc}\left(\frac{n}{2}\right) \left(e^{i\delta\varphi} + e^{i\pi n}\right) \end{aligned} \tag{5.2.4}$$

The intensities in the diffraction orders described by Eq. 5.2.4 increase with increasing etching error $\delta\varphi$. Thus, with increasing etching error more and more light is lost into noise orders. It is interesting to note that the diffraction amplitudes are only influenced by the absolute error in the etching depth rather than by the depth error relative to the desired phase depth. Similar results can be achieved for the other types of fabrication errors (see Exercises). We can conclude that the main effect of fabrication errors is a reduction of the diffraction efficiency in the desired signal orders. Since at the same time the light in the noise orders is increased we can also talk of a reduced signal-to-noise ratio.

5.3 Modelling of diffractive optics

This section is devoted to the theory of diffraction. The main goal is to outline the mathematical strategies for DOE modelling. We assume that the reader is familiar with the basics of the mathematical description of light. For the mathematical details, which are beyond the scope of this book, we refer to numerous books and review articles on this subject [5, 29, 30, 31, 32, 33]. Here, we rather outline the different approximations which can be applied for the description of different types of DOEs. Especially the range of validity of the various theoretical approaches is of interest. We first discuss the variety of approaches to "rigorous diffraction theory". We state the problem and outline the basic mathematical steps towards a solution of the diffraction problem. The validity of the different approaches is the topic of the following paragraph. There we will learn to distinguish between "thin and thick gratings". This paragraph will be followed by several sections dealing with "scalar diffraction theory". We will focus on "Fresnel and Fourier diffraction" and discuss "linear kinoform gratings" and "diffractive lenses" in this diffraction regime. Modelling of the aberrations of diffractive lenses through ray tracing is described in the section on "ray-tracing analysis of diffractive lenses". "Detour-phase diffractive optical elements" are another group of elements which we want to describe in the scalar approximation. The last two paragraphs are devoted to "holographic optical elements" which are generally described as Bragg gratings, and "zero-order gratings" which can easily be modelled through effective medium theory.

5.3.1 Approaches to rigorous diffraction theory

The mathematical description of light is based on Maxwell's equations. For propagation in a homogeneous, nonmagnetic, dielectric material the components of the electromagnetic wavefield (e.g., the electrical field vector $\boldsymbol{E}(x,z)$) are found as solutions to the Helmholtz equation:

$$\nabla^2 \boldsymbol{E}(x,z) + n^2 k^2 E(x,z) = 0 \tag{5.3.1}$$

The differential operator $\nabla^2$is defined as:

$$\nabla^2 = \frac{\partial^2}{\partial x^2} + \frac{\partial^2}{\partial z^2} \tag{5.3.2}$$

n is the refractive index of the medium and $\boldsymbol{k}$ denotes the wavevector:

$$|\boldsymbol{k}| = k = \frac{2\pi}{\lambda} \tag{5.3.3}$$

A plane wave $\boldsymbol{E}_{\varphi_i}(x,z)$ propagating at an angle φ_i is a classical solution of the Helmholtz equation:

$$\mathbf{E}_{\varphi_i}(x,z) = e^{ink[cos(\varphi_i)z + sin(\varphi_i)x]} \tag{5.3.4}$$

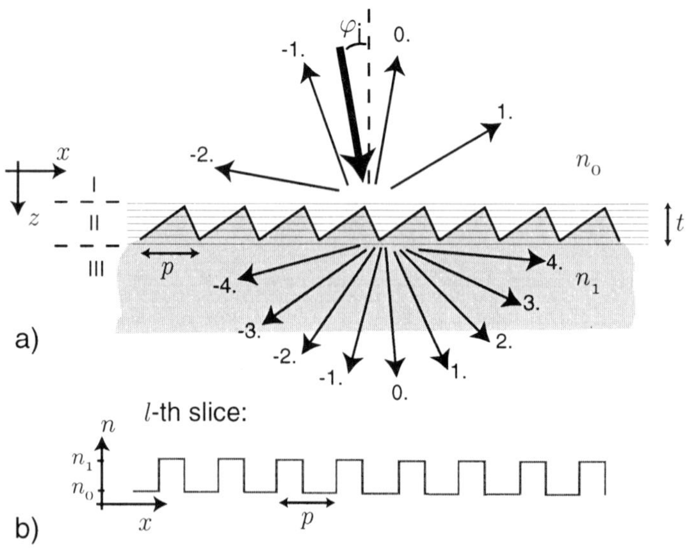

Fig. 5.8: Separation of a thick grating into different regions and layers.

In diffraction theory our goal is to find out how a diffraction grating influences such a plane wave. It is well known that a discrete set of propagating plane waves is generated behind the grating. More specifically we want to find ways to calculate the complex amplitudes of the individual diffraction orders. In order to analyse this problem we first have a look at a general example of a diffraction grating of finite thickness. As illustrated in Fig. 5.8a such a grating can be subdivided into 3 regions. In regions I and III the light waves are propagating in homegeneous media, e.g., with refractive indices n_0 and n_1, respectively. In the intermediate layer II of thickness t the dielectic features (e.g., the refractive index) of the medium may be varying periodically in the x-direction. In order to take into account the shape of the grating period this region is subdivided into "thin" slices. Within each slice, e.g., the refractive index variation along the x-direction assumes rectangular shape (Fig. 5.8b).

An electromagnetic field which solves this general diffraction problem needs to fulfil Maxwell's equations, i.e., the Helmholtz equations for all individual field components. The boundary conditions need to be satisfied at the numerous interfaces. Diffracted, reflected as well as transmitted field components need to be taken into account. Rigorous analytical solutions to this problem can be found only for specific diffraction screens. The classical example is Sommerfeld's solution for diffraction at a perfectly conducting half-plane [5, 34].

Lacking general analytical solutions for the rigorous diffraction problem, the focus is on numerical solutions. Without going into detail, we want to give the reader a flavour of some of the different approaches. We consider the incidence of a plane wave, described by Eq. 5.3.4, onto a grating as illustrated in Fig. 5.8. The Floquet theorem states that the solution $F(x)$ of a differential equation with periodic coefficients is pseudo-periodic [35], i.e.:

$$F(x+p) = \alpha F(x) \tag{5.3.5}$$

where p denotes the periodicity of the coefficients and α is a constant.

Combined with the boundary conditions of our diffraction problem this is the origin for the periodicity of the diffracted and reflected components in regions I and III. Using a Fourier expansion and by inserting the wave components into the Helmholtz equation (Eq. 5.3.1), we can derive the well known propagation angles φ_{r} and φ_{t} for the diffracted plane waves in regions I and III respectively:

$$\sin(\varphi_{\mathrm{r},m}) = \frac{m\lambda}{pn_0} + \sin(\varphi_i) \tag{5.3.6}$$

$$\sin(\varphi_{\mathrm{t},m}) = \frac{m\lambda}{pn_1} + \sin(\varphi_i)\frac{n_0}{n_1} \tag{5.3.7}$$

The transmitted electrical field component of the electromagnetic wave can be written as a plane-wave expansion of the transmitted diffraction orders. This expansion is often referred to as the Rayleigh expansion:

$$\mathbf{E}_{\mathrm{t}}(x,z) = \sum_{m=-\infty}^{m=+\infty} T_m e^{[\frac{2\pi i n_1}{\lambda}(\cos(\varphi_{\mathrm{t}/m})+\sin(\varphi_{\mathrm{t}/m}))]} \tag{5.3.8}$$

The Rayleigh expansion in plane waves is closely related to the Fourier expansion which we will use later on in scalar diffraction theory [5]. For a solution of the diffraction problem it is possible to calculate the diffraction amplitudes T_m from Maxwell's equations and the corresponding boundary conditions.

The most interesting aspect of the analysis of a grating, such as that depicted in Fig. 5.8, is the propagation of the electromagnetic field in region II with varying coefficients. In this region, e.g., the refractive index $n(x)$ varies periodically in lateral direction between n_0 and n_1. In the propagation direction z this region is separated into "thin" slices with a constant distribution $n(x)$ (Fig. 5.8b). In each of these slices the propagation is described by the Helmholtz equation for the electric field $\boldsymbol{E}(x, z)$, which reads:

$$\nabla^2 \boldsymbol{E}(x, z) = n_l^2(x) k^2 \boldsymbol{E}(x, z) \tag{5.3.9}$$

In the **rigorous coupled wave method** [36, 37, 38, 39] the periodic dielectric constant $\epsilon_l(x) = n_l^2(x)$ is developed as a Fourier series:

$$\epsilon_l(x, z_l) = \epsilon_1 + (\epsilon_2 - \epsilon_1) \sum_{j=-\infty}^{\infty} \hat{\epsilon}_{j,l} \cdot e^{2\pi i j x/p} \tag{5.3.10}$$

Here p denotes the period of the index grating.

The electric field in the l-th slice is expanded in a series, as follows:

$$\boldsymbol{E}_{2,l} = \sum_{q=-\infty}^{\infty} \boldsymbol{S}_{q,l}(z) e^{-i\left(\frac{2\pi\sqrt{\epsilon_{0,l}}}{\lambda} - q\frac{2\pi}{p}\boldsymbol{x}\right)\boldsymbol{r}} \tag{5.3.11}$$

The name "rigorous coupled wave method" stems from the interpretation of $\boldsymbol{S}_{q,l}(z)$ as amplitude functions of an inhomogeneous wave with the wavevector $\frac{2\pi\sqrt{\epsilon_{0,l}}}{\lambda} - q\frac{2\pi}{p}\boldsymbol{x}$, where $\epsilon_{0,l}$ is the arithmetic mean of the dielectric constant and $\boldsymbol{x}$ is the unity vector in the lateral direction. These waves are coupled so that only the complete set rather than any single wave fulfils the propagation equation. Inserting the expansions of the dielectric constant (Eq. 5.3.10) and the electric field (Eq. 5.3.11) into the propagation equation we obtain an infinite number of second-order differential equations. A solution is possible by restriction to a finite number Q of waves and taking into account the boundary conditions for each wave which are derived from Maxwell's equations.

A similar differential approach is taken in the **differential method** [40]. Here, both the dielectric constant and the electric field, which is pseudoperiodic, are developed as Fourier series. This converts the propagation equation into a coupled differential equation which can be solved by matrix methods.

In the **modal method** [41] the full set of eigenfunctions within the grating grooves which expand the electric field vector is calculated. This is only possible for perfectly conducting gratings with a selection of profiles, e.g., conducting binary gratings. For these specific grating profiles the modal method reduces the diffraction problem to the solution of a finite number of linear equations, and is thus one of the most efficient methods.

The various mathematical approaches for calculating the diffraction efficiencies of periodic gratings differ in their range of validity. The modal method is only applicable to perfectly conducting gratings. It is additionally restricted to binary gratings and superpositions of them. The rigorous coupled wave theory as well as the differential method, on the other hand, can be used for dielectric and metallic gratings, but do not allow a direct calculation for perfectly conducting gratings. There is no restriction for these methods as to the shape of the grating profile, however, for the differential method there is an upper limit on the thickness of the grating [42].

These brief remarks may be sufficient to give the reader a flavour of the different techniques and limitations of rigorous calculation methods for grating diffraction. For more information about this topic the reader is referred to the variety of publications in this field [30, 43].

5.3.2 Thin and thick gratings

Diffraction gratings are frequently described as "thin" or "thick". Before using these terms in the following sections, we want to define their meaning both intuitively and mathematically. Intuitively we call a diffraction grating a thin grating if an incoming wave interacts only once with the grating structure. This is the case if the diffraction screen is very thin or if the diffraction orders generated at the front surface of the grating propagate at small angles to the optical axis. To support our intuitive arguments, let us imagine a plane wave which is incident on an ideally absorbing screen with a transparent hole [44]. If the diameter of the hole is large enough, the diffraction angles will be small and all the light incident on the transparent area will pass through the screen (Fig.5.9a). In this case the complex light amplitude immediately behind the screen $U(x, z+0)$ is described by the product of the incident plane wave $U(x, z-0)$ and the amplitude screen $A(x) = \text{rect}[\frac{x}{D}]$:

$$U(x, z+0) = U(x, z-0)A(x) \tag{5.3.12}$$

Eq. 5.3.12 is often called the Kirchhoff approximation, which is valid for thin amplitude as well as thin phase gratings. We know that for gratings with sharp edges diffraction orders always exist, which propagate at large angles. If the screen has finite thickness, these orders are absorbed upon propagation through the hole. Nevertheless, the Kirchhoff approximation has a wide range of validity and is widely applied for DOE modelling.

We now imagine that the hole in the screen becomes small (Fig.5.9b), i.e., the diffraction angles increase. In this case an increasing number of orders diffracted at the front surface interact a second time with the screen. They are, e.g., absorbed during the propagation through the hole. The Kirchhoff approximation (Eq. 5.3.12) is not valid in this case. In order to calculate the effect of the grating rigorous approaches are necessary. The same effect can be observed if the thickness of the screen is increased while the hole diameter remains constant (Fig.5.9c).

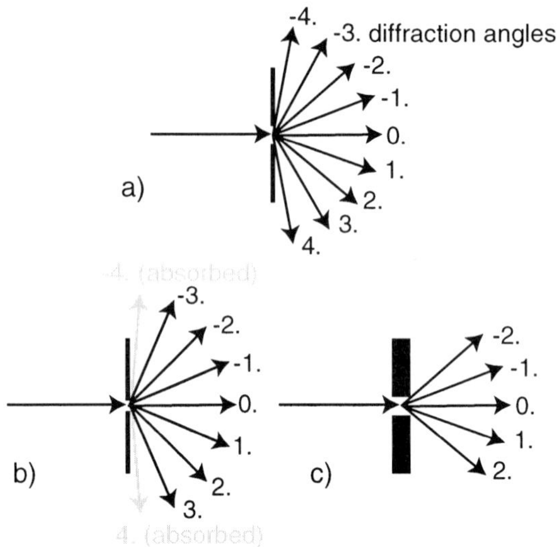

Fig. 5.9: Diffraction at "thin" and "thick" diffraction screens: a) "thin" diffraction screen with pinhole; b) "thick" screen with reduced pinhole diameter; c) "thick" screen with increased physical thickness.

The critical aspect, which decides whether a grating can be considered as being thick or thin is the ratio of the thickness to the period of the grating, rather than the physical thickness alone. In the following sections we will see that this relative thickness can have significant impact on the optical performance of the DOE. Depending on the underlying approximate diffraction theory a small finite thickness ($\approx \lambda$) of a phase grating may result in the best agreement with rigorous calculations [45].

A more precise distinction between thin and thick gratings can be found in the angular or wavelength sensitivity [46]. The angular or wavelength range within which the diffraction efficiency decreases to half of the maximum value can be described by the ratio between the physical thickness t of the grating and the grating period p. The sensitivity of a thick grating is very high. For infinitely thin diffraction gratings, the wavelength and angular sensitivity is much smaller. These considerations of wavelength sensitivity lead to the following classification of thin and thick gratings:

$$\left(\frac{t}{p}\right)_{\text{thick}} > 10 > \left(\frac{t}{p}\right)_{\text{thin}} \tag{5.3.13}$$

In the following sections we will discuss scalar diffraction theory and Bragg diffraction as useful approximations for modelling thin and thick gratings, respectively. Yet another mathematical definition of thin and thick gratings can be found by comparing the results gained from these approximations with rigorous diffraction theory [46]. Without proof, we show the results of these calculations. If we require the error between the approximate and the rigorous approaches to be $< 1\%$ we find for a thin grating:

$$\frac{\pi^2 n_1^2}{n_0^2}\frac{t^2}{p^2} < 1 \tag{5.3.14}$$

Similarly, we call the gratings "thick", if:

$$\frac{2\lambda^2}{p^2 n_1^2} > 10 \tag{5.3.15}$$

As we can see from these inequalities, the validity of the different approximations is only indirectly related to the thickness of the grating. Especially the definition for the thick grating rather depends on the diffraction angles than on the physical thickness of the grating. If none of the approximations in Eqs 5.3.14 and 5.3.15 is valid, the grating is neither thin nor thick in the sense that the diffraction efficiencies can be described with high precision neither by Bragg reflection nor by scalar approximation. Rigorous calculations are necessary to estimate the performance of such gratings.

5.3.3 Scalar diffraction theory

In the rigorous approaches to diffraction theory the vector characteristics of the electromagnetic wavefield are considered via the boundary conditions resulting from Maxwell's equations. They describe the coupling between the individual components of the electromagnetic wavefield. For a large variety of diffractive elements, however, it is possible to consider the wavefield as scalar. In this case we neglect the coupling between the electric and magnetic vector components upon interaction with inhomogeneous media. We only consider one transverse component of the electric or magnetic field. This approximation is justified for "thin" gratings, where the relevant diffraction orders concentrate around the optical axis. For a mathematical treatment we again start with the time independent Helmholtz equation:

$$\nabla^2 U(x,y,z) + \frac{2\pi}{\lambda} U(x,y,z) = 0 \tag{5.3.16}$$

In order to describe wave propagation, we decompose the scalar wavefield $U(x,y,z)$ into plane waves by means of a Fourier transformation:

$$U(x,y,z) = \int\!\!\int \tilde{U}(\nu_x,\nu_y,\nu_z)\, e^{2\pi i(\nu_x x+\nu_y y+\nu_z z)}\, d\nu_x d\nu_y d\nu_z \tag{5.3.17}$$

ν_x, ν_y, and ν_z are called spatial frequencies and denote the propagation directions of the components of the wavefield. Inserting Eq. 5.3.17 into Eq. 5.3.16 we derive the differential equation:

$$\frac{\partial \tilde{U}(\nu_x, \nu_y, \nu_z)}{\partial z^2} + k^2[1 - \lambda^2(\nu_x^2 + \nu_y^2)]\,\tilde{U}(\nu_x, \nu_y, \nu_z) = 0 \tag{5.3.18}$$

By solving this differential equation, we obtain the general equation for light propagation in a homogeneous medium:

$$U(x,y,z) = \tag{5.3.19}$$
$$\iiiint U(x_0, y_0, z_0)\, e^{2\pi i[\nu_x(x-x_0)+\nu_y(y-y_0)+\frac{z-z_0}{\lambda}\sqrt{1-\lambda^2(\nu_x^2+\nu_y^2)}]}\, d\nu_x d\nu_y dx_0 dy_0$$

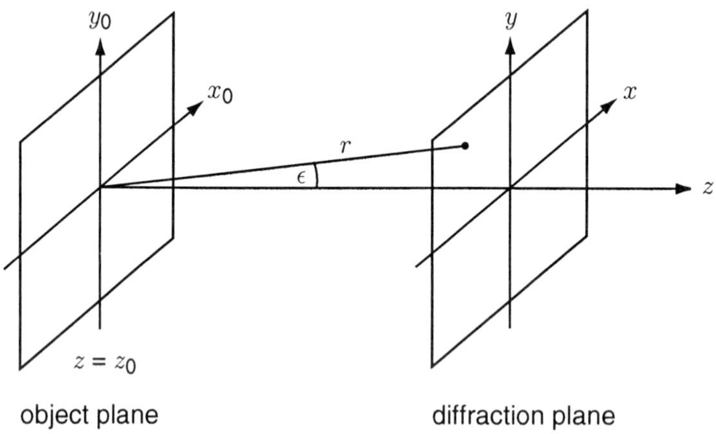

Fig. 5.10: Geometry for the description of light propagation by the diffraction integral.

According to the illustration in Fig. 5.10 this equation describes the development of the complex amplitude distribution upon propagation from an object plane z_0 to a plane z. In the paraxial approximation, i.e., for small propagation angles ϵ to the optical axis, this equation can be written as:

$$U(x,y,z) \propto \frac{i}{\lambda} \iint U(x_0, y_0, z_0)\, \frac{e^{ikr}}{r} \cos\epsilon\; dx_0 dy_0 \tag{5.3.20}$$

Eq. 5.3.20 is generally known as Kirchhoff's diffraction integral. It mathematically describes Huygens principle, which states that the wavefield at a location z behind a diffracting screen is described by a superposition of spherical waves ($\frac{e^{ikr}}{r}$) originating at the screen [5, 9].

5.3.4 Fresnel and Fraunhofer diffraction

Kirchhoff's diffraction integral can be approximated to describe the two most important regimes in scalar diffraction theory. In order to describe near-field diffraction or "Fresnel diffraction", we approximate the distance r between object and observation plane by [9, 44]:

$$\begin{aligned} r &= (z-z_0)\left[1+\left(\frac{x-x_0}{z-z_0}\right)^2+\left(\frac{y-y_0}{z-z_0}\right)^2\right]^{\frac{1}{2}} \\ &\approx (z-z_0)\left[1+\frac{1}{2}\left(\frac{x-x_0}{z-z_0}\right)^2+\frac{1}{2}\left(\frac{y-y_0}{z-z_0}\right)^2\right] \end{aligned} \tag{5.3.21}$$

In this case the diffraction integral is approximated by a Fresnel transform:

$$\begin{aligned} &U(x,y,z) \approx \\ &\quad \frac{i}{\lambda}\frac{e^{ik(z-z_0)}}{z-z_0}\iint U(x_0,y_0,z_0)\, e^{\frac{i\pi}{\lambda(z-z_0)}[(x-x_0)^2+(y-y_0)^2]}\, dx_0 dy_0 \end{aligned} \tag{5.3.22}$$

Eq. 5.3.22 is called the quadratic approximation of the Kirchhoff integral. By applying Eq. 5.3.21 we have effectively approximated Huygens spherical elementary waves by parabolic waves. This so-called paraxial approximation is valid in a region close to the optical axis. A well-known phenomenon of Fresnel diffraction is the "self-imaging" of periodic structures, the so-called Talbot effect. This will be discussed more closely in the context of "array illumination" (Chapter 8).

If we approximate Kirchhoff's integral for even larger distances $(z-z_0)$ we reach the regime of "far-field" or "Fraunhofer" diffraction [9, 44, 47]. This is valid if:

$$z-z_0 \gg \frac{1}{2}k(x^2+y^2) \tag{5.3.23}$$

In this case we can write the diffraction integral as

$$\begin{aligned} &U(x,y,z) \propto \\ &\quad \frac{i}{\lambda}\frac{e^{ik(z-z_0)}}{(z-z_0)} e^{\frac{i\pi}{\lambda(z-z_0)}(x_0^2+y_0^2)}\iint U(x_0,y_0,z_0)\, e^{-2\pi i\frac{xx_0+yy_0}{\lambda(z-z_0)}}\, dx_0 dy_0 \end{aligned} \tag{5.3.24}$$

This equation describes the Fourier transform of the input amplitude distribution $U(x_0, y_0, z_0)$. In other words, at large distances from a diffracting object the amplitude distribution can be described by the Fourier transform of the object distribution. In order to fulfil condition 5.3.23, very large ("infinite") distances between the diffracting object and the observation plane are required. Nevertheless, Fraunhofer diffraction is of significant importance in optics. The Fraunhofer approximation is, e.g., valid in the focal plane of a lens.

5.3.5 Linear kinoform grating

Linear diffractive gratings can be used in microoptical systems, e.g., as beam deflectors. In this paragraph we discuss linear kinoform gratings using a scalar approximation. We consider the setup illustrated in Fig. 5.11. The diffraction grating is illuminated by a plane wave. In the Kirchhoff approximation the wavefront immediately behind the "thin" object is described by the grating distribution $u(x_0) = g(x_0)$.

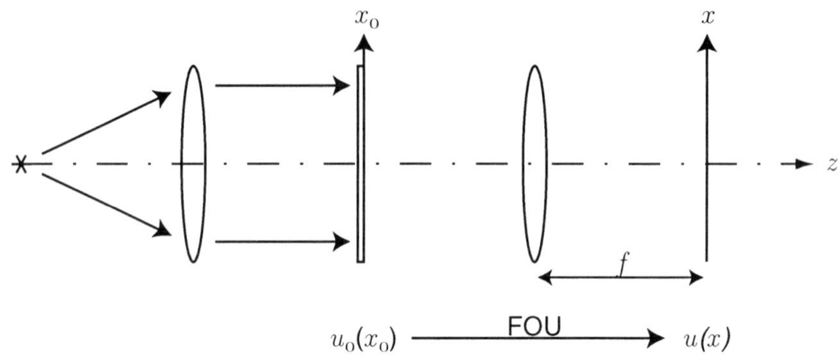

Fig. 5.11: The optical setup for reconstructing the Fraunhofer diffraction pattern of an object $u_0(x_0)$.

The directions α_m into which a wavefront incident at an angle α_{inc} is deflected can be calculated from the grating equation (see also Eq. 5.3.7):

$$\sin(\alpha_m) = m\frac{\lambda}{p} + \sin(\alpha_{\text{inc}}); \quad m = 0, \pm 1, \pm 2, \pm 3 \cdots \tag{5.3.25}$$

p denotes the grating period and the integer value of m gives the diffraction order. As we have seen, Eq. 5.3.25 is derived from Maxwell's equations without any approximation. For calculating the amplitude for various diffraction orders, we use the Fraunhofer approximation.

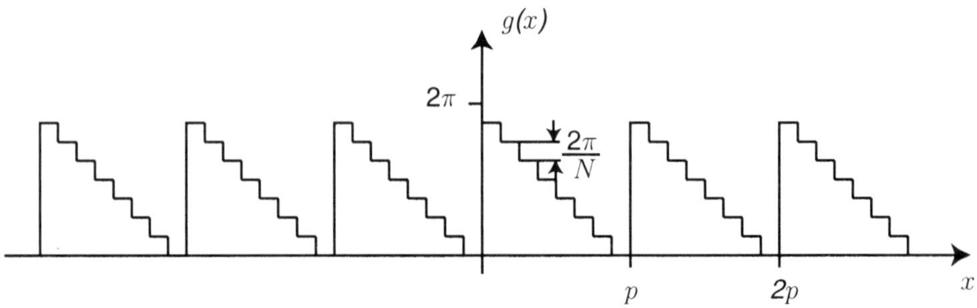

Fig. 5.12: The phase profile of a linear kinoform grating.

In order to optimize the light efficiency in the desired output direction (diffraction order), multilevel phase gratings are generally used. Linear multiple phase gratings can be considered as the diffractive version of beam deflectors optimized for lithographic fabrication.

We consider a one-dimensional linear grating (Fig. 5.12) with infinite extension and period p:

$$g(x) = g(x+p) \tag{5.3.26}$$

Because of the periodicity, the amplitude transmission function $g(x)$ can be expanded as a Fourier series:

$$g(x) = \sum_{m=-\infty}^{\infty} A_m \, e^{2\pi i m x/p} \tag{5.3.27}$$

The Fourier coefficients A_m give the values of the amplitudes of the diffraction orders (Fig. 5.13), which are calculated from:

$$A_m = \frac{1}{p} \int_0^p g(x) \, e^{-2\pi i m x/p} \, dx \tag{5.3.28}$$

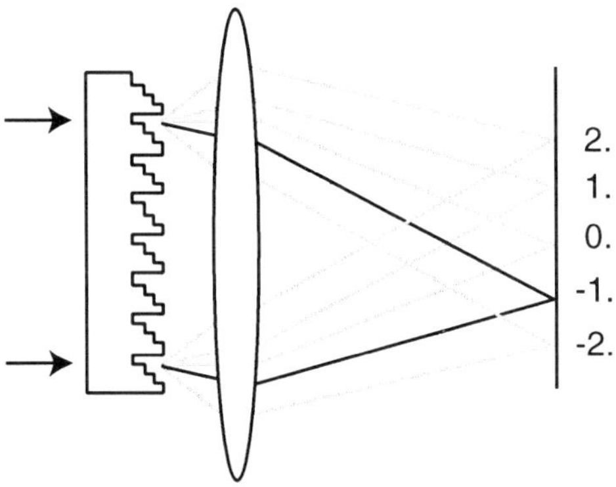

Fig. 5.13: Diffraction at a linear kinoform grating.

In what follows we shall set $p = 1$. Accordingly, from Parseval's theorem, we normalize the sum of intensities $I_m = |A_m|^2$ to unity:

$$\sum_{m=-\infty}^{\infty} I_m = 1 \tag{5.3.29}$$

We consider the more specific case of a grating with N discrete equidistant phase levels in the range $[0; 2\pi]$. Fig. 5.12 shows such a multilevel phase grating. The grating function can be described mathematically as:

$$g(x) = \sum_{k=-\infty}^{\infty} e^{-2\pi i \frac{k}{N}} \mathrm{rect}\left(\frac{x - \frac{k}{N} - \frac{1}{2N}}{\frac{1}{N}}\right) \tag{5.3.30}$$

Using Eq. 5.3.28, we calculate the amplitude of the m-th diffraction order [48]:

$$\begin{aligned} A_m &= \int_0^1 \sum_{k=0}^{N-1} e^{-2\pi i \frac{k}{N}} \ \mathrm{rect}\left(\frac{x - \frac{k}{N} - \frac{1}{2N}}{\frac{1}{N}}\right) e^{-2\pi i m x} dx \\ &= e^{i\pi m/N} \ \mathrm{sinc}\left(\frac{m}{N}\right) \frac{1}{N} \sum_{k=0}^{N-1} e^{2\pi i \frac{k(m+1)}{N}} \end{aligned} \tag{5.3.31}$$

The sum over k can be shown to be zero unless $m+1$ is a multiple of N:

$$\sum_{k=0}^{N-1} e^{2\pi i \frac{k(m+1)}{N}} = \begin{cases} N & : \ \text{if } m = jN - 1; \quad j \text{ integer} \\ 0 & : \ \text{else} \end{cases} \tag{5.3.32}$$

The blazed grating is designed to optimize the light intensity in a particular diffraction order. In our case, it is the -1st order. We therefore define the diffraction efficiency as the intensity of this order. From Eq. 5.3.31 we obtain for the efficiency of the quantized phase function:

$$\eta = I_{-1} = \mathrm{sinc}^2\left(\frac{1}{N}\right) \tag{5.3.33}$$

Eq. 5.3.31 corresponds to the result in Eq. 5.1.5, if $\varphi(x)$ is substituted by the linear phase slope of a prism. Fig. 5.14 shows the diffraction spectra of gratings with $N = 2$, 4 and 8 discrete phase levels. For $N = 2$, the symmetry of the grating profile causes a symmetrical spectrum with about 40.5% of the light being deflected into each of the two first orders. For $N > 2$, the grating profile becomes asymmetrical and, consequently, the spectrum becomes asymmetrical. The amount of light intensity in the -1st diffraction order is 81% for a four-level grating and 95 % for an eight-level grating [48].

It is important to point out that:

1. this calculation is true within the limits of scalar diffraction theory. If we violate the "thin" grating approximation, e.g., for features comparable to the wavelength of the light, the actual values for the diffraction orders can deviate significantly from the values predicted by scalar theory;
2. in practice, fabrication errors (e.g., etch depth errors and mask-to-mask misalignment) result in a reduction of the light efficiency.

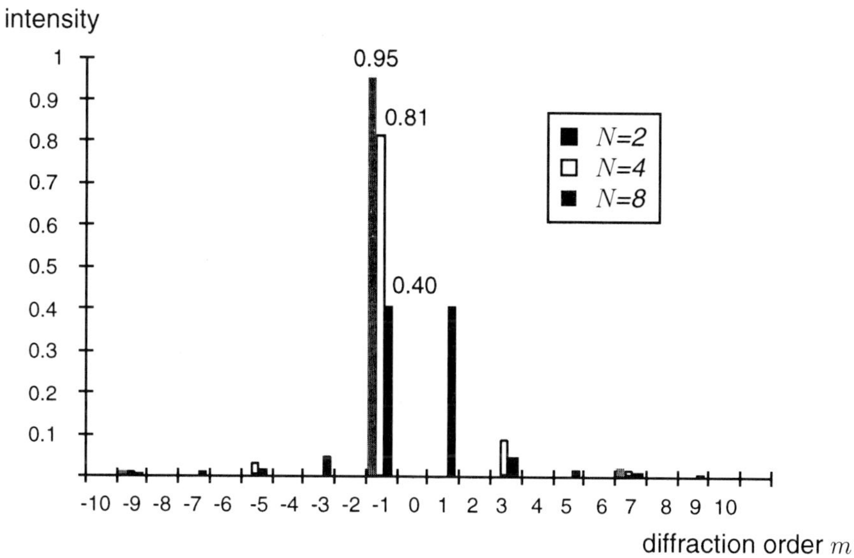

Fig. 5.14: The intensities in the diffraction orders of linear diffraction gratings with N phase levels.

5.3.6 Diffractive lenses

If the phase quantization procedure described above is applied to the circularly symmetrical phase profiles of microlenses, it yields a set of annular rings typical for diffractive lenses. If the structures within each of the annular zones consist of a continuous phase function the lenses are generally called Fresnel lenses. In Fresnel zone plates (FZPs) the ring system consists of diffracting binary phase or amplitude structures. In any case the annular rings are designed such that the optical path lengths for the light deflected from adjacent zones towards a common focal point differ by an integral multiple of a design wavelength. If r_j and f denote the radius of the j-th ring and the focal length, respectively, this condition is expressed mathematically as [2, 22, 49]:

$$r_j^2 + f^2 = (f + j\lambda)^2 \tag{5.3.34}$$

For the radius r_j of the j-th zone it follows that:

$$r_j^2 = 2j\lambda f + (j\lambda)^2 \tag{5.3.35}$$

For the paraxial approximation, where $f \gg j_{\max}\lambda$, we can approximate Eq. 5.3.35 by:

$$r_j^2 \approx 2j\lambda f \tag{5.3.36}$$

Eq. 5.3.36 describes the FZP pattern which is periodic in r^2 with period r_1^2. The focal length in this case is determined by:

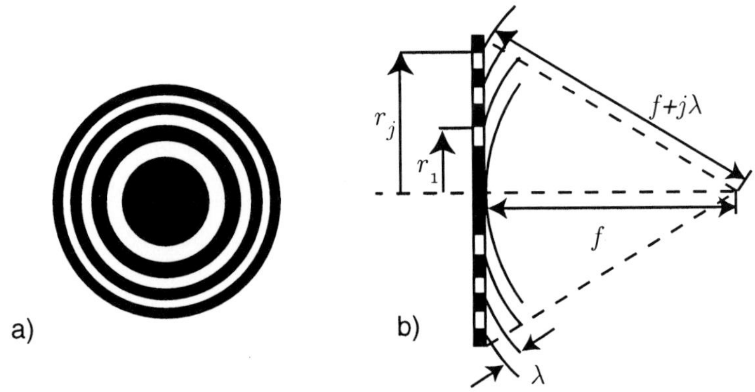

Fig. 5.15: Arrangement of the annular rings in Fresnel lenses: a) texture of the Fresnel zone plate; b) optical path differences between light rays from different zones.

$$f = \frac{r_1^2}{2\lambda} \tag{5.3.37}$$

Since it is very popular to approximate diffractive lenses by these paraxial expressions, we shall now discuss the validity of this approximation in more detail. To this end we calculate the maximum deviation of the paraxial phase profile from the ideal non-approximated phase profile. From Eqs 5.3.34 and 5.3.36 we obtain:

$$\Phi_{\text{ideal}}(r) = \frac{2\pi}{\lambda}(f - \sqrt{f^2 + r^2})$$
$$\Phi_{\text{parax}}(r) = -\frac{2\pi}{\lambda}(r^2/2f) = -\frac{\pi r^2}{\lambda f} \tag{5.3.38}$$

According to the Rayleigh criterion (Chapter 2) the lens can be considered as ideal if the maximum path length deviation is smaller than $\lambda/4$:

$$\left|\frac{\lambda}{2\pi}(\Phi_{\text{parax}}(r) - \Phi_{\text{ideal}}(r))\right| = \left|-\frac{r^2}{2f} - (f - \sqrt{f^2 + r^2})\right| < \frac{\lambda}{4} \tag{5.3.39}$$

With the aid of a series expansion of the square root in Eq. 5.3.39 we obtain for $r^6/(16f^6) \ll 1$:

$$r < \sqrt[4]{2\lambda f^3} \tag{5.3.40}$$

The f-number of the lens ($f/\# = f/(2r)$), for which the paraxial approximation can be applied safely, is calculated to be:

$$f/\# > \sqrt[4]{\frac{f}{32\lambda}} \quad \text{or} \quad r < 16\lambda(f/\#)^3 \tag{5.3.41}$$

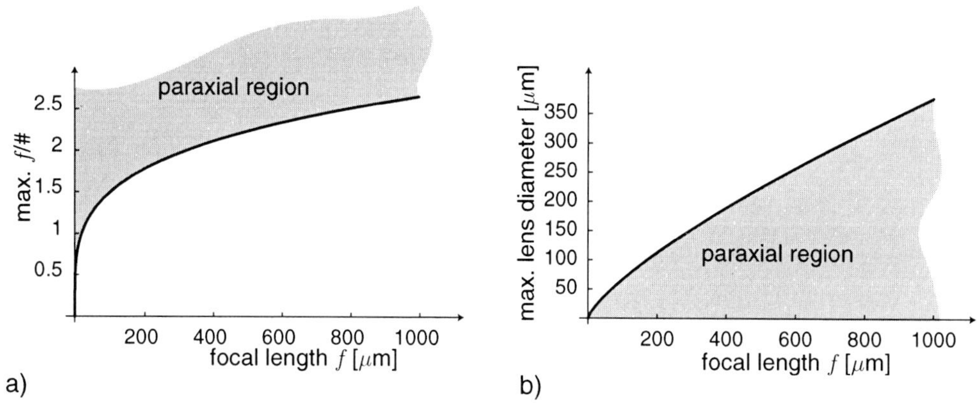

Fig. 5.16: The validity of the paraxial approximation for diffractive lenses ($\lambda = 0.633\ \mu$m): a) The minimum $f/\#$ of the diffractive lens for which the paraxial approximation is valid as a function of the focal length. b) The maximum lens diameter achieved in the paraxial regime as a function of the focal length.

These relations show some interesting aspects worth discussing. According to Eq. 5.3.41, the validity of the paraxial approximation for diffractive lenses not only depends on the $f/\#$ (Fig. 5.16) but also on the absolute value of the focal length. The consequences of this are illustrated best by some numerical examples. In the following examples we choose $\lambda = 0.633\ \mu$m. For a Fresnel lens with a relatively long focal length ($f = 12.5$ mm) to stay within the paraxial region the minimum f-number is $f/\# = 5$, corresponding to a maximum lens diameter of $2r = 2.5$ mm. If, however, we look at microlenses which typically have focal lengths in the range $50\ \mu\text{m} < f < 1000\ \mu$m, significantly smaller f-numbers are allowed without violating the paraxial approximation. A microlens with a focal length $f = 0.5$ mm, e.g., may have f-numbers as small as $f/\# > 2.2$. The radius of such a lens is restricted to $r < 112\ \mu$m.

In order to avoid large spherical aberrations in microlenses with f-numbers which do not obey the inequalities in Eq. 5.3.41, it is nessesary to avoid the paraxial approximation. In this case, Eq. 5.3.35 can be rewritten without approximation as [55]:

$$f = \frac{r_1^2 - \lambda^2}{2\lambda} \tag{5.3.42}$$

For the following mathematical treatment of diffractive lenses we will restrict ourselves to the paraxial case (Eq. 5.3.37). Then, one can use an analogy to linear blazed gratings for the mathematical treatment [22]. Due to the periodicity in r_1^2 the complex amplitude transmission of the diffractive lens can be expanded in a Fourier series of functions $e^{2\pi i m r^2/r_1^2}$:

$$g(x,y) = g(r^2) = \sum_{m=-\infty}^{+\infty} A_m \, e^{2\pi i m r^2/r_1^2} \tag{5.3.43}$$

In order to estimate the effect of such a lens on an incident wave field, the complex wave field $U(r, \Phi, z)$ at a distance z behind the lens will be calculated. To this end we calculate the Fresnel transform of the complex transmittance $g(r^2)$ of the grating. For a radially symmetrical object, $g(r^2)$, the Fresnel transform is expressed by the Bessel transform of $f(r) = g(r^2)e^{i\pi/(\lambda z)r^2}$ [22].

$$U(r, \Phi, z) = e^{2\pi i \frac{z^2+r^2}{\lambda z}} 2\pi \int_0^R J_0(2\pi \frac{rr'}{\lambda z}) f(r') r' dr' \tag{5.3.44}$$

R denotes the radius of the lens. Using Eq. 5.3.44 the complex amplitude at the focal plane $z_f = r_1^2/(2\lambda)$ is given by:

$$U(r, \Phi, z_f) = A_{-1} e^{2\pi i \frac{z_f^2+r_f^2}{\lambda z_f}} 2\pi R^2 \frac{J_1[2\pi(r_f R/(\lambda z))]}{2\pi(r_f R/(\lambda z))} + \sum_{m \neq -1} A_m u_m(r_f) \tag{5.3.45}$$

The first term in this equation is a good approximation of the light distribution in the focal plane. The shape of the focus is determined by the first-order Bessel function J_1 or, for rectangular apertures, by the sinc function (as defined in Chapter 2). The amount of light in the focus is determined by the amplitude factor $A_{-1} 2\pi R^2$. By analogy with the linear blazed grating, A_{-1} corresponds to the amplitude in the -1st order of $g(r^2)$, with the rest of the diffraction orders A_m contributing to the background noise. As for the linear blazed grating, we express the transmission function $g(r^2)$ as a "staircase" phase profile (Eq. 5.3.30). Thus, from the Fourier transform we obtain the same values for the diffraction efficiency of a diffractive lens as for the linear blazed grating. Fig. 5.17 shows the dependence of the diffraction efficiency of multilevel diffractive lenses on the number of phase levels. Although this was derived for the paraxial case, the results can be generalized to diffractive lenses with smaller f-numbers [22].

For describing a diffractive lens some additional parameters are of interest. The number $j_{\max}$ of zones in a diffractive lens with radius $R = r_{\max} = r_{j\max}$ and focal length f can be calculated from Eq. 5.3.36 as:

$$j_{\max} = \frac{R^2}{2\lambda f} = \frac{R}{4\lambda f/\#} = \frac{R}{2\lambda}\text{NA} \tag{5.3.46}$$

Here NA denotes the numerical aperture of the lens defined by NA $= R/f = 2/(f/\#)$. $j_{\max}$ is closely correlated to the Fresnel number N_F of the lens, which is defined as:

$$N_F = \frac{R^2}{\lambda f} = 2 j_{max} \tag{5.3.47}$$

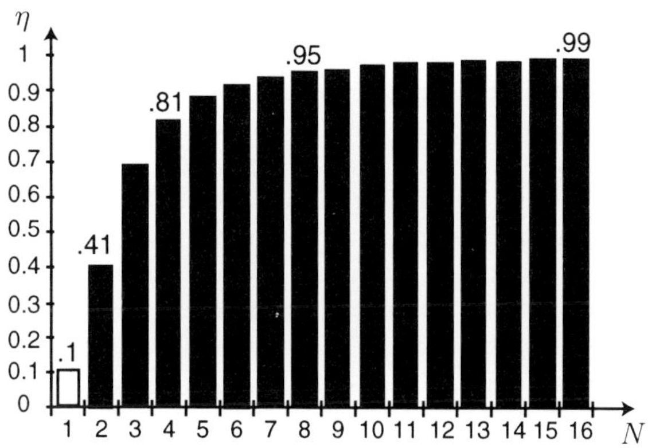

Fig. 5.17: Efficiency of diffractive lenses at the focal point vs. the number of phase levels [22]. Here N=1 refers to an amplitude Fresnel zone plate.

Several authors have pointed out the dependence of the lens performance on N_F. On nonuniformly illuminating the lens (e.g., with a Gaussian beam profile), a shift of the focal length can be observed, which depends on the Fresnel number N_F [51].

Another feature of diffractive lenses, which is especially interesting for fabrication considerations, is the minimum feature size $w_{\min}$ (i.e., the width of the narrowest line in the pattern). The minimum period of the FZP appears in the outermost zones. For a lens fabricated with N phase levels $w_{\min}$ can be calculated from $f/\#$ [22]:

$$\begin{aligned} w_{\min} &= \frac{r_1}{N}[\sqrt{j_{\max}} - \sqrt{j_{\max} - 1}] \\ &= \frac{r_1}{N}\left[\sqrt{j_{\max}} - \left(\sqrt{j_{\max}}\sqrt{1 - \frac{1}{j_{\max}}}\right)\right] \\ &= \frac{r_1}{N}\left[\sqrt{j_{\max}} - \left(\sqrt{j_{\max}}\left(1 - \frac{1}{2j_{\max}} + \frac{1}{8j_{\max}^2} + \ldots\right)\right)\right] \\ &\approx \frac{r_1}{2N\sqrt{j_{\max}}} = \frac{2\lambda f/\#}{N} = \frac{\lambda}{\mathrm{NA}\cdot N} \end{aligned} \tag{5.3.48}$$

We note that the minimum feature size only depends on the f-number ($f/\#$) of the lens. Generally, the available technology determines the minimum size of the features which can be generated. Thus, the achievable numerical aperture of diffractive lenses depends on the available fabrication process. As an example we consider a fabrication process with a minimum feature size of 1 μm. For lenses fabricated with 4 phase levels, at a wavelength of $\lambda = 0.633$ μm the achievable minimum f-number is $f/\# \approx 3.2$.

5.3.7 Ray-tracing analysis of diffractive lenses

From the considerations in the previous section we have seen that a diffractive lens generates an amplitude distribution at the first focal point which is described by a first-order Bessel function. Higher orders contributing to diffuse background light can be suppressed using multiple phase level profiles. For the on-axis case, ideal focusing behaviour (within the Rayleigh limit) can be achieved for both high and low f-numbers, if a suitable design of the radial zones is used (Eq. 5.3.42). If the lens is to be used for imaging an extended image field, we also have to take into account wavefront aberrations occurring for off-axis light beams. To this end we calculate the optical path difference (OPD) between the axial and the marginal ray for a wave incident at an angle α to the optical axis (Fig. 5.18) [52, 53].

$$\begin{aligned}\mathrm{OPD} &= R\sin(\alpha) + f\sqrt{1+\left(\tan(\alpha)-\frac{R}{f}\right)^2} - f\sqrt{1+\tan^2(\alpha)} \\ &\approx \frac{R^2}{2f} - \frac{R^4}{8f^3} + \underbrace{\frac{R^3\alpha}{2f^2}}_{\text{coma}} - \underbrace{\frac{3R^2\alpha^2}{4f}}_{\text{astigmatism}}\end{aligned} \tag{5.3.49}$$

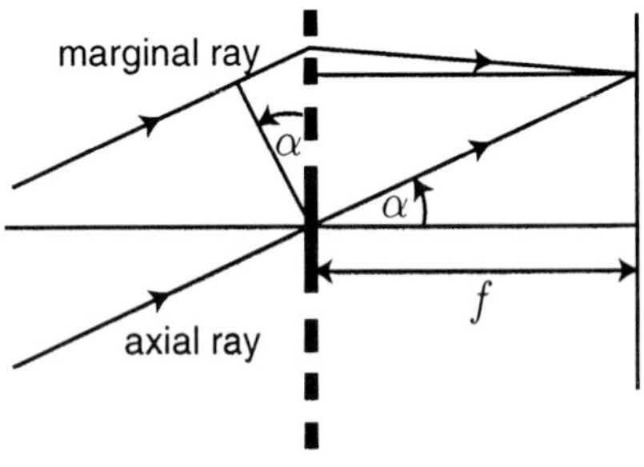

Fig. 5.18: Illustration of the optical path difference between the axial and the marginal ray for an off-axis angle α.

In Eq. 5.3.49 we recognize the first two terms from the expansion of the term $f - \sqrt{(f^2 + R^2)}$ in Eq. 5.3.39. The remaining terms form the third order or Seidel aberrations of the diffractive lens. The third term describes coma whereas the forth terms contains contributions of astigmatism as well as field curvature induced by the lens. We notice the absence of any term describing distortion. This illustrates that diffractive zone plates provide distortion-free imaging over extended image fields. If we again assume the Rayleigh limit for these aberration terms we find that for large field angles only the second aberration term will be significant. At the Rayleigh limit we find the maximum field angle:

$$\alpha[\mathrm{rad}] < \sqrt{\frac{\lambda f}{3R^2}} = \sqrt{\frac{4\lambda(f/\#)^2}{3f}} \tag{5.3.50}$$

As expected, the maximum field angle for aberration-free imaging is determined by the f-number. We further find that the imaging quality is also affected by the absolute value of the focal length. Thus, again, microlenses with relatively short focal lengths exhibit imaging properties which are superior to longer focal length lenses of larger dimensions. We shall now illustrate this by two numerical examples. For a microlens with a focal length of $f = 500\ \mu\text{m}$ and f-number $f/\# = 2.2$ we find a maximum total field angle for aberration-free imaging of $2\alpha = 10.4°$. For the $f/\# = 5$ lens considered before with a significantly longer focal length of $f = 12.4$ mm a total field angle of only $2\alpha = 4.7°$ results for which astigmatism remains smaller than the Rayleigh limit.

At small field angles, coma dominates the aberrations in Eq. 5.3.49. At the Rayleigh limit we find:

$$\alpha[\text{rad}] < \frac{\lambda f^2}{2R^3} = \frac{4\lambda(f/\#)^3}{f} \tag{5.3.51}$$

For the two examples, we calculate a maximum field angle of $2\alpha = 6.2°$ for the $f/\# = 2.2$ lens with short focal length and $2\alpha = 2.9°$ for the $f/\# = 5$ lens with longer focal length. Although the previous considerations only dealt with the case of infinite conjugate imaging, i.e., the object was considered to be at an infinite distance from the lens, a generalization to finite conjugates is straightforward and yields similar results.

Buralli et al. [55] extended the calculations above to the nonparaxial design of diffractive lenses. W. C. Sweat showed that the imaging behaviour of thin refractive elements converges to the behaviour of diffractive lenses if the refractive index of the lens material approaches infinity [53, 54]. This allows one to compare the relations derived for the primary aberrations of Fresnel zone lenses (Eqs 5.3.50 and 5.3.51) with those for the aberration of a classical thin refractive biconvex lens using a simple model. Such calculations show that the field extension for diffraction-limited performance tends to be slightly smaller for diffractive lenses than for refractive lenses. Exact values depend on the curvatures of the biconvex lens surfaces as well as on the exact values of the refractive indices.

The large refractive index model also allows one to use standard ray-tracing software to model the performance of diffractive Fresnel-type lenses. Based on such simulations it can be shown that it is possible to optimize the performance of diffractive lenses over large image fields by a simple shift of the aperture stop in the imaging setup. Such a shift of the aperture stop results in additional terms in the aberration functions of the lens. If the aperture stop is placed in the focal plane of a diffractive lens these additional terms cancel out some of the original aberration terms. Better overall imaging performance results. This fact has been used for the design of a diffractive landscape lens with large image field angles [55]. In order to compensate for the spherical aberration which is limiting the performance of such a lens an aspherical plate can be placed in the aperture stop. According to the principle of the Schmidt camera, with this plate in the front focal plane of the setup, the spherical aberration can be influenced without affecting the corrected aberration terms.

5.3.8 Chromatic aberrations of diffractive lenses

So far we focused on the monochromatic properties of diffractive microlenses. However, interesting possibilities also arise from the chromatic dispersion of these elements. For the wavelength dependence of the focal length we find according to Eq. 5.3.37:

$$\frac{\partial f_{\mathrm{d}}}{\partial \lambda} = -\frac{f_{\mathrm{d}}}{\lambda} \tag{5.3.52}$$

Here we have added the index d to denote the diffractive case. For a refractive lens with $f_{\mathrm{r}} = r_{\mathrm{c}}/(n-1)$ and $n \approx n_0 + (\partial n/\partial \lambda)(\lambda - \lambda_0)$ we obtain:

$$\frac{\partial f_{\mathrm{r}}}{\partial \lambda} = -\frac{(\partial n/\partial \lambda)_0}{n_0 - 1} f_{\mathrm{r}} \tag{5.3.53}$$

As an example we choose a lens made of fused silica glass with $n_0 = 1.45$ at $\lambda_0 = 1\ \mu\mathrm{m}$ and $(\partial n/\partial \lambda) = -0.0128\ \mu\mathrm{m}^{-1}$. Assuming two lenses with the same focal length, $f_{\mathrm{d}} = f_{\mathrm{r}}$, one obtains: $(\partial f/\partial \lambda)_{\mathrm{d}} \approx -35(\partial f/\partial \lambda)_{\mathrm{r}}$. The wavelength dispersion of a diffractive lens is considerably stronger than for a refractive lens and it has the opposite sign. In order to analyse the effect of a wavelength shift on the aberrations of the lens, we merely need to substitute the focal length f by $f \cdot \frac{\lambda_0}{\lambda}$ in Eqs 5.3.50 and 5.3.51. In this case we obtain the aberrations caused by a lens designed for the wavelength λ_0 but illuminated with λ [48].

The different signs of the chromatic dispersion of diffractive and refractive lenses are especially interesting for broadband applications. As we will see in Chapter 12 diffractive lenses are very useful for the achromatisation of refractive lenses.

5.3.9 Detour-phase diffractive optical elements

In diffractive optical elements generally phase functions are generated by microlithographic profiling of dielectric materials. A completely different approach to encode phase values in diffractive elements is the concept of the detour-phase. Phase values are not generated directly through dielectric layers of variable thickness. Instead, in detour-phase elements local shifts of the period of a carrier grating are used to generate phase shifts in the diffraction orders of the grating. The detour-phase φ_{d} is based on the shift theorem of Fourier optics, which states that a lateral shift of a grating causes a phase shift of the corresponding diffraction orders [9]. Similarly local shifts of a grating cause relative phase shifts between the wavefronts diffracted into one specific order (Fig. 5.19). A spatially sampled 2D phase function can be encoded by pixelwise applying local shifts to the carrier grating. In the reconstruction, each of these shifts transforms into the corresponding phase value. Fig. 5.20a shows a 1D schematic of three pixels of a detour-phase element fabricated on a binary amplitude carrier grating. From geometrical considerations we obtain the relation for the detour-phase for a local shift of Δp of a carrier grating with period p:

$$\begin{aligned} \varphi_{\mathrm{d}} &= \sin(\alpha_m)\frac{2\pi}{\lambda}\Delta p \\ &= m 2\pi \frac{\Delta p}{p} \end{aligned} \qquad (5.3.54)$$

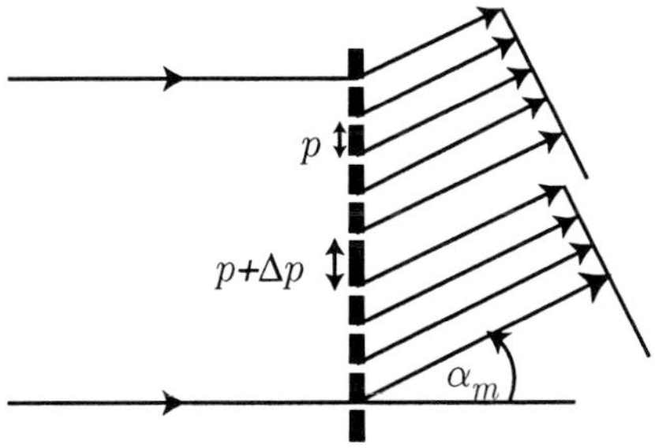

Fig. 5.19: The detour-phase implemented through a local shift of the period of a carrier grating.

Suggested by A. W. Lohmann [56], the detour-phase approach marks the beginning of the interest in computer-generated diffractive optical elements [57, 58, 59]. Much of the success of detour-phase elements is due to the fact that at the time of its discovery lithographic techniques only started to find access to the fabrication of optical elements. Etching and thin film techniques were not developed for the fabrication of dielectric profiles with lateral extension in the micrometre range. The detour-phase was the first technique which allowed the implementation of complex amplitudes with multilevel phase values. In addition to phase encoding, a modification of the local aspect ratio of the carrier grating allows one to influence the diffraction amplitude locally. It is thus possible to generate a diffractive element which reconstructs phase as well as amplitude information in the diffraction order of the carrier grating. The detour-phase approach, therefore, became famous under the name "computer-generated holography" [60]. Detour-phase holography has found numerous applications, e.g., in spatial filtering and correlation applications. Its potential to reconstruct an arbitrary complex amplitude distribution designed in a computer triggered most of the interest. For application as components in microoptical setups, however, the low efficiency of detour-phase elements is a severe obstacle. The diffraction efficiency is determined by the amplitude which is diffracted into the desired signal order. Due to the implementation on a carrier grating the diffraction efficiency of the detour-phase element is composed of the efficiency of the implemented phase function η_{phase} and the efficiency of the carrier grating η_{carrier} [58, 62]:

$$\eta_{\mathrm{element}} = \eta_{\mathrm{phase}} \cdot \eta_{\mathrm{carrier}} \qquad (5.3.55)$$

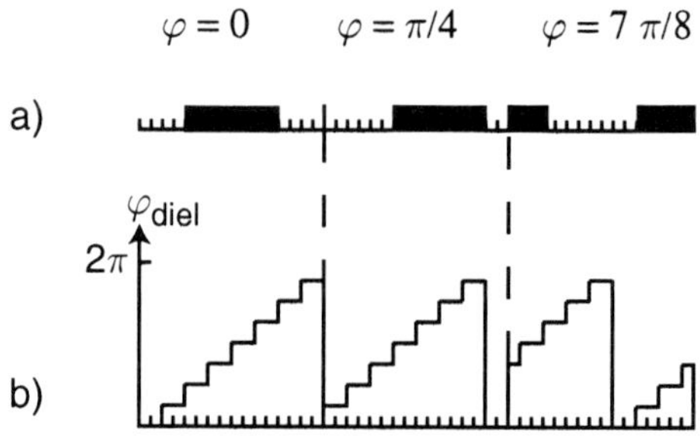

Fig. 5.20: Three pixels of a detour-phase element implementing three different phase values as local detours to a) a binary amplitude grating and b) a kinoform phase grating [62].

In so far as only binary gratings with efficiencies of $\eta_{carrier} < 40.5\%$ are used as carrier gratings (Fig. 5.20a), this efficiency is not sufficient for implementation in microoptical systems. Although the use of a variety of different carrier gratings with higher efficiencies has been indicated very early on, only recently have elements been investigated which combine the detour-phase concept with high efficiency carrier gratings (e.g., kinoform gratings) (Fig. 5.20b). These elements are called kinoform detour-phase holograms (KDPHs) [61, 62].

In KDPHs the detour-phase can be exploited as an additional degree of freedom for the design of efficient DOEs [63]. This can be advantageous for the implementation of a very complex phase function, where the number of phase levels L necessary to achieve the desired performance is very high (e.g., $L > 16$). Due to the limited resolution of the fabrication process it becomes difficult or impossible to fabricate the element as a conventional kinoform element, where all phase values are implemented as dielectric layers of different thickness. In KDPHs only a limited number (i.e., 8 or 16) of dielectric phase levels is necessary to fabricate a high efficiency linear carrier grating. The phase levels of the DOE are achieved by locally shifting this carrier grating. The available number of detour-phase levels depends on the number of positioning steps Δp per carrier period p. This number can be very high for large carrier periods. The detour-phase concept can be understood as a method for trading phase resolution for spatial resolution. For 1D phase functions it is possible to change the orientation of the carrier grating in order not to sacrifice spatial resolution [64]. If the carrier grating is oriented perpendicular to the phase function, very large grating periods are possible, which allow extremely high phase resolution without a reduction of the spatial resolution [65]. In a similar way the detour-phase element can be optimized in the case of 2D phase functions with different complexity in the two lateral directions (e.g., off-axis elements). In conclusion, combined with kinoform carrier gratings, the detour-phase implementation of phase functions is an interesting alternative to conventional multilevel DOEs. KDPHs have potential for application in microoptical setups.

5.3.10 Polarisation-selective diffractive optical elements

Generally, lithographically fabricated diffractive optical elements are insensitive to the incident light polarisation. Exceptions are high frequency gratings and thick holographic elements, where rigorous diffraction theory predicts a strong dependence of the diffraction efficiencies on the polarisation (e.g., form birefringence of zero-order gratings). In order to maintain the design flexibility of DOEs it is advantageous to be able to fabricate polarisation-selective computer-generated holograms. This is possible with conventional fabrication techniques, if anisotropic substrate materials (e.g., lithium niobate) are used. A conventional DOE etched into a lithium niobate substrate already shows polarisation dependent properties since the etching depths lead to different relative phase shifts for the two orthogonal polarisations. In order to be able to build an optical element with arbitrary functionality for each of the polarisations, it is necessary to apply independent phase functions for each of the polarisations. This is possible in a compound element consisting of two binary DOEs fabricated in substrates with different birefringence. The etch depths in the two substrates are calculated to provide the necessary phase delays for the two polarisation directions. This approach has been demonstrated for the fabrication of polarising beam splitters, polarisation-selective lenses and beam-shaping components [66]. For interconnect applications polarisation-selective DOEs are of particular interest since they provide an additional degree of freedom for the design of the system [67, 68].

5.3.11 Holographic optical elements as thick Bragg gratings

Yet another approach to the fabrication of microoptical elements is based on optical holography. Holography was invented in the 1950s by Dennis Gabor who was looking for ways to improve the quality of images gained from electron microscopy. He demonstrated that it is possible to record and reconstruct the full information content of a complex optical wavefront (i.e., the amplitude as well as the phase information) by recording the interference pattern of the wavefront with a reference wavefront. He called this process holography (from the greek phrase "holos" meaning "complete", referring to the recording of the entire complex wavefront) [69]. Although the holographic approach was not successful for the application Gabor had in mind, his invention had an impact in many areas of optics. A significant amount of research on holographic techniques has been performed aiming at improvements of the quality and efficiency of the reconstructed images as well as on specific applications. Consequently, holographically recorded elements are also considered for microoptical applications. Such elements are generally called holographic optical elements (HOEs) [70] .

The holographic recording process is based upon the interference of an object wave with a reference wave. The complex amplitude of the object wave can be reconstructed from the recorded interferogram by illumination with the reference wave. Fig. 5.21 shows the case of two plane wavefronts propagating at different angles. The total angle between the propagation directions is denoted by 2θ. In the

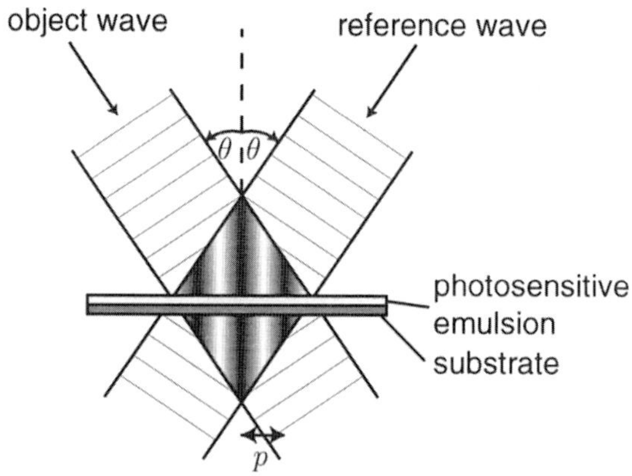

Fig. 5.21: Holographic formation of a linear grating by interference of two tilted plane waves.

overlap area of the plane waves we can observe an interference pattern consisting of linear fringes. The period of the this grating is determined by θ:

$$p = \frac{\lambda}{2\sin(\theta)} \tag{5.3.56}$$

If one of the incident plane waves in Fig. 5.21 is replaced by a more complex object wavefront, the linear interference fringes are distorted to form a complex interference pattern in the overlap area with the reference wave. This hologram pattern can be recorded by placing a photographic plate or any other photosensitive material in the region of the interferogram. The interference fringes cause a grey-level illumination of the photoemulsion. Depending on the material chosen for the recording process an amplitude or a phase pattern is generated in the photoemulsion. Surface profiles can also be generated using photoresist materials followed by an analog developing process similar to that used, e.g., in direct laser or e-beam writing or grey-scale mask illumination. If the developing parameters are chosen properly it is also possible to fabricate binary gratings. A holographic grating recorded with this process is structured such that the reference wave incident on the grating is diffracted into the object wave. In other words, the object wave is reconstructed by the incident reference wave.

The interferometric technique was suggested very early on as being suitable for the fabrication of blazed gratings [71]. Its main advantage is the possibility of fabricating continuous phase profiles with a single illumination step. The minimum periods of holographic gratings are achieved for $\theta = 90°$ (Eq. 5.3.56). In this case the interfering waves propagate in opposite directions and the fringe period is $p = \lambda/2$. Any other period can be achieved by changing the relative angle between the interfering beams or by tilting the recording photoplate. Holographic recording has the potential for creating very high frequency gratings.

So far we have considered HOEs merely as blazed diffractive elements, fabricated with the interferometric recording technique. With these elements it is possible to achieve high grating frequencies (corresponding to large diffraction angles). If the interference patterns are recorded in thin photosensitive layers, the resulting gratings are considered as thin gratings. These thin gratings with sinusoidal fringes exhibit low diffraction efficiencies (see Exercises). This is due to the large number of diffraction orders generated by a sinusoidal phase grating. The sinusoidal amplitude grating, on the other hand, generates only 3 diffraction orders (-1; 0; 1) but absorption significantly reduces the efficiency in these orders.

More frequently, holographic recording techniques are used for the fabrication of volume gratings [35]. For this purpose, thick ($t > 10$ μm) layers of photoemulsion are used to record the interference fringes. Frequently used materials for the recording are silver-halide photoemulsions, dichromatic gelatine or photopolymers. For silver-halide photoemulsions (e.g., Agfa-Gevaert 8E75HD) the processing is similar to standard photographic developing and fixing, followed by a bleaching process to convert the grey-level profiles into phase gratings. For thick layers in particular, this chemical processing is rather complex and expensive. Due to the finite grain size scattering effects inside the bleached photographic layers reduce the achievable resolution and efficiencies [72, 73].

Up to the present time, dichromatic gelatine (DCG) is the most widely used material for volume holographic recording for scientific applications [74, 75]. In particular, the low scattering effects inside the material make it an interesting candidate for volume holography. DCG had already been discovered in the 19th century for photographic processes. The basic material for this process is gelatine, which is a transparent organic protein-like molecule. The recording of thick volume holograms in DCG is based upon the capacity of the gelatine to absorb significant amounts of water. The sensitivity to the illuminating light stems from doping with dichromate molecules. The gelatine layer is immersed into a liquid solution of, e.g., ammoniumdichromat $(NH_4)2Cr_2O_7$. Subsequently, the DCG layer can be illuminated, e.g., using the 488 nm or 514 nm line of an Argon ion laser. Upon illumination, the dichromate ions in the gelatine are reduced and initiate crosslinks between the gelatine molecules. This local manipulation of the physical structure of the gelatine layer causes local changes of the refractive index. A subsequent complex chemical development process involves hydration followed by dehydration of the gelatin layer. During this development the very small refractive index change after illumination is increased significantly to typical values of $\Delta n > 0.1$. This explains the high efficiency of the recorded holograms. A major drawback is the complexity of the DCG processing. Since formation of the volume index grating is based upon the hydration and dehydration of the gelatine layer, the whole process is very sensitive to temperature and the humidity of the processing environment. In order to achieve good reproducibility, precise control of the processing environment is even more critical for DCG than for binary photoresist processing. Furthermore, after fabrication, the HOEs need to be sealed to be protected against degradation due to environmental influences such as humidity [74]. With sufficient care high diffraction efficiencies $\eta_{\mathrm{DCG}} > 80\%$ can be achieved with holographic gratings in DCG.

Due to the significant improvement in quality of photopolymer materials in recent years, they have also become interesting materials for volume holographic recording [76]. The processes which lead to the formation of the refractive index profile are similar to the processes used for the generation of refractive elements in polymer materials (Chapter 4). The polymer layer (e.g., DuPont HRF 150) contains polymerized PMMA and photoinitiators, as well as monomer material. Upon illumination the polymer chains are partly broken down so that additional bonds can be formed. This structural change triggers an index variation. The processing is very simple, since the developing takes place automatically during the illumination process. Fixing of the generated index profile can be performed easily through flash illumination with UV light which passivates the photoinitiator. Major drawbacks so far are the relatively low refractive index change, relatively high absorption and scattering which cause a significant drop of efficiencies especially for large Bragg angles (i.e., high grating frequencies). The efficiencies for holograms recorded in the DuPont photopolymer are $\eta_{\mathrm{polymer}} \approx 60$–$70\%$ [77, 78].

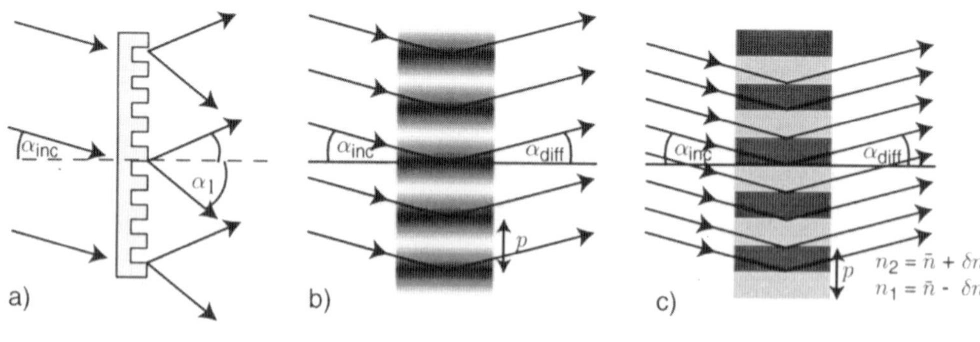

Fig. 5.22: Beam diffraction at a) a thin planar grating and b) a thick volume grating (grey-levels illustrate the refractive index profile) c) the stacked layer model of the thick grating.

HOEs are thick volume gratings, which cannot be described by scalar diffraction theory. Instead, HOEs are approximated as thick Bragg gratings [35]. Fig. 5.22b illustrates the beam diffraction at a volume grating. The thick grating can be approximated as a stack of layers with alternating refractive indices (Fig. 5.22c, where $\delta n \ll \bar{n}$). The thickness of each of the layers is $p/2$. The diffracted (or better Bragg-reflected) wavefront is considered to be composed of the multiple reflected beams interfering in the output plane. Due to a phase shift of π occurring on reflection at the higher refractive index layer, the condition for constructive interference of the reflected beams is given by ($\alpha_{\mathrm{inc}} = \alpha_{\mathrm{diff}}$):

$$2\frac{p}{2}\sin(\alpha) = (2j+1)\frac{\lambda}{2}; \quad j = 0, 1, 2, 3, ... \tag{5.3.57}$$

Comparing Eq. 5.3.57 with Eq. 5.3.25 which describes the general grating equation we find that for Bragg diffraction at thick gratings only one out of the large number of possible diffraction orders is selected. This can be understood with the help of the stacked layer model. The orders which are directed in transmission through the layer stack are eliminated by destructive interference with the multiple reflected beams travelling in the same direction. For the other orders we find from Eq. 5.3.57 and Eq. 5.3.25 that $(2j+1) = 2m$. Since m and j are integers, this condition can only be fulfilled for j=0. Thus, Eq. 5.3.57 becomes:

$$\sin(\alpha) = \frac{\lambda}{2p}; \quad j = 0, 1, 2, 3, ... \tag{5.3.58}$$

The observation that Eq. 5.3.58 is equivalent to Eq. 5.3.56 proves that holographically recorded thick gratings fulfil the Bragg condition if illuminated with either of the recording waves.

The fact that thick holographic gratings diffract light with high efficiency into a single diffraction order is their most important property. It is also possible to achieve large diffraction angles due to the holographic recording process. These are obviously attractive features for applications in microoptics. Using a variety of object beams it is possible to fabricate HOEs to perform a variety of different tasks such as beam deflection, beam focusing and beam splitting [79, 80, 81, 82]. Let us assume, e.g., that an interference pattern between a converging spherical wave and a plane reference wave is recorded. When this HOE is reconstructed with the plane reference beam the light behind the element will be converging. If illuminated with a diverging beam the output will be the collimated plane reference beam. The HOE performs as a focusing or collimating lens, respectively. In a step and repeat process it is also possible to fabricate facetted holograms which, e.g., perform as microlens arrays [83, 84]. The potential of HOEs can be further enhanced by multiplexing of multiple holograms into a single element.

One of the main problems in the application of HOEs is due to the high angular and wavelength selectivity for their performance. As described, a HOE is optimized to yield nearly 100% efficiency for reconstruction under the Bragg condition. If, however, the Bragg condition is violated because of angular or wavelength detuning, the efficiency is reduced dramatically [85, 86]. For practical applications this puts high constraints on the alignment of the optical setups as well as the control of the wavelength. In microoptics, semiconductor lasers are often used as sources which exhibit relatively wide spectra and suffer from wavelength variations. In combination with these sources the use of HOEs is difficult. It is, however, possible to use combined elements or suitable optical setups which are relatively insensitive to wavelength variations [87].

5.3.12 Effective medium theory of zero-order gratings

In the previous paragraphs we discussed scalar diffraction theory for thin DOEs and Bragg theory for thick HOEs. Now we focus on yet another family of DOEs which are called zero-order grating. A grating with period p is called a zero-order grating if the condition:

$$p \ll \lambda \tag{5.3.59}$$

is fulfilled.

Using Eq. 5.3.7 we try to calculate the propagation angles for the diffraction orders of such a grating. We find that all but the zeroth order are evanescent, i.e., they yield diffraction angles $> 90°$. Hence these gratings are called zero-order gratings. The transmitted zeroth order is influenced by the structure of the period. From the boundary conditions of Maxwells equations, we know that for gratings with sufficiently small periods the electric field component $\boldsymbol{E}_\perp$ orthogonal to the grating period is continuous across the grating period [5]. For the dielectric constant $\epsilon_\perp$ experienced by this field component we can derive an average value:

$$\epsilon_\perp = \epsilon_1 \frac{w}{p} + \epsilon_2 (1 - \frac{w}{p}) \tag{5.3.60}$$

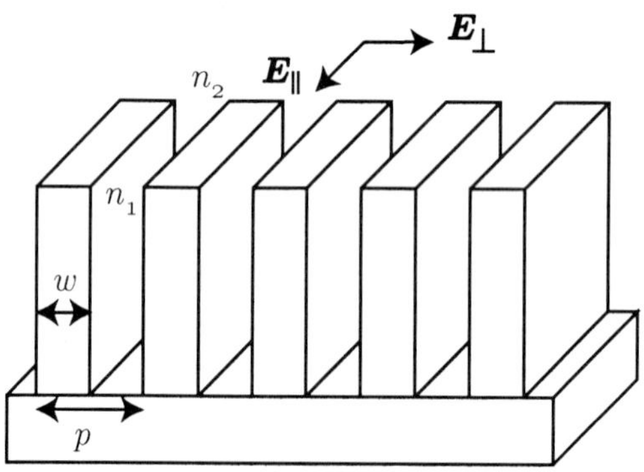

Fig. 5.23: Schematic of a laminar zero-order grating (p $\ll \lambda$) with the orientations of the incident electric wavefield.

Here ϵ_1 and ϵ_2 are the dielectric constants of the grating bar and the environment respectively and the duty cycle of the grating is $\frac{w}{p}$ (Fig. 5.23). For the electric field component $\boldsymbol{E}_\|$ oriented parallel to the grating bars, we obtain an effective dielectric constant $\epsilon_\|$:

$$\epsilon_\| = \frac{\epsilon_1 \epsilon_2}{\epsilon_2 \frac{w}{p} + \epsilon_1 \frac{w}{p}} \tag{5.3.61}$$

With the refractive index $n = \sqrt{\epsilon}$ we find that the sub-wavelength grating exhibits birefringence, i.e., different refractive indices for different polarisation angles of the incident light. Since the amount of birefringence, described by $\Delta n = \sqrt{\epsilon_\perp} - \sqrt{\epsilon_\|}$ depends on the grating profile, this phenomenon is called "form birefringence" [5].

The approximations of the refractive indices described by Eqs 5.3.60 and 5.3.61 are the foundations of the effective medium approximation to zero-order gratings. They describe the fact that upon transmission through a sub-wavelength grating the zeroth order experiences an effective refractive index resulting from the averaging of the dielectric constants of the grating media. Zero-order gratings acting as anti reflection coatings [88], wave plates [89] and artificial distributed index gratings have been demonstrated [90, 91, 92].

5.4 Design of diffractive optical elements

After discussing various possibilities for describing the effect of diffractive components on an incident light wave we now focus on the inverse problem. The issue we address in this section is how to find the structure of a diffraction grating which performs a specific task. We specifically address two widely used design techniques for DOEs. In the first section we describe the design procedure for diffractive lenses optimized for specific imaging configurations based on ray tracing. The second paragraph describes the application of iterative techniques for DOE optimization.

5.4.1 DOEs optimized for imaging along a tilted optical axis

Diffractive lenses offer diffraction-limited performance over extended image fields. For microlenses this is also true for relatively small f-numbers. One of the major advantages of diffractive lenses over refractive lenses is related to the fabrication process. Due to lithographic multilevel fabrication, high optical quality, almost perfect reproducibility and ease of replication can be achieved for DOEs. Although related to the fabrication, we wish to point out one further advantage which might turn out to be the most significant for application to microoptical systems. As mentioned before in combination with phase quantization, lithographic fabrication allows one to generate any desired phase profile. Thus, it is possible to fabricate lenses which are optimized for specific imaging configurations (e.g., finite conjugate imaging, etc). As an important example we address the design of a lens, optimized for imaging along a tilted optical axis. This design is of special importance for microoptical systems which are folded in order to increase the packaging density, e.g., in planar optics [93].

In order to find the optimum phase profile for an off-axis diffractive lens we use the ray-tracing approach (Chapter 2). We start by calculating the $\boldsymbol{k}$-vector of the diffractive lens $\boldsymbol{k}_{\text{lens}}$ necessary to focus the obliquely incident plane wave described by the $\boldsymbol{k}$-vector $\boldsymbol{k}_{\text{inc}}$ [94] (Fig. 5.24).

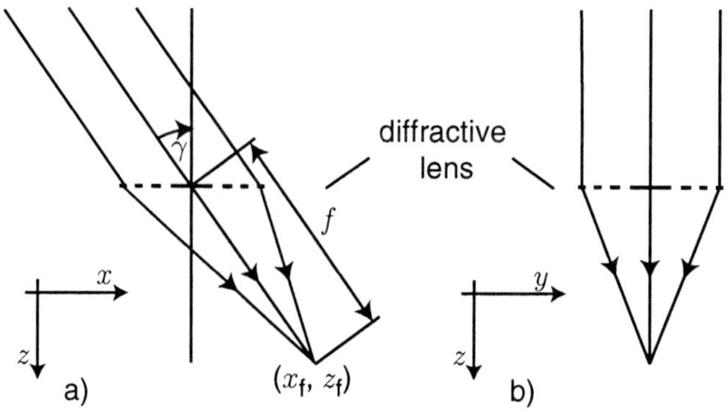

Fig. 5.24: Illustration of the optical path difference necessary for collimation a) along a tilted optical axis and b) on-axis.

$$\boldsymbol{k}_{\mathrm{inc}} = \frac{2\pi}{\lambda} \begin{pmatrix} \sin(\gamma) \\ 0 \\ \cos(\gamma) \end{pmatrix} \tag{5.4.1}$$

In order to form a focused beam, the $\boldsymbol{k}$-vector $\boldsymbol{k}_f$ is required behind the lens, where:

$$\begin{aligned} \boldsymbol{k}_{\mathrm{f}} &= \frac{2\pi}{\lambda} \frac{\boldsymbol{r}_{\mathrm{f}} - \boldsymbol{r}_{\mathrm{l}}}{|r_{\mathrm{f}} - r_{\mathrm{l}}|} \\ &= \frac{2\pi}{\lambda} \begin{pmatrix} f\sin(\gamma) - x \\ -y \\ f\cos(\gamma) \end{pmatrix} \frac{1}{\sqrt{(f\sin(\gamma) - x)^2 + y^2 + f^2\cos(\gamma)^2}} \end{aligned} \tag{5.4.2}$$

Here $\boldsymbol{r}_{\mathrm{f}}$ and $\boldsymbol{r}_{\mathrm{l}}$ denote the coordinates of the focal point and the lens respectively:

$$\boldsymbol{r}_{\mathrm{f}} = \begin{pmatrix} f\sin(\gamma) \\ 0 \\ f\cos(\gamma) \end{pmatrix} \qquad \boldsymbol{r}_{\mathrm{l}} = \begin{pmatrix} x \\ y \\ 0 \end{pmatrix} \tag{5.4.3}$$

For the $\boldsymbol{k}$-vector components of the lens we then obtain:

$$k_{\mathrm{L}x} = k_{\mathrm{f}x} - k_{\mathrm{inc}x} = \frac{2\pi}{\lambda} \left(\frac{f\sin(\gamma) - x}{\sqrt{(f\sin(\gamma) - x)^2 + y^2 + f^2\cos(\gamma)^2}} - \sin(\gamma) \right) \tag{5.4.4}$$

and

$$k_{\mathrm{L}y} = k_{\mathrm{f}y} - k_{\mathrm{inc}y} = \frac{2\pi}{\lambda} \frac{-y}{\sqrt{(f\sin(\gamma) - x)^2 + y^2 + f^2\cos(\gamma)^2}} \tag{5.4.5}$$

From our knowledge of these $\boldsymbol{k}$-vectors we can calculate the necessary phase profile of the lens by using Eq. 2.1.9:

$$\varphi_C(x,y) = -\frac{2\pi}{\lambda}\left[\sqrt{(x - f\sin(\gamma))^2 + y^2 + f^2\cos^2(\gamma)} - f + x\sin(\gamma)\right] \qquad (5.4.6)$$

This lens profile resulted for the focusing of a collimated beam along the oblique axis. If used for finite conjugate 1:1 imaging along the tilted axis, an optimized lens has to be symmetrical relative to the central axis. The corresponding phase distribution can be found from [95]:

$$\begin{aligned}\varphi_S(x,y) &= \frac{1}{2}[\varphi_C(x,y) + \varphi_C(-x,y)] \qquad (5.4.7)\\ &= -\frac{\pi}{\lambda}\left[\sqrt{x^2 + y^2 + 2xf\sin(\gamma) + f^2} + \sqrt{x^2 + y^2 - 2xf\sin(\gamma) + f^2} - 2f\right]\end{aligned}$$

In order to fabricate a diffractive lens which effectively performs like the refractive lens with this phase profile, we may calculate the shape of the Fresnel rings from:

$$|\varphi_S(x,y)| = 2\pi m \qquad \text{where } m: \text{ integer} \qquad (5.4.8)$$

Solving Eq. 5.4.8 we find elliptical contour lines for the Fresnel zones of this lens. The different radii of the ellipses denote the different focal lengths ($f_x = f\cos(\gamma)$, $f_y = f$) of the lens along the two lateral directions. The optimized lenses of Eqs 5.4.6 and 5.4.7 can be fabricated lithographically just like conventional diffractive microlenses. The fabrication of the corresponding refractive phase profile, if possible, would require significant additional effort. This example illustrates the design freedom of diffractive optics.

5.4.2 Iterative design techniques for DOEs

In addition to the ray-tracing approach, which we have just discussed for the example of optimized diffractive lenses, iterative optimization algorithms are also very popular for the design of phase functions with specific diffraction patterns [96]. An iterative Fourier transform algorithm is widely applied for the far-field diffraction regime [97]. The basic design algorithm is illustrated in Fig. 5.25. This algorithm is adopted from the reconstruction of image phase information from knowledge of the intensity distribution of the diffraction pattern in electron microscopy [98]. The algorithm is started with either a predesigned or a random phase distribution. The diffraction pattern for this distribution is calculated by means of a Fourier transform and compared to the desired output. Very often, the goal of the design is a specific amplitude distribution, while the phases of the diffraction amplitudes are free parameters. Thus, the next step in the algorithm is to introduce the desired amplitude distribution while leaving the phases as they are. After an inverse Fourier transformation we are back in the element plane. Here, for efficiency reasons, we require the element to be a pure phase element. We thus adjust the amplitudes of all element pixels to be unity. After another Fourier transform we find another diffraction pattern. This iteration is continued until a phase distribution is found which generates a diffraction pattern which is sufficiently close to the desired output. The advantage of this iterative design procedure is that any desired output can be optimized. The algorithm is often applied to the design of beam splitter gratings, in which case it is not restricted to symmetrical spot arrays or equal intensity diffraction orders. It is also possible to generate diffraction orders with predetermined phases if a sufficient number of free design parameters is available.

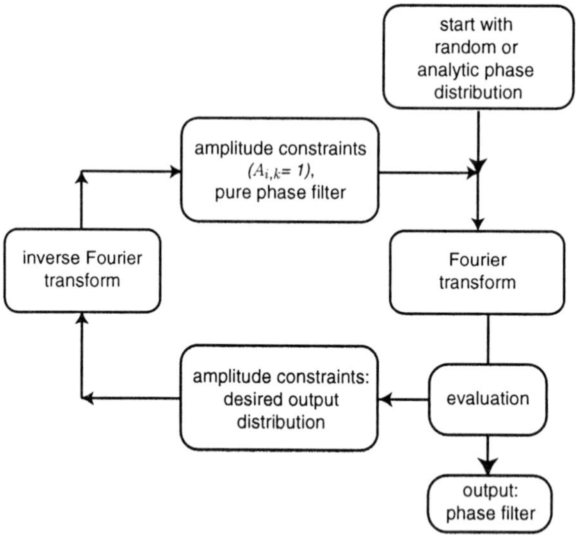

Fig. 5.25: Illustration of the iterative Fourier transform algorithm (Gerchberg-Saxton algorithm) for the design of diffractive optical elements.

The basic iterative Fourier transform approach is the direct numerical approach to diffractive element design. The most important problems which might occur during the iterative design process are due to the large number of possible solutions. The problem here is not to find a solution but to find the best possible solution. A large number of modifications to the basic algorithm have been tested [99]. The choice of the starting phase distribution is very critical. Using analytically predesigned starting distributions results in better solutions. Other modifications focus on the converging behaviour of the algorithm. A stepwise adjustment of the constraints in both domains as well as two separate optimization loops for amplitude and efficiency improves the convergence in many cases [100].

Apart from the iterative Fourier transform technique described above, iterative techniques can also be applied in other diffraction regimes to design DOEs with specific diffraction patterns. In the regime of Fresnel diffraction, an iterative Fresnel algorithm can be applied [101]. Also more rigorous propagation calculations between the diffraction pattern and the corresponding DOE can be used [102].

In Chapter 9 on multiple beam splitters we will learn about the design of Dammann gratings. This is an example of yet another design approach for DOEs, which might be called a direct approach. To this end the diffraction pattern of the DOE is calculated as a function of the fabrication parameters. This calculation can be performed for various different approximations, e.g., Fourier diffraction, Fresnel diffraction or rigorous approaches. In order to adjust the DOE design to the application in question, a suitable cost function has to be found, which allows optimization of the diffraction pattern, depending on the fabrication freedom in the element. With such a cost function, a variety of algorithms, such as, e.g., simulated anhealing or steepest descent, can be applied for the optimization.

5.5 Glossary

- **blazed grating**
  diffraction grating resulting from the 2π quantization of an analog phase profile. Each period of the grating has a continuous phase profile with phase steps of $N2\pi$ at the edges of the periods.

- **detour-phase**
  phase values in the nonzero diffraction orders of a carrier grating, which result from a shift (detour) of the grating. Mathematically, the detour-phase results from the shift theorem of Fourier optics.

- **effective medium theory**
  an approximation to the diffraction theory of sub-wavelength diffraction gratings. The transmitted zeroth diffraction order experiences an effective refractive index of the grating, which results from an averaging of the dielectric constants of the grating material and the environment, weighted with the duty cycle of the grating. For laminar 1D gratings, effective medium theory predicts different refractive indices for light with different polariations (form birefringence).

- **Fraunhofer diffraction**
  far-field diffraction. In scalar diffraction theory, the mathematical relation between the complex amplitude transmission of a DOE and its diffraction pattern results from a Fourier transformation.

- **Fresnel diffraction**
  near-field diffraction. In scalar theory, the mathematical relation between the complex amplitude of a DOE and its diffraction pattern is decribed by a Fresnel transform. A typical phenomenon of Fresnel diffraction is the Talbot self imaging of periodic objects.

- **Kirchhoff approximation**
  describes the wavefront behind a DOE through a multiplication of the incident wavefront and the complex amplitude transmission of the grating. This is only valid for "thin" diffraction gratings.

- **zero-order grating**
  diffraction grating with sub-wavelength periods, which results in the zeroth diffraction order to be the only nonvanishing order.

5.6 List of new symbols

A_0:	amplitude of an optical wavefront
A_m:	diffraction amplitudes
α:	constant to describe the Floquet theorem; field angle supported by a lens
α_{inc}:	propagation angle of the incident plane wave
α_m:	diffraction angles
$\boldsymbol{E}$:	electric field vector
$\epsilon_l(x, z_l)$:	dielectric constant in the respective layer
η:	efficiency of a DOE
f:	focal length
$f/\#$:	the f-number of the lens, defined as: $f/\# = \frac{f}{2r}$
$\hat{\mathcal{G}}(\nu_x)$:	Fourier transform of the quantized phase function $\hat{g}(x)$
$\hat{G}_m(\nu_x)$:	ghost imaged resulting for the phase quantization
$\tilde{g}(\nu_x)$:	Fourier transform of the non-quantized phase function $g(x)$
$g(x)$:	phase function
$\hat{g}(x)$:	quantized phase function
γ:	angle of incidence of a plane wave onto a lens
I_m:	intensity in the m-th diffraction order
I_{inc}:	intensity of the incident wave
$\boldsymbol{k}$:	wavevector
k:	length of the wavevector: $k = \frac{2\pi}{\lambda}$
N:	number of phase quantization values
NA:	numerical aperture of a lens, defined as : $\text{NA} = \frac{R}{f} = \frac{2}{f/\#}$
n_0, n_1:	refractive indices
n, m, q:	integer values; indices for the summations
p:	period of a grating
Δp:	shift of the period of a grating
r_j:	radius of the j-th ring in a Fresnel zone pattern
t_N:	etching depth for the N-th mask layer of a kinoform DOE
θ:	angle of inclination between two interfering waves
$U(x)$:	complex optical wavefront
$U(x, z-0)$:	complex amplitude of the light wave just in front of the plane $z = 0$
$U(x, z+0)$:	complex amplitude of the light wave just behind the plane $z = 0$
$\varphi(x)$:	phase of a wavefront
φ_d:	detour-phase value
$\varphi_{\text{error}}, \delta\varphi$:	deviations from the ideal phase
$\varphi_{\text{qu}}(x)$:	quantized phase of a wavefront
$\Delta\varphi(x)$:	the height of a phase step
$\varphi_{r,m}, \varphi_{t,m}$:	diffraction angles
$w, \delta w$:	width of the features in a DOE, and deviations from the ideal width

5.7 Exercises

1. **Etching depth for the fabrication of multilevel phase gratings**

   For the fabrication of a multilevel phase grating a series of binary etching steps is used. Calculate the etching depths necessary to fabricate an 8-phase level linear kinoform grating in a fused silica glass substrate

   a) for the case of a transmission grating fabricated for $\lambda = 633$ nm (use for the refractive index of fused silica: $n_{633} = 1.457$).

   b) What are the etching depths if the grating is optimized to be used in reflection? (2 possibilities!)

2. **The validity of the paraxial approximation for the design of diffractive lenses**

   A typical value for the minimum feature size of lithographic fabrication processes is $w_{\text{min}} \approx 1\ \mu$m. For this value calculate the minimum f-numbers of lenses which can be fabricated with 2, 4 and 8 phase levels (wavelength: $\lambda = 633$ nm). What are the minimum focal lengths for these lenses if the paraxial approximation is still valid? What are the corresponding maximum lens diameters?

3. **Effect of fabrication errors on the performance of kinoform gratings**

   In Fig. 5.7 the effect of a lateral misalignment of a mask level on the phase profile of a kinoform grating is illustrated.

   a) Express the phase error introduced by a lateral misalignment δs of the N-th mask as a function of the period p of the kinoform grating and the etching depth for the N-th etching step $t_N = \frac{\lambda}{2^N(n-1)}$, where n is the refractive index of the substrate material.

   b) Estimate the effect of this misalignment in the output plane by calculating the Fourier transform.

4. **Efficiency of holographically recorded blazed gratings**

   Calculate the light efficiency in the first diffraction order of a thin blazed grating recorded holographically in a setup shown in Fig. 5.21. For the calculation assume that the developing process after illumination generates

   a) a cosinusoidal amplitude grating:

   $$g_1(x) = \frac{1}{2}[1 + \cos(2\pi \frac{x}{p})]$$

   b) after bleaching a cosinusoidal phase grating:

   $$g_2(x) = e^{i \cos(2\pi x/p)}$$

References

[1] O. E. Miller, J. H. McLeod and W. T. Sherwood,"Thin sheet plastic Fresnel lenses of high aperture", *JOSA* **41** (1951), 807–815.

[2] K. Miyamoto, "The phase Fresnel lens", *JOSA* **17** (1961), 17–21.

[3] R. W. Wood, "Improved diffraction gratings and replicas", *JOSA* **34** (1944), 509.

[4] M. C. Hutley, "Diffraction Gratings", Academic Press, London (1982), Chap. 5, p. 129.

[5] M. Born and E. Wolf, "Principles of Optics", Pergamon Press, London (1980), 6th ed.

[6] L. B. Lesem, P. M. Hirsch and J. Jordan, "The kinoform: a new wavefront reconstruction device", *IBM J. Res. Dev.* **13** (1969), 150–155.

[7] H. Dammann, "Blazed Synthetic Phase-Only Holograms", *Optik* **31** (1970), 95–104.

[8] J. A. Jordan, P. M. Hirsch, L. B. Lesem and D. L. Van Roy, "Kinoform lenses", *Appl. Opt.* **9** (1970), 1883–1887.

[9] J. W. Goodman and A. M. Silvestri, "Some effects of Fourier-domain phase quantisation", *IBM J. of Res. Develop.* **14** (1969), 478.

[10] W. J. Dallas, "Phase quantization — a compact derivation", *Appl. Opt.* **10** (1971), 673.

[11] W. J. Dallas, "Phase quantization in holograms — a few illustrations", *Appl. Opt.* **10** (1971), 674.

[12] D. A. Buralli, G. M. Morris, "Effects of diffraction efficiency on the modulation transfer function of diffractive lenses", *Appl. Opt.* **31** (1992), 4389–4396.

[13] J. Futhey and M. Fleming, "Superzone diffractive lenses", *OSA Techn. Digest* **9**, Diffractive optics: design, fabrication, and applications, New Orleans, LA, USA (1992), 4–6.

[14] M. W. Farn, "Agile beam steering using phased-arraylike binary optics", *Appl. Opt.* **33** (1994), 5151–5158.

[15] J. A. Davis, D. M. Cottrel, R. A. Lilly and S. W. Connely, "Multiplexed phase encoded lenses written on spatial light modulators", *Opt. Lett.* **14** (1989), 420–422.

[16] E. C. Tam, S. Zhou and M. R. Feldman, "Spatial-light-modulator-based electro-optical imaging system", *Appl. Opt.* **31** (1992), 578–580.

[17] E. Carcolé, J. Campos and S. Bosch, "Diffraction theory of Fresnel lenses in low resolution devices", *Appl. Opt.* **33** (1994), 162–174.

[18] M. Kuittinen and H.-P. Herzig, "Encoding of efficient diffractive microlenses", *Opt. Lett.* **20** (1995), 2156–2158.

[19] J. J. Clair and C. I. Abitbol, "Recent advances in phase profile generation", in Progress in Optics **XVI** (1978), E. Wolf (ed.), North Holland, Amsterdam, 71–117.

[20] W.-H. Lee, "High efficiency multiple beam gratings", *Appl. Opt.* **18** (1979), 2152–2158.

[21] H. Zarschitzky, A. Stemmer, F. Mayerhofer, G. Lefranc and W. Gramann, "Binary and multilevel diffractive lenses with submicrometre feature sizes", *Opt. Eng.* **33** (1994), 3527–3536.

[22] J. Jahns and S. J. Walker, "Two-dimensional array of diffractive microlenses fabricated by thin film deposition", *Appl. Opt.* **29** (1990), 931–936.

[23] B. Goebel, L. L. Wang and T. Tschudi, "Multilayer technology for diffractive optical elements", *Appl. Opt.* **35** (1996), 4490–4493.

[24] U. Krackhardt, N. Streibl and J. Schwider, "Fabrication errors of computer-generated multilevel phase-holograms", *Optik* **95** (1994), 137–146.

[25] V. Arrizón and S. Sinzinger, "Modified quantization schemes for array illuminators", *Opt. Comm.* **140** (1997), 309–315.

[26] W. B. Veldkamp and T. J. McHugh, "Binary Optics", *Scientific American* (1992) No. 5, 50–55.

[27] J. A. Cox, T. Werner, J. Lee, B. Fritz and J. Bergstrom, "Diffraction efficiency of binary optical elements", *Proc. SPIE* **1211**, Computer and optically generated holographic optics (1990), 116–124.

[28] J. A. Cox, B. Fritz and T. Werner, "Process error limitations on binary optics performance", *Proc. SPIE* **1555**, Computer and optically generated holographic optics (1991), 80–88.

[29] J. W. Goodman, "Introduction to Fourier Optics", McGraw Hill, New York (1968).

[30] R. Petit (ed.), "Electromagnetic theory of gratings", Springer Verlag, Berlin (1980).

[31] H. Römer, "Theoretische Optik", VCH, Weinheim (1994).

[32] H. Hoenl, A. W. Maue and K. Westfahl, "Theorie der Beugung", in Handbuch der Physik, **25** Springer Verlag, Berlin (1961).

[33] J. Turunen, "Diffraction theory of microrelief gratings", in Micro-optics, elements, systems, and applications, H.-P. Herzig, Taylor & Francis, London (1997), 31–52.

[34] A. Sommerfeld, "Optics", Academic Press, San Diego (1949).

[35] L. Solymar and D. J. Cooke, "Volume holography and volume gratings", Academic Press, London (1981).

[36] K. Knop, "Rigorous diffraction theory for transmission phase gratings with deep rectangular profiles", *JOSA* **68** (1978), 1206–1210.

[37] M. G. Moharam and T. K. Gaylord, "Diffraction analysis of dielectric surface-relief gratings", *JOSA* **72** (1982), 1385–1392.

[38] P. Lalanne and M. Morris, "Highly improved convergence of the coupled wave method for TM polarization", *JOSA A* **13** (1996), 779–784.

[39] L. Li, "Use of Fourier series in the analysis of discontinuous periodic structures", *JOSA A* **10** (1996), 1184–1189.

[40] M. Nèviere and P. Vincent, "Differential theory of gratings: answer to an objection on its validity for TM polarization", *JOSA* **5** (1988), 1522–1524.

[41] J. R. Andrewartha, J. R. Fox and I. J. Wilson, "Resonance anomalies in the lamellar grating", *Opt. Act.* **26** (1979), 69.

[42] H. Haidner, "Theorie der Subwellenlängenstrukturen für diffraktive optische Elemente", PhD. thesis, Universität Erlangen/Nürnberg (1994).

[43] R. Morf, "Diffraction theory", in Perspectives for parallel optical interconnects, P. Lalanne and P. Chavel (eds), Springer Verlag, Berlin (1993).

[44] A. W. Lohmann, "Optical information processing", Optik + INFO, Uttenreuth, Germany (1986).

[45] M. Totzeck and B. Kuhlow, "Validity of the Kirchhoff approximation for diffraction by weak phase objects", *Opt. Comm.* **78** (1990), 13–19.

[46] T. K. Gaylord and M. G. Moharam, "Thin and thick gratings: terminology clarification", *Appl. Opt.* **20** (1981), 3271–3273.

[47] W. Stößel, "Fourieroptik", Springer Verlag, Berlin, Germany (1993).

[48] J. Jahns, "Diffractive Optical Elements for Optical Computing", in Optical Computing Hardware, J. Jahns and S. H. Lee (eds), Academic Press, San Diego (1994), Chap. 6, pp. 137–167.

[49] H. Nishihara and T. Suhara, "Micro Fresnel lenses", in Progress in Optics **XXIV** (1987), E. Wolf (ed.), North-Holland, Amsterdam, 3–37.

[50] D. A. Buralli, G. M. Morris and J. R. Rogers, "Optical Performance of Holographic Kinoforms", *Appl. Opt.* **28** (1989), 976–983.

[51] C. J. R. Sheppard, "Imaging in optical systems of finite Fresnel number", *JOSA A* **3** (1986), 1428–1432.

[52] M. Young, "Zone plates and their aberrations", *JOSA* **62** (1972), 972–976.

[53] W. A. Kleinhans, "Aberrations of curved zone plates and Fresnel lenses", *Appl. Opt.* **16** (1977), 1701–1704.

[54] W. C. Sweat, "Describing holographic optical elements as lenses", *JOSA* **67** (1977), 803–808.

[55] D. A. Buralli and M. G. Morris, "Design of a wide field diffractive lenscape lens", *Appl. Opt.* **28** (1989), 3950–3959.

[56] D. Hauk and A. W. Lohmann, "Minimumstrahlkennzeichnung bei Gitterspektrographen", *Optik* **15** (1958), 275–277.

[57] B. R. Brown and A. W. Lohmann, "Complex spatial filtering with binary masks", *Appl. Opt.* **5** (1966), 967–969.

[58] B. R. Brown and A. W. Lohmann, "Computer-generated binary holograms", *IBM J. Res. Dev* **13** (1969), 160–168.

[59] A. W. Lohmann and D. P. Paris, "Binary Fraunhofer holograms, generated by Computer", *Appl. Opt.* **6** (1967), 1739–1748.

[60] W. H. Lee, "Computer-generated holograms: techniques and applications", in Progress in Optics **XVI** (1978), E. Wolf (ed.), North Holland, Amsterdam, 119–132.

[61] J. Turunen, J. Fagerholm, A. Vasara and M. R. Tagizadeh, "Detour-phase kinoform interconnects: the concept and fabrication considerations", *JOSA A* **7** (1990), 1202–1208.

[62] S. Sinzinger and V. Arrizón, "High efficiency detour-phase elements", *Opt. Lett.* **22** (1997), 928–930.

[63] S. Sinzinger and V. Arrizón, "The detour-phase as design freedom for kinoform diffractive optical elements", *EOS meetings digest* **12**, Diffractive optics, Savonlinna (1997), 44–45.

[64] A. W. Lohmann, J. Ojeda-Castañeda and A. Serano-Heredia, "Synthesis of 1D complex amplitudes using Youngs experiment", *Opt. Comm.* **101** (1993), 17–20.

[65] V. Arrizón, S. Kinne and S. Sinzinger, "Efficient detour-phase encoding of 1D multi-level phase diffractive elements", *Appl. Opt.* **37** (1998), 5454–5460.

[66] J. E. Ford, F. Xu, K. Urquhart and Y. Fainman, "Polarisation-selective computer-generated holograms", *Opt. Lett.* **18** (1993), 456–458.

[67] A. V. Krishnamoorthy, F. Xu, J. E. Ford and Y. Fainman, "Polarization-controlled multistage switch based on polarization selective holograms", *Appl. Opt.* **36** (1997), 997–1010.

[68] A. Kirk, N. Nieuborg, P. Heremans, M. Kuijk, R. Vounckx and I. Veretennicoff, "Anisotropic diffractive optical logic inverter for pnpn optical thyristors", *Techn. Digest.* , **1**, Optical Computing, Sendai, Japan (1996), 162–163.

[69] D. Gabor, "A new microscope principle", *Nature* **161** (1948), 177.

[70] D. H. Close, "Holographic optical elements", *Opt. Eng.* **14** (1975), 408–419.

[71] N. K. Sheridon, "Production of blazed holograms", *Appl. Phys. Lett.* **12** (1968), 316–318.

[72] D. J. Cooke and A. A. Ward, "Reflection-hologram processing for high efficiency in silver-halide emulsions", *Appl. Opt.* **23** (1984), 934–941.

[73] R. K. Kostuk, J. W. Goodman and L. Hesselink, "Volume reflection holograms with multiple gratings: an experimental and theoretical evaluation", *Appl. Opt.* **25** (1986), 4362–4369.

[74] B. J. Chang and C. D. Leonard, "Dichromated gelatine for the fabrication of holographic optical elements", *Appl. Opt.* **18** (1979), 2407.

[75] D. Meyerhofer, "Dichromated gelatin", in Holographic recording materials, H. M. Smith (ed.), Springer Verlag, Berlin, Germany (1977).

[76] A. A. Friesem, Z. Rav-Noy and S. Reich, "Photodielectric polymer for holographic recording", *Appl. Opt.* **16** (1977), 427–432.

[77] J. M. Wang, L. Cheng and A. A. Sawchuk, "Light efficient two dimensional perfect shuffles with DuPont photopolymer holograms", *Appl. Opt.* **32** (1993), 7148–7154.

[78] K. Curtis and D. Psaltis, "Characterization of the DuPont photopolymer for three-dimensional holographic storage", *Appl. Opt.* **33** (1994), 7425–7428.

[79] R. K. Kostuk, J. W. Goodman and L. Hesselink, "Design considerations for holographic optical interconnects", *Appl. Opt.* **26** (1987), 3947–3953.

[80] J. Schwider, W. Stork, N. Streibl and R. Völkel, "Possibilities and limitations of space-variant holographic optical elements for switching networks and general interconnects", *Appl. Opt.* **31** (1992), 7403–7410.

[81] H. Kobolla, J. Schmidt, E. Gluch and J. Schwider, "Holographic perfect shuffle permutation element for miniaturized switching element", *Appl. Opt.* **34** (1995), 2844–2847.

[82] H.-P. Herzig, P. Ehbets, D. Prongué and R. Dändliker, "Fan-out elements recorded as volume holograms: optimized recording conditions", *Appl. Opt.* **31** (1992), 5716–5723.

[83] S. K. Case, P. R. Haugen and O. J. Løkberg, "Multifacet holographic optical elements for wave front transformation", *Appl. Opt.* **20** (1981), 2670–2675.

[84] K.-H. Brenner and F. Sauer, "Diffractive reflective optical interconnects", *Appl. Opt.* **28** (1988), 386–388.

[85] R. Kostuk, Y.-T. Huang, D. Hetherinton and M. Kato, "Reducing alignment and chromatic sensitivity of holographic optical interconnects with substrate-mode holograms", *Appl. Opt.* **28** (1989), 4939–4944.

[86] T. Stone and N. George, "Wavelength performance of holographic optical elements", *Appl. Opt.* **24** (1985), 3797–3810.

[87] J. Schwider, "Achromatic design of holographic optical interconnects", *Opt. Eng.* **35** (1996), 826–831.

[88] R. C. Enger and S. K. Case, "Optical elements with ultrahigh spatial frequency surface corrugation", *Appl. Opt.* **22** (1983), 3220–3228.

[89] L. H. Cescato, E. Gluch and N. Streibl, "Holographic quaterwave plates", *Appl. Opt.* **29** (1990), 3286–3290.

[90] W. Stork, N. Streibl, H. Haidner and P. Kipfer, "Artificial distributed-index media fabricated by zero-order gratings", *Opt. Lett.* **16** (1991), 1921–1923.

[91] H. Haidner, P. Kipfer, W. Stork and N. Streibl, "Zero-order gratings used as an artificial distributed indes medium", *Optik* **89** (1992), 107–112.

[92] M. W. Farn, "Binary gratings with increased efficiency", *Appl. Opt.* **31** (1992), 4453–4458.

[93] J. Jahns and A. Huang, "Planar Integration of Free-Space Optical Components", *Appl. Opt.* **28** (1989), 1602–1605.

[94] T. Shiono and H. Ogawa, "Reflection aspherical microlenses for planar optics fabricated by electron beam lithography", *Opt. Lett.* **17** (1992), 565–567.

[95] J. Jahns and B. Acklin, "Integrated Planar Optical Imaging System with High Interconnection Density", *Opt. Lett.* **18** (1993), 1594–1596.

[96] V. Soifer, V. Kotlyar and L. Doskolovich, "Iterative methods for diffractive optical elements computation", Taylor & Francis, London (1997).

[97] R. W. Gerchberg and W. O. Saxton, "A practical algorithm for the determination of phase from image and diffraction plane figures", *Optik* **35** (1972), 237–246.

[98] J. R. Fienup, "Phase retrieval algorithms: a comparison", *Appl. Opt.* **21** (1982), 2758–21769.

[99] O. Bryngdahl and F. Wyrowski, "Digital holography — computer-generated holograms", in Progress in Optics **XXVIII** (1990), E. Wolf (ed.), North-Holland, New York, USA, 1–86.

[100] F. Wyrowski, "Design theory of diffraction elements in the paraxial domain", *JOSA A* **10** (1993), 1553–1561.

[101] R. Dorsch, A. W. Lohmann and S. Sinzinger, "Fresnel ping-pong algorithm for two-plane computer-generated hologram display", *Appl. Opt.* **33** (1994), 869–875.

[102] R. Bräuer and O. Bryngdahl, "Design strategy for diffractive elements with prescribed diffraction angles in non-paraxial region", *Opt. Com.* **115** (1995), 411–416.

6 Integrated waveguide optics

In this chapter we digress briefly from our main topic of free-space microoptics and introduce the field of integrated waveguide optics. We now leave the discussion of individual microoptical components and technological aspects, and start our discussion of microoptical system aspects, which will be supplemented by the following chapters on "microoptical systems" and "optoelectronic devices and smart pixel arrays".

Waveguide optics is an important field of optics which, just as free-space microoptics, has been motivated by the desire to build stable and rugged optical systems using planar fabrication techniques. The potential for integrating waveguiding structures onto planar substrates led to the development of waveguide optical systems for applications in optical communications and sensing [1, 2, 3, 4]. We give an overview of the field and try to point out the potential of and the differences between waveguide optics and free-space optics. To this end we first focus on the propagation of the light wave in guiding media and learn about the development of "modes in optical waveguides". These are the fundamentals which are necessary in order to understand the mechanisms of "waveguide couplers and beam splitters" and "waveguide modulators", which are discussed in subsequent sections. At the end of this chapter we will discuss several "integrated waveguide optical systems" for specific applications.

6.1 Modes in optical waveguides

The fundamental principle of "waveguiding" in dielectric media is quite simple. Total internal reflection is responsible for the confinement of the electromagnetic field inside a waveguide. The core of an optical fiber or the waveguide consists of a medium with higher refractive index n_2 than the surrounding material n_1. All the light propagating at an angle larger than α_c relative to the normal of the material interface is reflected back into the high refractive index material due to total internal reflection (Fig. 6.1). α_c is called the **critical angle**, which is given by:

$$\sin(\alpha_c) = \frac{n_1}{n_2} \tag{6.1.1}$$

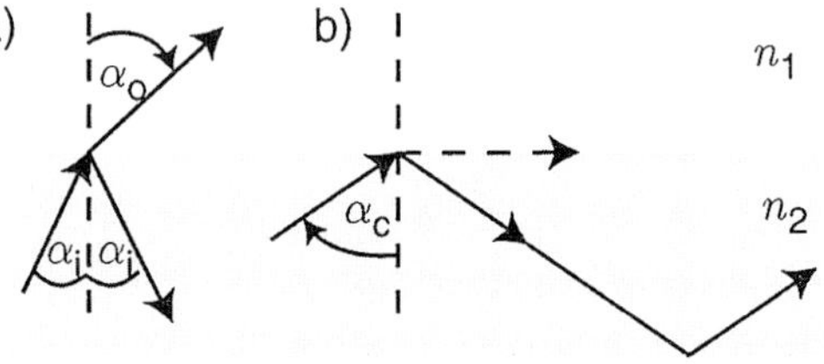

Fig. 6.1: Illustration of the reflection of an optical beam at a medium with lower refractive index ($n_1 < n_2$): a) the angle of incidence is smaller than the critical angle ($\alpha_i < \alpha_c$), so that a refracted and a reflected beam exist; b) angle of incidence is equal to the critical angle ($\alpha_i = \alpha_c$).

"Light guiding" based on total internal reflection in a jet of water was demonstrated in the 19th century by John Tyndall. However, it took until the 1960s, when lasers became available as powerful and efficient light sources, for the potential of guided wave optics to become realized, e.g., in optical communication. Initially fiber optics set the pace for this development, by rapidly developing techniques for the fabrication of fibers with low signal attenuation of less than 0.2 dB/km [5]. It was soon realized that optical waveguides can also be fabricated on planar substrates using lithographic thin film techniques. This directly leads to the concept of integrated optics, where small strips of dielectric material form channels for the optical beams, which can be integrated with active devices, e.g., in InP [6]. Coupling and modulation of these optical channels can be achieved in integrated optical couplers.

6.1.1 Discrete waveguide modes

Firstly we shall discuss light propagation in a waveguide. We consider the situation depicted in Fig. 6.2. A plane wave is propagating in a thin layer of dielectric material. A typical thickness d for a waveguide is in the order of several microns O(μm). As long as the propagation angle is smaller than the critical angle, most of the incident wave is refracted and continues to propagate outside the dielectric. Small fractions of the light wave, which are reflected back into the dielectric, can be neglected since they are attenuated further with each reflection. If, however, the angle of propagation exceeds the critical angle, no refracted wave exists and all the light is reflected into the dielectric. The incident and reflected beams interfere inside the waveguide. For constructive interference, the beams have to fulfil the self-consistency condition [7], i.e., after two reflections the wave has to reproduce itself. In this case the interference pattern does not change in z-direction. Here, we ignore phase shifts which occur upon reflection at the material interface. We calculate the path difference between the reflected wave and the original wave (Fig. 6.2):

$$\overline{AC} - \overline{AB} = \overline{AC}\,[1 - \cos(2\theta)] = 2\sin^2(\theta)\overline{AC} = 2d\sin(\theta) \qquad (6.1.2)$$

For self-consistency we require the phase differences between these waves to be integral multiples of 2π:

$$\frac{2\pi}{\lambda}(\overline{AC} - \overline{AB}) = \frac{2\pi}{\lambda}[2d\sin(\theta)] = m2\pi; \quad m = 1, 2, 3, ... \qquad (6.1.3)$$

$\lambda = \frac{\lambda_0}{n_2}$ denotes the wavelength in the medium. Eq. 6.1.3 results in a set of discrete propagation angles θ_m, for which the self-consistency criterion is fulfilled:

$$\sin(\theta_m) = m\frac{\lambda}{2d} \qquad (6.1.4)$$

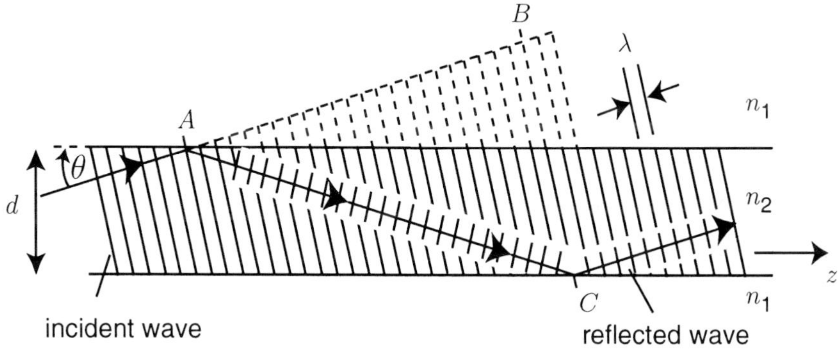

Fig. 6.2: Development of discrete modes in a waveguide [7].

The propagating waves are restricted to this discrete set of propagation angles inside the substrate. These propagation directions are called the modes of the waveguide. We calculate the wavevectors (the k-vectors) of the modes using:

$$k_m = \frac{2\pi}{\lambda}\sin(\theta_m) = m\frac{\pi}{d} \qquad (6.1.5)$$

From Eq. 6.1.5 we find that the origin of the discrete modes is the lateral confinement of the wave in the waveguide. The angular distance between the individual modes is inversely proportional to the thickness d of the waveguide. For large thicknesses a quasi-continuous angular spectrum of propagating waves results. This corresponds to the case of free-space optics.

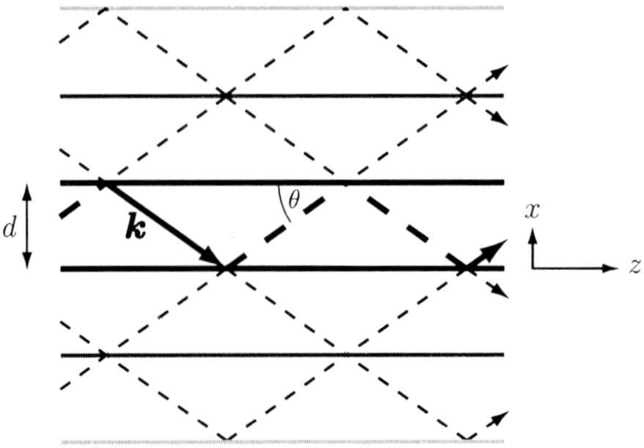

Fig. 6.3: Explanation of the discrete waveguide modes by analogy with grating diffraction.

The mode spectrum of a waveguide can also be understood by analogy with grating diffraction. Total internal reflection at the material interfaces causes a virtual lateral periodicity (Fig. 6.3) [8]. Thus, propagation in the waveguide corresponds to the situation of laterally periodic wavefields in free-space optics. Since the waves propagate in opposite x-directions in adjacent stripes, the period of the virtual structure is $2d$. The propagation directions, as described by Eq. 6.1.4, are found from diffraction due to this grating structure (see grating diffraction in Chapter 5). Propagating modes in the waveguide also have to fulfil the condition of total internal reflection, i.e., the propagation angles must be smaller than $\theta_m < \frac{\pi}{2} - \alpha_c$ (Fig. 6.1). Eqs 6.1.1 and 6.1.4 yield the maximum number of modes M propagating in the waveguide as the smallest integer value greater than:

$$\begin{aligned} M &> 2\sin\left(\frac{\pi}{2} - \alpha_c\right)\frac{d}{\lambda} = 2\cos(\alpha_c)\frac{d}{\lambda} \\ &= 2\sqrt{\frac{n_2^2 - n_1^2}{n_2^2}}\frac{d}{\lambda} = 2\sqrt{n_2^2 - n_1^2}\frac{d}{\lambda_0};\ M:\ \text{integer} \end{aligned} \tag{6.1.6}$$

For $M = 1$ we find the so called cut-off wavelength.

$$\lambda_{\text{co}} = 2d\sin\left(\frac{\pi}{2} - \alpha_c\right) = 2d\sqrt{\frac{n_2^2 - n_1^2}{n_2^2}} \tag{6.1.7}$$

For all wavelengths larger than λ_{co} only a single mode can propagate in the waveguide [7]. In this case we call it a monomode waveguide.

6.1.2 Field distribution of the modes

We shall now discuss the field distributions of the different modes in the waveguide. Mathematically they can be derived as solutions of the Helmholtz equation inside the different layers. In the guiding layer, the field distribution for each mode is composed of two plane waves travelling at angles θ_m and $-\theta_m$. Assuming propagation in the $x - z$ plane, the wavevectors of these plane waves can be written as $(\pm n_2 k_0 \sin(\theta), 0, n_2 k_0 \cos(\theta))$. The z-components of the k-vectors are called the propagation constants β_m of the modes:

$$\beta_m = n_2 k_0 \cos(\theta_m) \tag{6.1.8}$$

For the field distributions in the guiding layer we obtain:

$$\boldsymbol{E}(x, z) \propto U_m(x) e^{i\beta_m z} \tag{6.1.9}$$

where $U_m(x)$ denote the transverse field distributions:

$$U_m(x) = \begin{cases} \cos\left(\frac{2\pi \sin(\theta_m)}{\lambda} x\right), & m: \text{ even} \\ \sin\left(\frac{2\pi \sin(\theta_m)}{\lambda} x\right), & m: \text{ odd} \end{cases} \tag{6.1.10}$$

For the field outside the guiding dielectric, we know from the boundary conditions that it has to match the field inside at every point of the boundary. Thus, the z-dependence is again described by $e^{-i\beta_m z}$. From energy considerations we conclude that the transversal field distribution can be described by an exponential decay:

$$U_m(x) = \begin{cases} e^{-\gamma_m x}, & x > \frac{d}{2} \\ e^{\gamma_m x}, & x < -\frac{d}{2} \end{cases} \tag{6.1.11}$$

γ_m, which describes the exponential decay of these so-called evanescent waves, is called extinction coefficient:

$$\gamma_m = n_1 \frac{2\pi}{\lambda_0} \sqrt{\frac{\cos^2(\theta_m)}{\cos^2(\frac{\pi}{2} - \alpha_c)} - 1} = n_1 \frac{2\pi}{\lambda_0} \sqrt{\frac{\cos^2(\theta_m) n_2^2}{n_1^2} - 1} \tag{6.1.12}$$

Fig. 6.4 summarizes the field distributions of several TE modes in a symmetrical waveguide structure. The evanescent tails of the fields depend on the extinction coefficient. From Eq. 6.1.12 we find that γ decreases with increasing θ_m. Higher-order modes thus penetrate deeper into the adjacent dielectric layers. Furthermore, with increasing refractive index step, the extinction coefficient increases and the field penetration is reduced.

In the above discussion we assumed an ideally symmetrical waveguide with equal refractive index materials in the layers above and below the waveguiding layer. An asymmetry in the layer profile will cause assymmetrical field distributions. Specifically the penetration depth of the evanescent field is different for different waveguide coatings. Fig. 6.5 shows the geometries of different types of waveguide structures. Fabrication of each of these types is possible using lithographic thin layer deposition and structuring techniques. The technological efforts necessary for the different types varies significantly. For a detailed discussion of the properties of the various profiles, the reader is referred to the literature [1, 2].

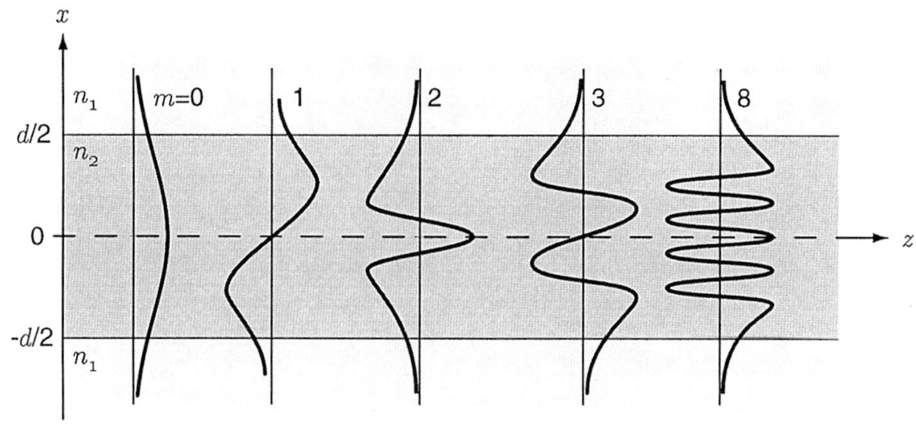

Fig. 6.4: Field distributions for the TE modes guided in a symmetrical planar waveguide structure [8].

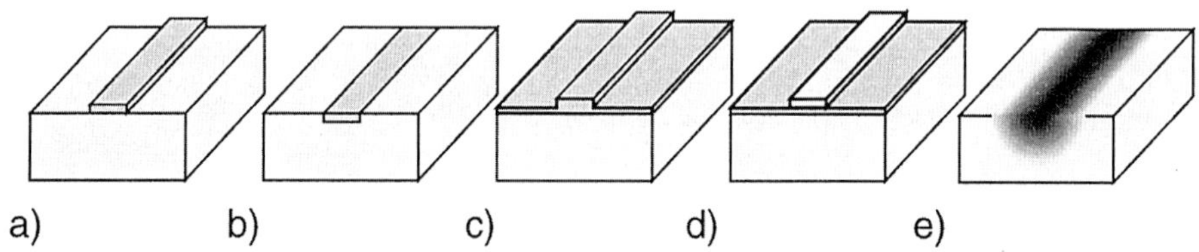

Fig. 6.5: Geometries of different types of waveguides: a) strip, b) embedded strip, c) rib or ridge, d) strip loaded and e) GRIN waveguide.

6.2 Waveguide couplers and beam splitters

Coupling elements are amongst the most important passive optical elements in integrated optics. We distinguish between external coupling (i.e., coupling from free-space or other light-guiding elements into a waveguide) and coupling between waveguides, e.g., directional couplers.

6.2.1 External coupling

Coupling into the waveguide is a critical task, where the goal is to convert the light from free-space or other waveguiding structures, e.g., optical fibers or semiconductor lasers, into modes which can be guided by the waveguide. The light can be introduced into the waveguide from the edge ("transverse coupling") or from the planar surface of the waveguide ("surface coupling").

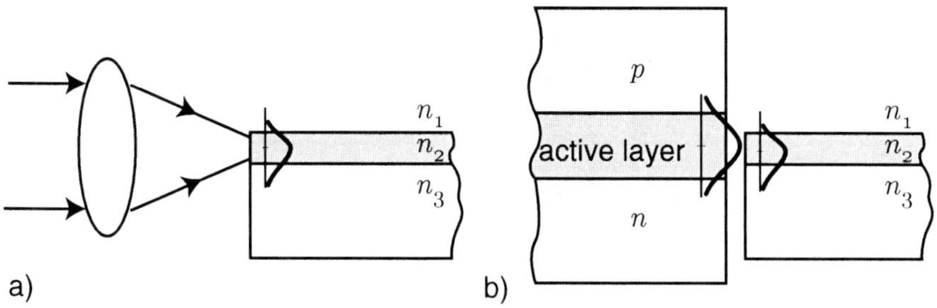

Fig. 6.6: Transverse coupling of light into a waveguide a) from free-space through direct focusing; b) from a semiconductor light source (e.g., laser or light-emitting diode) through butt-end coupling ($n_2 > n_1$; $n_2 > n_3$).

Transverse coupling

Fig. 6.6 illustrates direct focusing from free-space and butt-end coupling from a semiconductor light source. These are the most common types of transverse coupling. For most efficient coupling through direct focusing (Fig. 6.6 a), the numerical aperture of the lens has to match the numerical aperture NA of the waveguide, which is defined as (see Exercises):

$$\mathrm{NA} = \sqrt{n_2^2 - n_1^2} \tag{6.2.1}$$

In this case the coupling efficiency is determined by the ratio of the size of the focus to the thickness of the waveguide. For many types of lasers the coupling efficiency can be close to 100% since lasers have a high degree of spatial coherence, i.e., they are good point sources. This good coupling behaviour, as well as the potential for fast modulation, is the reason for the great impact the laser has had on the breakthrough in guided-wave optics, for example, in communication applications. However, high alignment precision is necessary for this coupling technique since the waveguide layers are often very thin (i.e., $\mathrm{O}(\mu\mathrm{m})$).

The second transverse coupling technique is butt coupling. To this end the waveguiding layers of two components are brought into close contact so that the waveguide modes continue to propagate directly into the waveguide. This is often used for coupling the waveguiding layers of semiconductor lasers or fibers to waveguides. Best coupling efficiencies are achieved if the thicknesses of the waveguiding layers are adjusted. Since the shape of the fundamental modes of the waveguides are similar, good coupling effiencies can be achieved (e.g., efficiency $\eta \approx 68\%$).

Surface coupling

Besides transverse coupling through the edges of the waveguide, surface coupling can also be applied. Here the light is coupled into the guiding layer through the planar surfaces of the waveguide. By definition, it is not possible to achieve angles of incidence which are large enough to couple the lightwave directly through the surface into the waveguide. After refraction at the higher refractive index waveguide dielectric, the beam cannot fulfil the phase matching condition for total internal reflection anymore. Therefore, one has to use optical components, such as a grating or a prism, to convert the light into the desired guided modes.

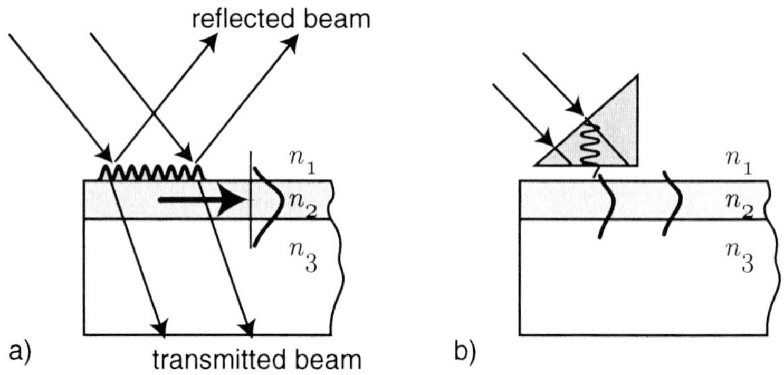

Fig. 6.7: Waveguide coupling by a) a grating and b) a prism coupler.

Fig. 6.7a shows the principle of a grating coupler, schematically. The incident beam is diffracted at the periodic grating structure. For the diffracted beam it is possible to match the phase condition for guided waves. Thus, energy can couple from the incident beam into waveguide modes. By careful determination of the angle of incidence, grating couplers allow one to couple energy into a specific waveguide mode. Another way to look at the grating coupler is to observe a perturbation of the waveguide modes due to the overlaying dielectric grating. Each waveguide mode is split into a number of modes in the grating regions. For some of these modes the phase matching condition can now be fulfilled by the incident beam.

The prism coupler depicted in Fig. 6.7b works according to the same principle as the grating coupler. The incident lightwave experiences total internal reflection at the interface between the prism material and the gap between prism and waveguide. Interference between the incident and reflected wave results in a standing wave inside the prism. As described in the previous section, outside the prism the standing wave has evanescent tails which extend into the region between prism and waveguide. The standing wave propagates in the z-direction with a propagation constant β_p. If the gap between prism and waveguide is small enough, an overlap between the evanescent tails of the prism mode and the waveguide modes occurs. Coupling is possible into the waveguide mode with a propagation constant $\beta_m = \beta_p$.

Again, by variation of the angle of incidence, it is possible to excite specific modes in the waveguide. For obvious reasons the optical coupling through the evanescent tails is often referred to as optical tunneling (see Exercises).

6.2.2 Coupling between waveguides

Optical tunneling through the evanescent tails of guided modes can also be directly applied to the coupling between waveguides. Such waveguide couplers are generally called directional couplers since the direction of propagation is maintained during the coupling. The modes of two parallel optical waveguides which run close enough to each other are coupled through the evanescent portions of their fields. Consequently, during propagation through one of the waveguides the light amplitude couples over into the adjacent waveguide. An analogous situation can be observed in mechanics with two coupled pendulums [8]. If pendulum 1 is excited, energy is transferred to pendulum 2 via the coupling, i.e., it also starts to oscillate. After some time all the oscillation energy is found in the second pendulum while pendulum 1 is at rest. The process then takes place in the other direction until all the oscillation occurs in pendulum 1 again. Thus, the energy alternates between the two pendulums.

For two coupled waveguides the situation is analogous. If a light wave propagates in waveguide 1, due to the overlap of the evanescent tails of the modes, energy is coupled into waveguide 2 (Fig. 6.8). After a well defined propagation length the light wave is completely transferred into waveguide 2 and the inverse process starts. Mathematically, the situation is described by coupled mode theory [9]. To this end the electric field of the waveguide mode is separated into a complex amplitude $A(z)$ containing the propagation term $e^{i\beta z}$ and the normalized lateral field distribution $E(x, y)$. In the following we assume identical waveguides. The complex amplitudes in the coupled waveguide system can be described by two symmetrical coupled differential equations [7]:

$$\begin{aligned} \frac{dA_1(z)}{dz} &= -i\beta A_1(z) + i\kappa A_2(z) \\ \frac{dA_2(z)}{dz} &= -i\beta A_2(z) + i\kappa A_1(z) \end{aligned} \tag{6.2.2}$$

For the more general case it would be necessary to assume different propagation constants β in each waveguide and possibly different coupling constants κ for the coupling between the modes of waveguide 1 and waveguide 2.

We assume the following initial condition:

$$A_1(0) = 1; \; A_2(0) = 0 \tag{6.2.3}$$

This means, at first all the light amplitude is propagating in waveguide 1. In this case the solutions of Eq. 6.2.2 read:

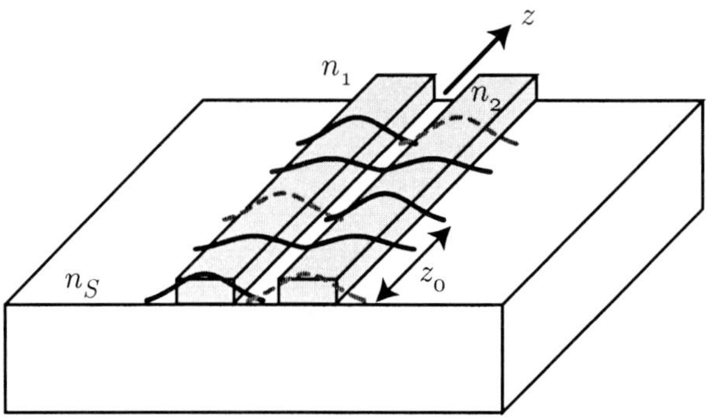

Fig. 6.8: Mode coupling between two neighbouring waveguides.

$$A_1(z) = \cos(\kappa z)e^{i\beta z} \tag{6.2.4}$$

and

$$A_2 = -i\sin(\kappa z)e^{-\beta z} \tag{6.2.5}$$

If we take into account the wave attenuation α upon propagation through the waveguide we write for β:

$$\beta = \beta_r + i\frac{\alpha}{2} \tag{6.2.6}$$

For the optical power in the waveguides we obtain:

$$\begin{aligned} P_1(z) &= \cos^2(\kappa z)e^{-\alpha z} \\ P_2(z) &= \sin^2(\kappa z)e^{-\alpha z} \end{aligned} \tag{6.2.7}$$

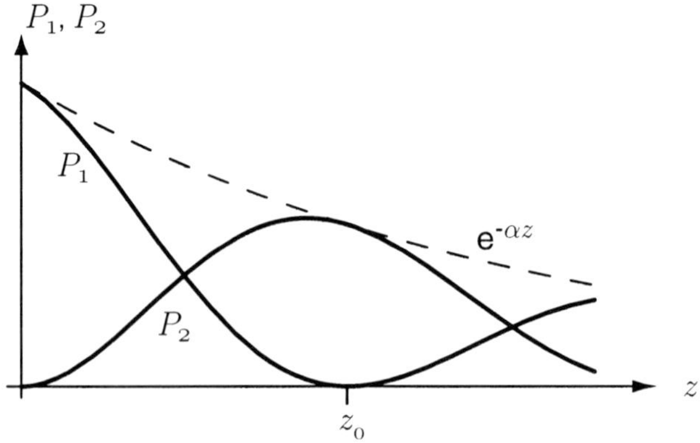

Fig. 6.9: Power transfer from one waveguide to the adjacent one by mode coupling [8].

Fig. 6.9 illustrates the power distribution in both waveguides. The coupling length z_0, after which the whole energy is transferred from one waveguide to the other, is calculated from:

$$\kappa z_0 = \frac{\pi}{2} \quad \longrightarrow \quad z_0 = \frac{\pi}{2\kappa} \tag{6.2.8}$$

The coupling constant κ between the waveguides is calculated from the overlap integral of the waveguide modes:

$$\kappa = \frac{1}{2}(n_2^2 - n_1^2)\frac{4\pi^2}{\lambda^2 \beta} \iint\limits_{\mathcal{A}} E_1 E_2 \, dxdy \tag{6.2.9}$$

Here the integration region $\mathcal{A}$ is the area covered by one waveguide.

6.2.3 3 dB couplers for beam splitting

Directional couplers can also be used to distribute the light signal by splitting the waveguide channel (Fig. 6.10a). If we want to achieve equal field amplitudes in the two resulting data channels, we choose for the coupling distance z_S :

$$P_1 = P_2 \quad \longrightarrow \quad \sin(\kappa z_S) = \cos(\kappa z_S) \tag{6.2.10}$$

We find for z_S:

$$z_S = \frac{\pi}{4\kappa} = \frac{z_0}{2} \tag{6.2.11}$$

Each of the output channels of such a beam-splitting coupler carries approaximately one half of the input beam energy. The couplers are, therefore, often called 3 dB couplers.

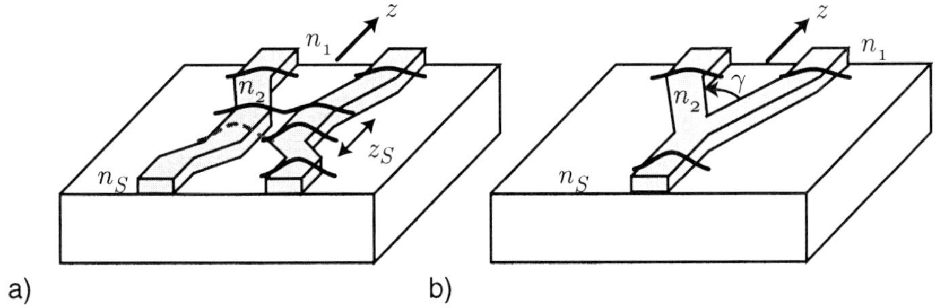

Fig. 6.10: Beam-splitting waveguide components a) 3 dB directional coupler; b) Y-branch

6.2.4 Branching waveguides

As an alternative to 3 dB couplers, beam splitting in waveguide optics can be achieved through branching waveguides (Fig. 6.10b). The concept here is to achieve maximum overlap of the modes of the two output channels in a small area at the center of the branch. Here, the coupling coefficient reaches a maximum and coupling can take place over very short coupling distances. However, the mode coupling depends on the angle γ between the output waveguide channels. Thus, with increasing angle, the coupling becomes less efficient. For low insertion loss the angle must not exceed $\gamma \leq 0.5°$. Consequently, in order to achieve reasonable channel separation, e.g., to be able to access both channels separately, the Y-branch has to be very long. The limit in the branching angle also limits the splitting ratio for the branch to low numbers (e.g., 2-3) [1].

6.3 Waveguide optical modulators

6.3.1 The electro-optic effect

Most active waveguide modulator devices are based on the electro-optic effect, i.e., the change of the refractive index of a material induced by an electric field. The refractive index n of most optical materials varies with the application of an electric field $\boldsymbol{E}$. Generally n is only a weak function of $\boldsymbol{E}$ so that it can be expanded in a Taylor serie about $\boldsymbol{E} = 0$:

$$n(\boldsymbol{E}) = n - \frac{1}{2} r_P n^3 \boldsymbol{E} - \frac{1}{2} r_K n^3 \boldsymbol{E}^2 + ... \tag{6.3.1}$$

Here r_P and r_K are material coefficients which describe the strength of the electrooptic effects in the material. From Eq. 6.3.1 we see that two different kinds of dependencies of the refractive index on the electric field can be distinguished. The linear dependence on the electric field is called the Pockels effect (coefficient r_P), while the quadratic effect is called the Kerr effect (coefficient r_K). In many materials r_P is significantly stronger than r_K so that we can neglect the third term in Eq. 6.3.1 [7].

$$n(E) \approx n - \frac{1}{2} r_P n^3 E \tag{6.3.2}$$

Such media are called Pockels media or Pockels cells. The Pockels coefficient r_P, is very small and ranges from 10^{-12}–$10^{-10} \frac{m}{V}$. An electric field as strong as $|\boldsymbol{E}| = E = 10^6 \frac{V}{m}$ only generates an index change of $\Delta n = 10^{-6}$–10^{-4}. Typical media for Pockels cells are crystalline materials such as KDP (KH_2PO_4), ADP ($NH_4H_2PO_4$), $LiNbO_3$ or $LiTaO_3$.

In centrosymmetrical materials, such as gases, liquids and a few specific crystals, the Pockels coefficient vanishes since $n(\boldsymbol{E})$ has to be a symmetrical function. In this case a quadratic dependence of the refractive index on the electric field can be observed (Kerr effect).

$$n(\boldsymbol{E}) \approx n - \frac{1}{2} r_K n^3 E^2 \tag{6.3.3}$$

The Kerr effect is just as weak as the Pockels effect, with typical values of $r_{\mathrm{K}} = 10^{-22}$–$10^{-12}$. An electric field of $E = 10^6 \ \frac{V}{m}$ generates an index change of $\Delta n = 10^{-6}$–10^{-2}.

Despite the fact that both the Kerr and the Pockels effect are rather weak, it is possible to exploit them in optical modulator devices. Waveguide devices are well suited since the length of the device can be adjusted in order to generate a sufficiently strong impact on the propagating mode despite the weak physical effect.

6.3.2 The electro-optic phase modulator

From our knowledge of the electro-optic effects it is now straightforward to "build" an optical phase modulator. To this end we use a waveguide of length L from a Pockels material. If an electric field is applied to this waveguide structure, the refractive index of the waveguide is changed slightly. Consequently, the lightwave travelling through the waveguide experiences a phase retardation with respect to the case without electric field. This phase shift $\Delta\varphi$ depends on the Pockels coefficient r_P, the electric field $\boldsymbol{E}$ and the length L of the waveguide.

$$\Delta\varphi = \varphi_0 - \frac{2\pi}{\lambda_0} n(\boldsymbol{E}) L = \frac{\pi L}{\lambda_0} r_P n^3 \boldsymbol{E} \tag{6.3.4}$$

Here $\varphi_0 = \frac{2\pi n L}{\lambda}$ is the phase generated by the waveguide of length L with no voltage applied. Assuming a thickness d of the waveguide, we apply the electric field perpendicular to the waveguide. We find for the electric field $\boldsymbol{E} = \frac{V}{d}$, where V denotes the voltage applied perpendicular to the waveguide. The voltage V_π necessary to cause a phase shift of π in the modulator can be calculated as:

$$V_\pi = \frac{d\lambda_0}{L r_P n^3} \tag{6.3.5}$$

6.3.3 Polarisation modulator — dynamic phase retarder

In anisotropic birefringent crystals the electro-optic effect can also be used for polarisation switching and the implementation of dynamic wave retarders. In birefringent materials such as $LiNbO_3$, different polarisation directions of the transmitted wave experience different refractive indices. Additionally, the anisotropy has the consequence that the Pockels coefficients are different for the different polarisations. By application of an electric field we can change the birefringence of the crystal. We calculate the voltage necessary to change the relative phase between the perpendicular linear polarisation directions by π from:

$$V_{\mathrm{bi}/\pi} = \frac{d}{L} \frac{\lambda_0}{n_1^3 r_{P1} - n_2^3 r_{P2}} \tag{6.3.6}$$

Here n_1, n_2 and r_{P1}, r_{P2} denote the refractive indices and the Pockels coefficients for the two crystal axes, respectively [7].

6.3.4 Integrated intensity modulators

The electro-optic phase retarder as well as the electro-optic phase modulator can be converted into intensity modulating devices. For this purpose, the dynamic phase retarder is placed between two crossed polarisers with suitable orientation to the crystal axes of the birefringent waveguide. On applying a voltage $V_{\mathrm{bi}/\pi}$ (Eq. 6.3.6), the polarisation is switched. At the same time, due to the polarisers, the transmission of the device is switched. Losses due to reflection, absorption and scattering cause a reduced maximum transmission value in this device. Misalignment of the polarisers relative to the optical axes causes the minimum transmission values to be larger than zero. Nevertheless, high contrast ratios of 1000:1 are possible with this type of intensity modulator. Full integration of such modulators is difficult because of the difficulties involved in producing integrated polarisers [1].

In order to convert the electro-optical phase modulator into an intensity modulator we use an integrated waveguide optical interferometer. To this end two Y-branches are used for beam splitting and beam combination, respectively. Thus, a structure results which can be understood as a Mach-Zehnder interferometer integrated in waveguide optics (Fig. 6.11a). The incident wave I_i is split at the first Y-branch into two equal intensity waves I_1 and I_2. After travelling in separate channels over some distance, these optical waves are recombined at the second Y-branch. If the optical path length for both interferometer arms is identical the combined waves interfere constructively in the output waveguide channel. A relative phase shift $\Delta\varphi$ in one of the channels, however, results in a decreasing intensity of the interfering waves. The intensity I_O in the output channel of the interferometer is described by:

$$I_O = I_1 + I_2 + 2\sqrt{I_1 I_2}\cos(\Delta\varphi) = I_i \cos^2(\frac{\Delta\varphi}{2}) \tag{6.3.7}$$

where: $I_1 = I_2 = \frac{I_i}{2}$.

The phase shifts necessary to switch the intensity in the output waveguide can be generated with an electro-optical phase modulator integrated into one branch of the interferometer. Using Eq. 6.3.4, we find for the dependence of the intensity of the output wave on the voltage V applied to the waveguide with thickness d:

$$I_o = I_i \cos^2\left(\frac{\pi L}{\lambda_0 d} r_P n^3 V\right) \tag{6.3.8}$$

For maximum intensity contrast, we choose a modulation between 0 and the voltage V_π for a π phase shift. Linear intensity modulation is also possible if the voltage is modulated at relative phase shifts around $\Delta\varphi = \frac{\pi}{2}$, where the cosine function is approximately linear. Integrated interferometric intensity modulators of this kind which can operate at modulation frequencies of > 1 GHz are commercially available. Modulation speeds as high as 25 GHz have already been demonstrated [1].

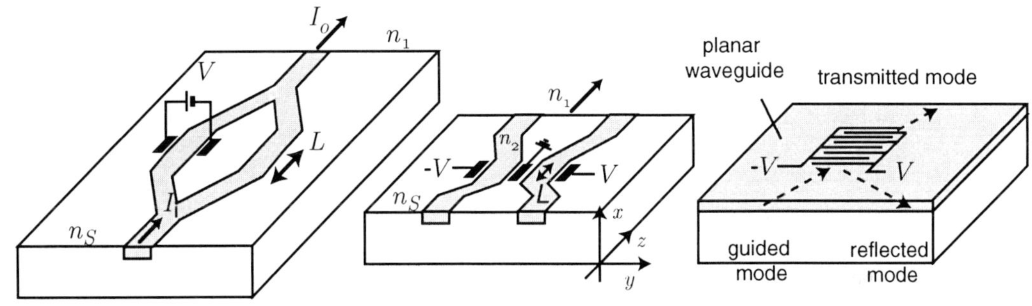

a) b) c)

Fig. 6.11: Integrated optical modulators: a) Mach-Zehnder interferometer with electro-optic phase modulator for intensity modulation; b) electro-optically controlled directional coupler, c) electro-optically generated Bragg grating.

6.3.5 Electro-optic directional couplers

The electro-optic effect can also be applied to control the energy transfer in directional couplers. As we have seen in the previous section of this chapter, the coupling between identical waveguides in directional couplers depends on the coupling coefficient κ which is described by the overlap integral in Eq. 6.2.9. If the waveguides are not identical, the coupling efficiency also depends on the mismatch of the propagation constants $\Delta\beta = \beta_1 - \beta_2$. In this case, we have to take into account that the coupling coefficients for the coupling from waveguide 1 to waveguide 2 and vice versa are different $\kappa_{12} \neq \kappa_{21}$. We can derive a power transfer ratio $\mathcal{T}$ using the coupled mode approach described in section 1.2.2. [7]:

$$\mathcal{T} = \frac{I_2(z_0)}{I_1(0)} = \left(\frac{\pi}{2}\right)^2 \mathrm{sinc}^2\left(\frac{1}{2}\sqrt{\left[1+\left(\frac{\Delta\beta z_0}{\pi}\right)^2\right]}\right) \qquad (6.3.9)$$

Here, sinc(x) is used according to the definition in Chapter 2. From Eq. 6.3.9, we find that $\mathcal{T}$ is maximum for $\Delta\beta = 0$, and reaches a first zero transition at $\Delta\beta = \frac{\sqrt{3}\pi}{z_0}$. At this point no power is transferred from waveguide 1 to waveguide 2.

It is possible to control the phase mismatch $\Delta\beta$ using the electro-optic effect. To this end an electric field is applied to a waveguide in the coupling region (Fig. 6.11b). The index change Δn generates a proportional change of the propagation constants:

$$\Delta\beta = \frac{2\pi\Delta n}{\lambda_0} = -n^3 r_P \boldsymbol{E} \frac{\pi}{\lambda_0} \tag{6.3.10}$$

In order to increase the index change, often voltages with opposite signs are applied to both waveguides in the coupler. Thus, the phase mismatch can be doubled [1].

Yet another kind of electro-optic modulator can be incorporated into an integrated optical system using specifically shaped electrodes. For example, using a periodic grating-like arrangement of electrodes, we can generate a diffraction grating inside a waveguide (Fig. 6.11c). The periodic shape of the electrodes generates a refractive index grating. These gratings are similar to thick Bragg gratings in holographic optical elements (Chapter 5) and are thus capable of efficiently deflecting incident lightwaves. The electro-optic implementation allows one to modulate the grating profile in order to distribute energy dynamically between the transmitted and the reflected waveguide mode. In the planar waveguide illustrated in Fig. 6.11c the light is confined only in the x-direction, while it is unguided in the y-direction. This situation is a combination of free-space optics and integrated waveguide optics [1].

This brief discussion of the basic principles of waveguide optical modulators based on the electro-optic effect may suffice in this context. Acousto-optical modulators can be used as an alternative. Their functionality is based on a change of the refractive index of the waveguide medium due to strain. In acousto-optics, a periodic strain pattern is generated by an acoustic wave travelling through the medium. As a result of diffraction at the resulting index grating, an incident or guided wave in the medium can be modulated in the desired fashion.

6.4 Applications of waveguide optics

Integrated waveguide optical systems have applications in a large variety of fields [2]. In the following we want to discuss examples from the field of optical interconnects and optical sensing.

6.4.1 Waveguide optics in optical interconnects

Optical communications form the most important application of guided-wave optics [10, 11]. The breakthrough of fiber optics for long distance communications also causes a lot of interest in optics for switching and routing fabrics. Integrated waveguide optics is compatible to fiber optics with respect to the wave guiding and to electronics with respect to the fabrication techniques. In many applications it, therefore, forms the interface between fiber optics and electronics.

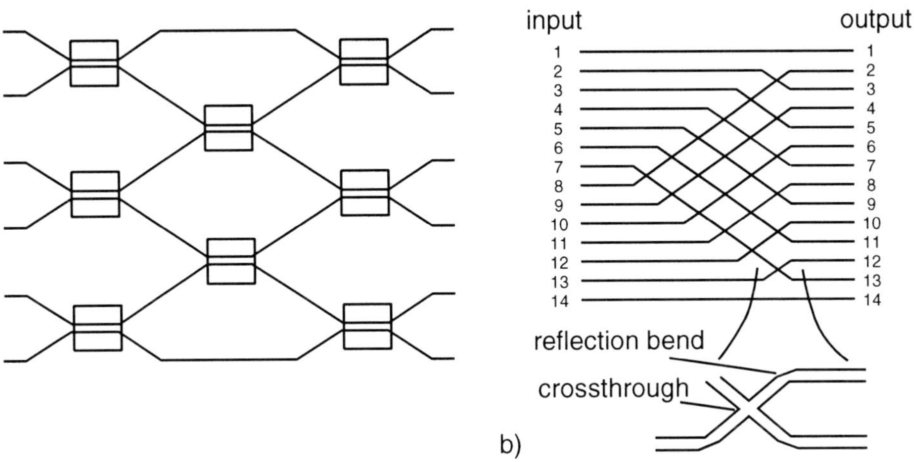

a) b)

Fig. 6.12: Integrated optical systems for optical interconnect applications: a) a matrix of integrated optical switches for signal routing; b) integrated optical Perfect Shuffle interconnect using crossthrough circuits and reflection bends.

Waveguide optics for interconnection networks

Using Y-branches or directional couplers it is straightforward to implement beam splitters for broadcasting interconnects as well as point-to-point interconnection networks. In combination with electro-optic materials it is even possible to realize switching fabrics with variable point-to-point interconnects. Fig. 6.12a shows the schematic of a matrix of integrated optical switches implemented, e.g., as electro-optic directional couplers for signal routing. For larger interconnect networks such as Perfect Shuffles or Crossbars, 45° bends as well as crossthrough interconnects can be realized (Fig. 6.12b). Reflection bends can be used to reduce power losses in large angle waveguide bends. For multimode waveguides, bend losses as low as 0.4 and 0.15 dB can be achieved for 90° and 45° bends, respectively. For single mode waveguides, losses are generally higher, around 1.5–3 dB for 60° and 90°-bends. Crossthrough interconnects are useful for shuffle interconnect schemes. They can be implemented very efficiently with 0.006 dB for a 90° single mode crossthrough [11, 12].

These components provide the necessary flexibility for implementing a variety of shuffle point-to-point interconnects. Broadcast interconnects, e.g., for optical clock distribution have also been suggested for an implementation in waveguide optics. With an H-tree arrangement of the waveguide channels it is possible to achieve equal optical path lengths to a large number of clock receivers [11]. In recent years plastic optical waveguides and fibers have been suggested and demonstrated for optical interconnects. Their main advantage is the cheap fabrication, e.g., using replication techniques such as injection moulding [13, 14].

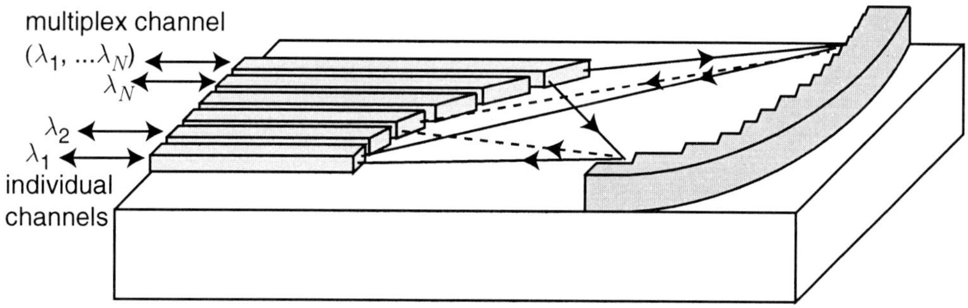

Fig. 6.13: An integrated wavelength de/multiplexing system using waveguides for the I/O channels and a free-space optical phase grating for spectral splitting.

Guided-wave optical systems for wavelength division multiplexing (WDM)

Wavelength multiplexing or demultiplexing systems are of great significance for optical data links. A variety of integrated waveguide optical systems which are capable of multiplexing and demultiplexing have been suggested [11, 2]. Common to most of these systems is that they combine waveguide components with quasi-free-space components in order to achieve the multiplexing functionality [15, 16]. Fig. 6.13 shows an integrated optical system for λ-multiplexing or demultiplexing. The input and output channels are formed by waveguides, while the multiplexing or demultiplexing is performed in free-space by a blazed phase grating, which is etched vertically into the substrate [17].

Another system for WDM was suggested by Dragone (Fig. 6.14a). He uses a pair of optical star couplers and a set of waveguides with different lengths for the wavelength de/multiplexing [18, 19]. The star couplers combine waveguide optics and one-dimensional free-space propagation in a planar waveguiding layer. The shape of the free-space optical section is chosen such that light from one of the input channels is distributed equally among the output waveguide channels. After a section of guided wave propagation they form the input channels for the next star coupler. Since the optical path length varies from channel to channel, the waveguide modes experience relative phase shifts. Upon mixing of the modes from the different waveguides in the second star coupler, this results in a diffraction pattern. The variable lengths waveguide structure effectively acts like a phase grating or quantized prism ("phased array" quantization). In order to understand the analogy between the integrated system and a free-space optical system, Fig. 6.14b shows the corresponding free-space optical system.

These few examples are sufficient to illustrate the variety of possible waveguide optical systems for application in optical communications. We will now focus on a second field of applications for integrated optics, namely optical sensors.

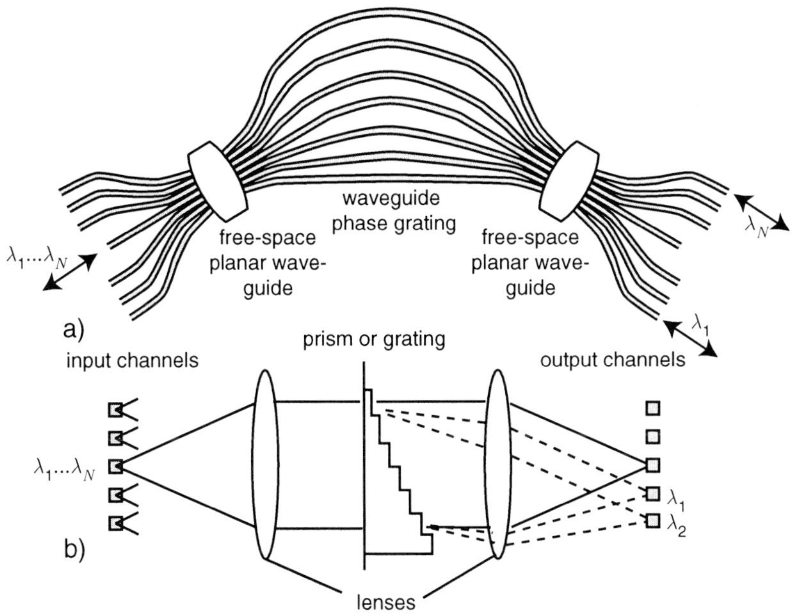

Fig. 6.14: Integrated optical system for wavelength division de/multiplexing: a) integrated system using waveguide optics and 1D-free-space propagation; b) analogous system using free-space optical components.

6.4.2 Waveguide optical sensors

Voltage and temperature sensing

Sensing is a second important field of application for optical systems. In guided-wave optics, sensors can be built very compactly and with high sensitivity [4]. As we have seen during our discussion of the electro-optic effect, very weak physical effects can have significant effects on the waveguiding properties of the system. The reasons for this are the long interaction lengths of the optical wave and the waveguide. For example, a system for voltage sensing is straightforward from our discussion of the electro-optic effect. For this purpose we can use the very same configuration as for the intensity modulator decribed previously (Fig. 6.11a). There, we varied the intensity throughput of the system by applying a certain voltage. Similarly we now calibrate the system such that by measuring the intensity in the output of the Mach-Zehnder interferometer we can determine the voltage applied to the electro-optic waveguide material [1].

Along the same line it is possible to build temperature sensors. A temperature variation generates an index change as well as a change in the length of a waveguide. In a Mach-Zehnder configuration with a specific difference ΔL in the optical arm lengths, this generates a sinusoidal intensity variation [1]:

$$\frac{P_{\text{out}}}{P_{\text{in}}} \propto 1 + \cos\left(\frac{2\pi}{\lambda_0} b\, \Delta L\, T\right) \tag{6.4.1}$$

where b determines the temperature dependence of the refractive index and the length ΔL:

$$b = \frac{dn}{dT} + \frac{n}{\Delta L}\frac{d\Delta L}{dT} \tag{6.4.2}$$

Due to the periodicity of the intensity ratio as described by Eq. 6.4.1, unambiguous temperature measurements are only possible in a range of width $\frac{\lambda_0}{2b\Delta L}$. The sensitivity of the sensor, on the other hand, is proportional to ΔL. Consequently, the temperature range over which measurements can be performed is reduced if the sensitivity is improved. This problem can be overcome by integrating several interferometer setups with different values of ΔL onto one substrate [20]. In this case, the configuration with the smallest difference in arm length allows one to determine the temperature range with coarse precision. The long ΔL system provides the high precision measurements. With such a combination, temperature measurements with a resolution of $\delta T = 2 \cdot 10^{-3}$°C have been demonstrated over a range of $\Delta T > 700$°C. Similar configurations can also be used for measuring pressure, strain or electric and magnetic field strengths.

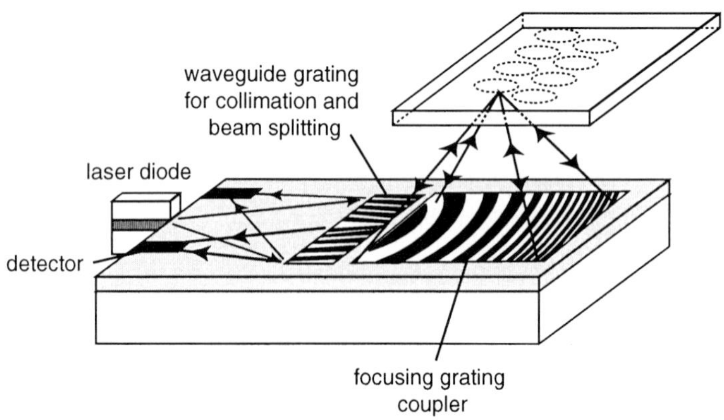

Fig. 6.15: Integrated confocal sensor using waveguide and free-space optics [21].

Waveguide-optical disc pick-up

Fig 6.15 shows another type of integrated optical sensor. Again planar waveguide optics is combined with 1D and 2D free-space optics. The system shown is a confocal sensor which can be used, e.g., as optical disc pick-up system or as depth sensor [21]. A laser diode is butt coupled to a planar waveguiding layer. Integrated waveguide gratings are used for collimating the laser mode. A focusing waveguide grating coupler deflects the light out of the waveguide vertically and focuses onto an object, for example, an optical disc. The scattered or reflected light is collected again by the grating coupler and focused onto position-sensitive detectors which allow tracking and focus detection. Obviously, the system is implemented in a very compact fashion compared to assembled free-space systems. The critical issue of this configuration is the quality of the waveguide coupler and the waveguide grating. Lateral as well as depth resolution in the micrometre regime have been demonstrated with such a sensor [22].

Integrated waveguide-optical systems can also be used for signal processing. For this purpose, the signal, e.g., can be applied to the periodic electrodes integrated on a planar waveguide in order to generate a Bragg-grating in the waveguide (Fig. 6.11c). Diffraction of a waveguide mode at the Bragg grating allows one to perform a spectral analysis of the input signal (Fig. 6.16).

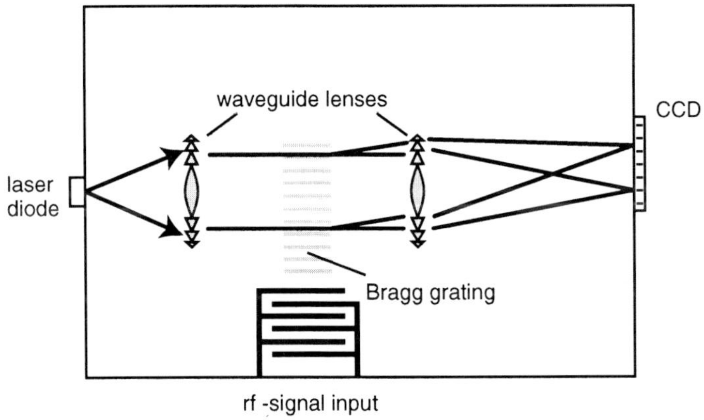

Fig. 6.16: Integrated waveguide optical system for spectrum analysis of the input signal [8].

6.5 Glossary

- **directional coupler**
  waveguide optical component which allows one to couple light between two parallel waveguides through coupling of the evanescent tails of the waveguide modes. In the area where the coupling takes place, the waveguides are arranged close to each other and in parallel.

- **Kerr effect**
  quadratic dependence of the refractive index (dielectric constant) on an applied electric field. Generally, the Kerr coefficient is significantly weaker than the Pockels coefficient. The Kerr effect is thus observed in media which do not show the Pockels effect.

- **Pockels effect**
  the linear dependence of the refractive index of a dielectric material on an applied electric field.

- **star coupler**
  waveguide optical component which distributes the light from one input waveguide (or fiber) channel equally over a number of output waveguide (or fiber) channels. Multiple beam splitter in waveguide optics.

- **waveguide**
  thin layer or strip of dielectric medium with higher refractive index than the surrounding layers. The light waves propagating at angles larger than the critical angle of total internal reflection are guided inside the medium.

- **waveguide mode**
  propagation direction of a wave in the lateral confinement of a waveguiding dielectric, for which the self-consistency condition is fulfilled, i.e., for which the twice-reflected wave and the incident wave interfere constructively. The modes in the waveguides are characterized by the z-component of their k-vectors which are called the propagation constants β.

6.6 List of new symbols

A_1, A_2	lateral amplitude distribution of waveguide modes
$\mathcal{A}$	the area covered by a waveguide; integration area to calculate the coupling between the waveguide modes
α_c	critical angle of total internal reflection
α_i	angle between the incident wavefront and the substrate normal
b	coefficient which describes the effect of temperature on the optical path length in an integrated interferometer
β, β_r	propagation constants of waveguide modes, real part of the constant
$\boldsymbol{E}$	electric field vector
E_1, E_2	electric field distributions of modes in different waveguides
$\kappa, \kappa_{12}, \kappa_{21}$	coupling coefficients of coupled waveguides
L, ΔL	length of a waveguide or difference in the waveguide length in the arms of an integrated interferometer
$\lambda_0; \lambda$	wavelength of the lightwave in vacuo and a dielectric, respectively
λ_{co}	cut-off wavelength of the monomode waveguide in the dielectric
M	maximum number of mode in a waveguide
n_1, n_2, n_3, n_S	refractive indices of different areas of a waveguide
$n(\boldsymbol{E})$	refractive index as a function of the electric field
P_1, P_2	optical powers in waveguide couplers
r_P, r_K	material constants describing the Pockels and the Kerr effect.
SAW	surface acoustic wave
T	temperature
$\mathcal{T}$	power transfer ratio in a waveguide coupler
θ_a	aperture angle of a waveguide
θ, θ_m	propagation angles of the modes with respect to the substrate surface
V_π	voltage necessary to generate a phase shift of π in an electro-optic waveguide
$V_{\mathrm{bi}/\pi}$	voltage necessary to generate a phase shift of π between the orthogonal linear polarisation directions in an electro-optic birefringent waveguide
WDM	wavelength division multiplexing
z_0, z_S	length of waveguide couplers for full energy transfer and beam splitting, respectively

6.7 Exercises

1. **The numerical aperture of a symmetrical waveguide**
   The numerical aperture of a waveguide determines the maximum angle of incidence for which a light beam can propagate in the waveguide. Show that for a symmetrical waveguide, as illustrated in Fig. 6.2, the NA is given as:

$$\mathrm{NA} = \sqrt{n_2^2 - n_1^2}$$

2. **Number of modes in slab waveguides**
   Calculate the number of modes in two waveguide layers with n=1.457 (quartz glass) surrounded by a vacuum at a wavelength of $\lambda_0 = 633$ nm. Assume the thickness of the waveguides to be $d = 1$ μm and $d = 3$ mm, respectively. What is the angular separation between the modes in both cases?

3. **Cut-off wavelength**
   What is the cut-off wavelength λ_{co} of a waveguide? Calculate λ_{co} for the thin waveguide in the previous problem.

4. **Optical tunneling**
   What is optical tunneling? Explain the basic principles and how it is used for optical couplers.

References

[1] R. G. Hunsperger, "Integrated optics", Springer Verlag, Berlin (1995).

[2] H. Nishihara, M. Haruna and T. Suhara, "Optical integrated circuits", McGraw-Hill, New York (1987).

[3] T. Tamir (ed.), "Integrated optics", Springer Verlag, Berlin (1982).

[4] W. Bludau, "Lichtwellenleiter in Sensorik und optischer Nachrichtentechnik", Springer Verlag, Berlin (1998).

[5] J. M. Senior, "Optical fiber communications: principles and practice", Prentice Hall, Cambridge (1992), 2nd ed.

[6] S. E. Miller, "Integrated optics: an introduction", *Bell Systems Techn. Journal* **48** (1969), 2059–2068.

[7] B. E. A. Saleh and M. C. Teich, "Fundamentals of Photonics", John Wiley & Sons, New York (1991).

[8] J. Jahns, "Optische Nachrichtentechnik", FernUniversität Hagen, Hagen (1997).

[9] A. Yariv, "Coupled-mode theory for guided-wave optics", *IEEE J. Quant. El.* **QE-9** (1973), 919.

[10] A. E. Willner, "Optical communications", in Handbook of Photonics, M. C. Gupta (ed.), CRCPress LLC, Boca Raton, Fl, USA (1997), 624–718.

[11] M. Kobayashi, "Guided-wave optical interconnect techniques", in Photonics in switching II, J. E. Midwinter (ed.), Academic Press, San Diego, CA, USA (1993), 271–317.

[12] A. Guha, J. Bristow, C. Sullivan and A. Husain, "Optical interconnects for massively parallel architectures", *Appl. Opt.* **29** (1990), 1077–1093.

[13] M. Jöhnck, B. Wittmann and A. Neyer, "64 channel two dimensional POF based optical array inter chip interconnect", *EOS meetings digest* **3490**, Optics in Computing, Brugge (1998), 285–288.

[14] Y. Li and J. Popelek, "Clock delivery using laminated polymer fiber circuits", *EOS meetings digest* **3490**, Optics in Computing, Brugge (1998), 278–281.

[15] E. Acosta and K. Iga, "Design of a wavelength multiplexer-demultiplexer by the use of planar microlenses", *Appl. Opt.* **33** (1994), 3415–3419.

[16] D. Intani, T. Baba and K. Iga, "Planar microlens relay optics utilizing lateral focusing", *Appl. Opt.* **31** (1992), 5255–5258.

[17] J. B. D. Soole, A. Scherer, H. P. LeBlanc, N. C. Andreadakis, R. Bhat and M. A. Koza, "Monolithic InP/InGaAsP/In grating spectrometer for the 1.48–1.56 μm wavelength range", *Appl. Phys. Lett.* **58** (1991), 1949–1951.

[18] C. Dragone, C. H. Henry, I. P. Kaminow and R. C. Kistler, "Efficient multichannel integrated optics star coupler on silicon", *IEEE Phot. Techn. Lett.* **1** (1989), 241–243.

[19] C. Dragone, "An $N \times N$ optical multiplexer using a planar arrangement of two star couplers", *IEEE Phot Techn. Lett.* **3** (1991), 812–814.

[20] L. M. Johnson, F. J. Leonberger and G. W. Pratt, "Integrated optical temperature sensor", *Appl. Phy. Lett.* **41** (1982), 134–136.

[21] S. Sheard, T. Suhara and H. Nishihara, "Integrated-optic implementation of a confocal scanning optical microscope", *IEEE J. of lightwave technology* **11** (1993), 1400–1403.

[22] S. Ura, M. Shinohara, T. Suhara and H. Nishihara, "Integrated-optic grating-scale-displacement sensor using linear focusing grating couplers", *IEEE Phot. Techn. Lett.* **6** (1994), 239–241.

7 Microoptical systems

In previous chapters we have discussed a variety of techniques for the fabrication and optimization of microoptical components or component arrays. For most applications, integration of these components into fully functional systems is of similar importance. Systems integration is crucial, in order to reduce size and cost, as well as to increase the robustness and stability of the optical systems. The possibility for integrating optical systems was, therefore, part of our definition of microoptics in Chapter 1. While in macro- and miniature optics discrete components are mounted to form optical systems using mechanical alignment equipment (i.e., optomechanics), for microoptics new integration approaches are necessary. This becomes obvious when looking at the precision requirements for microoptical systems. For the alignment of components with lateral dimensions of the order of milimetres O(mm), alignment precision in the micrometre or submicrometre regime $O(\mu m)$ and corresponding stability are required. In this chapter we shall discuss possible techniques for microoptical systems integration.

We address three different approaches: micro-opto-electro-mechanical systems (MOEMS), stacked optics and planar integrated free-space optics. MOEMS are microsystems which combine mechanical, optical and electronic functionality. In optical systems integration, MOEMS have been suggested to use micromechanics for the alignment, generating an optical microbench. Rather than by micromechanics, in "stacked optics", components or arrays of components are assembled in a multilayer stack using alignment structures and bonding techniques. In order to simplify the assembly, a variety of lithographic techniques is applied. Hybrid integration techniques as well as deep lithography for the fabrication of alignment slots are applicable. Finally we discuss an approach towards microoptical systems integration, which is called "planar optical integration" or "planar optics". Here the whole optical system is integrated monolithically, i.e., into a single substrate (e.g., fused silica), where the light propagates within the substrate. Thus, the difficult and expensive mechanical assembly of discrete components can be completely avoided in planar optics. After discussing the different approaches towards systems integration, we address the specific example of an imaging system for optical interconnection applications. We focus on optimization of the imaging system as well as integration in planar optics.

7.1 Systems integration

7.1.1 MOEMS for optical systems integration

Progress in micromachining techniques (mostly of silicon) has led to a variety of new micromechanical components. In Chapter 3 we already learned about the basics of "bulk silicon micromachining" by anisotropic etching. We also mentioned the possibility of increasing the variety of feasible structures by doping of the bulk silicon substrate, which influences the etch rate. For more complex 3D microstructures, a different approach called "surface micromachining" adds to the potential of silicon etching (see e.g., [1]). In surface micromachining, thin layer deposition techniques (e.g., chemical vapour deposition) are applied to coat the bulk silicon substrate with layers of SiO_2 or polycrystalline silicon (poly silicon), respectively. Each of these layers can be structured with appropriate etching techniques (e.g., RIE). Finally the substrate is immersed into an etching solution, e.g., based on hydrofluoric acid (HF). For this solution the etch rates of SiO_2 are significantly higher than for poly Si. The SiO_2 layers act as sacrificial layers which are removed during HF etching, leaving behind the 3D poly silicon structures. With this approach it is possible to fabricate truely 3D structures in polysilicon, such as free-standing cantilevers, beams or bridges (Fig. 7.1). With an increasing number of layers, the complexity of the fabricated structures can be increased. This technique has been applied to the fabrication of mobile parts such as microgears and microactuators [2]. The potential to integrate the micromechanical parts with CMOS electronics leads the way to micro-electro-mechanical systems (MEMS) [3].

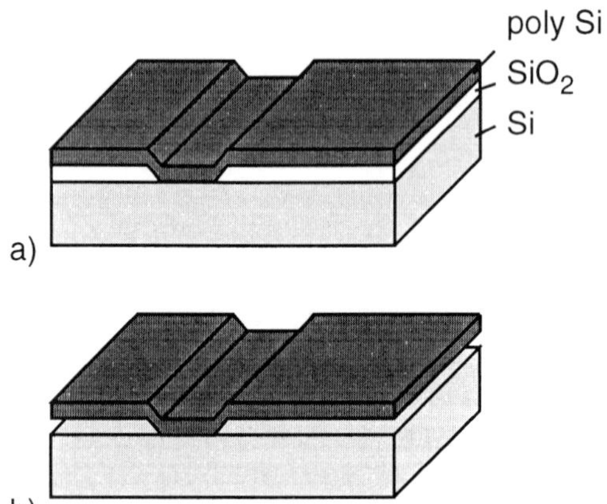

Fig. 7.1: Fabrication of a cantilever in silicon. a) Layer structure before HF etching with a sacrificial layer of SiO_2. b) Cantilever after etching.

Although truly 3D structures can be fabricated, the profiling depth, i.e., the number of subsequent sacrificial and polycrystalline layers is limited. For the integration of 3D free-space optics the use of hinges has been suggested, which can be fabricated and structured in the substrate plane and afterwards folded into the third dimension (Fig. 7.2) [4]. Based on these microhinges a microoptical bench has been suggested and demonstrated [5]. The optical components are fabricated on the microhinges using planar fabrication techniques as described in Chapters 4 and 5. Finally the components are flipped out of plane into the appropriate orientation in the 3D system. In order to improve the stability and angular alignment, pairs of spring and side latches are used. With these latches it is possible to reduce the error in the angular position of the microlens to less than 0.1°. The lateral position of the hinges is determined with submicron precision by the planar fabrication process.

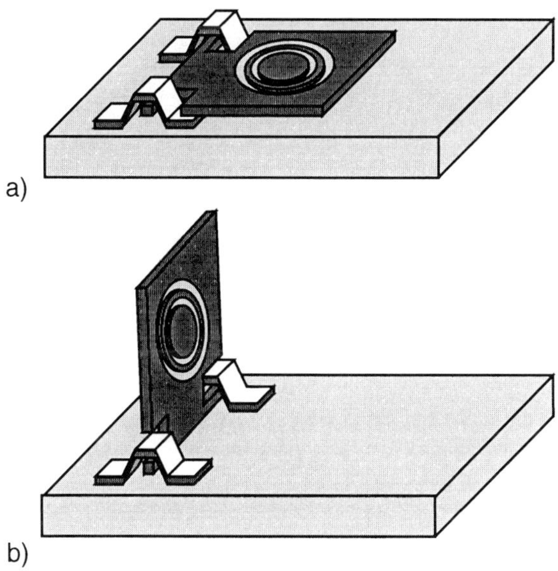

Fig. 7.2: Microhinges fabricated by surface micromachining using two sacrificial layers: a) microhinges after fabrication and structuring in the substrate plane; b) microhinges after folding into the third dimension.

A micromachined optical bench using the microhinge approach has been demonstrated with the example of an optical disc pick-up system [6]. The integrated free-space optical system consists of three micro Fresnel lenses on microhinges. A semiconductor laser is aligned with respect to the optical system using alignment latches. The system is completed by a semitransparent mirror for beam splitting and two mirror structures oriented at ±45° to the substrate normal, for coupling to the optical disc and the detector, respectively. For the integrated optical disc pick-up system, full-width-half-maximum (FWHM) focus diameters of 1.8 μm on the optical disc and 4 x 8 μm^2 on the detector have been demonstrated.

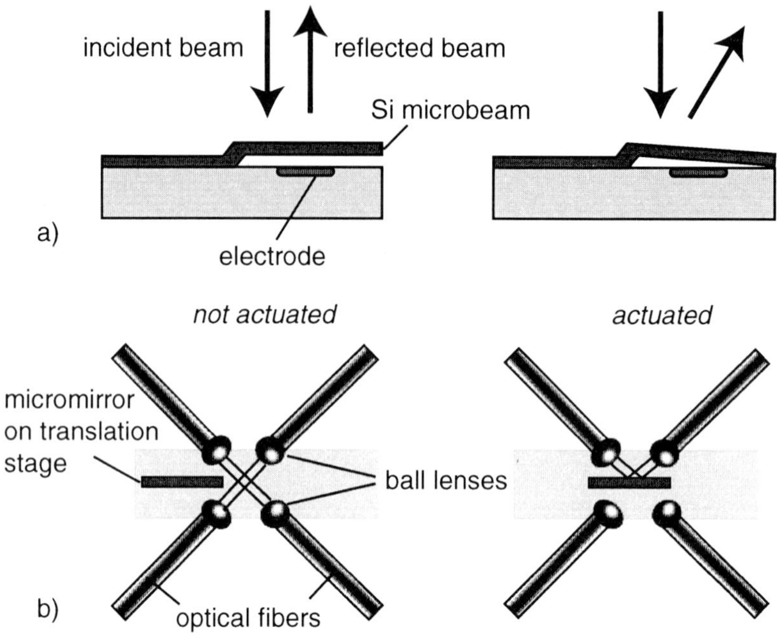

Fig. 7.3: Two examples for micromechanics used for optical switches: a) a flexible microbeam used as switchable mirror; b) fiber-optical exchange-bypass switch, integrated using miniature optics and micromechanics.

One of the major drawbacks of optical systems integrated on a micromechanical bench is the limited stability and the sensitivity to environmental conditions. The sensitivity of the microbench is obvious with the thin microhinges aligned vertical to the substrate plane. Due to the complex layer structure, replication techniques cannot be applied to the integrated optical system. Fabrication costs remain relatively high for systems fabricated on the microoptical bench. The variety of optical components integrated with this technique is limited due to constraints in the materials which can be processed. Amplitude gratings and Fresnel-type microlenses have been integrated on poly-Si microhinges. More efficient refractive reflow microlenses in photoresist have also been demonstrated.

The greatest potential for the application of MOEMS lies in the integration of electronically controlled actuators with the integrated optical system. This is possible using moveable microtranslation or rotation stages for positioning and tilting the integrated microoptical elements. A variety of different principles for microactuators has been demonstrated. Examples are comb-drive actuators, linear microvibromotors, stepper motors, microengines, scratch-drive actuators and thermal actuators. Such actuators have been used for the demonstration of optical switching and beam-steering systems [7]. Fig. 7.3 schematically shows two examples of optical switches using micro-electro-mechanical actuators. Large arrays of switchable micromirrors are already commercially available for video projection systems (DMD^{TM} by Texas Instruments) [8].

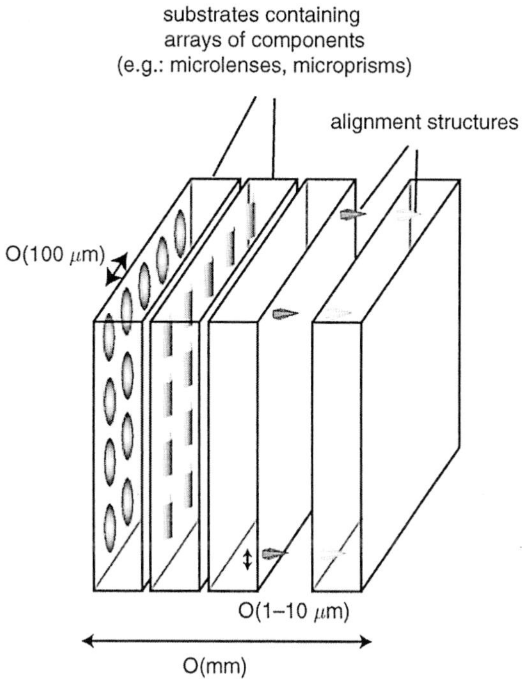

Fig. 7.4: The concept of stacked microoptics for the integration of microoptical systems: system dimensions in the order of mm, O(mm); component dimensions: O(10–500 μm); dimensions of the alignment structures: O(10 μm) with alignment precision O(1 μm).

7.1.2 Stacked optics

The term stacked optics was coined by Iga et al. in 1982 [9]. They suggested using 2D arrays of optical components separately fabricated on planar optical substrates to build integrated imaging systems. After positioning, these substrates are attached to one another using adhesives in order to fix the relative alignment. For a number of parallel optical systems the alignment task is reduced to the alignment of the 2D component arrays rather than the individual components (Fig. 7.4). The systems are arranged such that each planar substrate only contains one type of optimized elements. Iga et al. suggested using arrays of GRIN microlenses fabricated by ion-exchange in glass in combination with spatial or aperture filter planes and circular hole arrays to hold optical fibers for the input. Since the GRIN profile is contained within the substrate, the surface of these microlens arrays is flat. Assuming good substrate quality and flatness, parallel alignment of component layers can be achieved by stacking the substrates on top of each other.

Meanwhile the concept of stacked optics has been extended to an integration technique for an arbitrary optical system. To this end, integration with deflecting optical components has been suggested by Brenner [10]. With suitable technology for the fabrication of microprisms and beam-splitting components, stacked optics is a versatile integration concept for microoptical systems. In order to alleviate the alignment of the component planes further, the use of mechanical alignment and fixing structures in combination with optical alignment marks is necessary [11]. For the fabrication of such assembling grooves and structures, micromachining (e.g., by diamond turning or silicon etching) as well as deep lithographic methods have been considered. Hybrid bonding techniques, such as flip-chip bonding, can be used as an alternative for alignment and fixing of the various component planes. Nevertheless, a considerable amount of alignment is still necessary in order to build compact optical systems. This is the reason why stacked integration has only been demonstrated so far for microoptical systems with small complexity. For example, a stacked system using an optical biprism and two planar GRIN lenses to perform a shift and interlace of two data planes (Fig. 7.5) has beed demonstrated by Moisel et al. [12, 13].

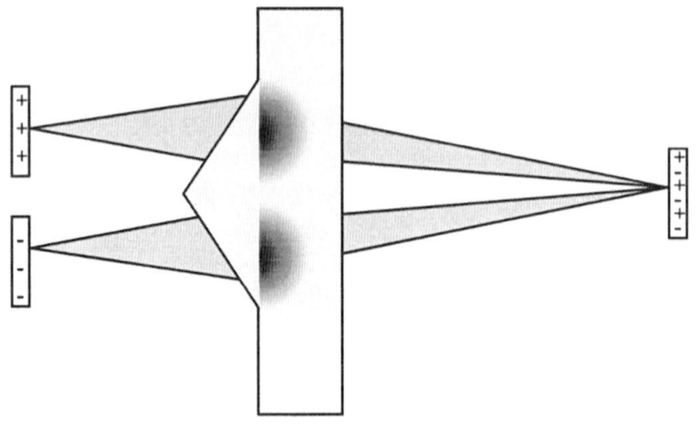

Fig. 7.5: Schematic of a stacked microoptical system integrating two planar GRIN microlenses and a biprism for interlacing two input data planes .

In principle, in stacked optics a variety of different substrates can be used for assembly in the system. It is possible to use optimized techniques for the fabrication of each class of components. For example, GRIN lenses fabricated by ion-exchange in glass substrates can be integrated with microprisms fabricated in PMMA substrates. However, for many applications the different behaviour of the substrates under changing environmental conditions (e.g., different thermal expansion coefficients) may cause difficulties.

7.1.3 Planar optics

The concept of planar optics is one step further along the integration scale for microoptical systems. Here the whole optical system is integrated monolithically in one single substrate. To this end the system is folded in such a way that the light propagation takes place along a zig-zag path inside the substrate (Fig. 7.6) [14, 15]. The light wave is coupled into and out of the substrate by transmissive components which deflect the light into the propagation direction. During propagation through the substrate, the wave hits the optical components which are integrated as reflective elements on the surfaces of the substrate. Due to the 2D layout of the whole free-space optical system, lithographic techniques can be used for fabrication as well as alignment of the optical components. Submicron alignment precision of the components can already be achieved during the fabrication process. No subsequent optomechanical alignment of discrete components or component arrays is necessary.

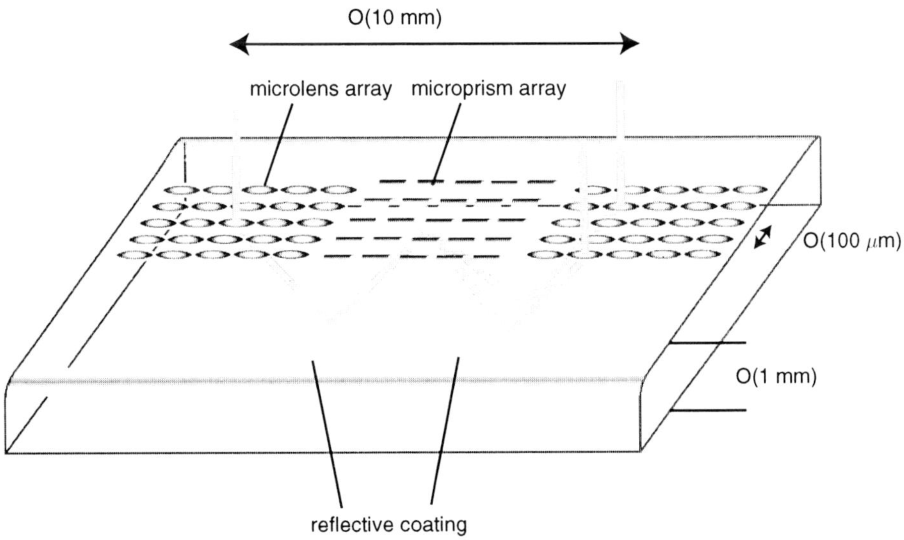

Fig. 7.6: Planar optical integration: all optical elements are integrated on the surfaces of a single substrate. The light propagates inside the substrate along a zig-zag path.

The folding of the optics into a two-dimensional geometry makes planar optics compatible to the planar fabrication techniques which are used for the processing of integrated circuits. It also allows one to use surface-mount approaches to place optoelectronic chips on the substrate surfaces. The chips are, therefore, accessible for handling, cooling, testing and repair purposes. The substrate may either be glass, plastic or a semiconductor material such as silicon or gallium arsenide, provided it is transparent at the wavelength used. Substrates with a thickness of several millimetres are used, in order to allow for laterally unguided free-space optical light propagation along a zigzag path as indicated in Fig. 7.6. It is important to note that planar optics is free-space optics. Although the light is confined to propagation inside the substrate, it is not confined in a direction perpendicular to its propagation direction.

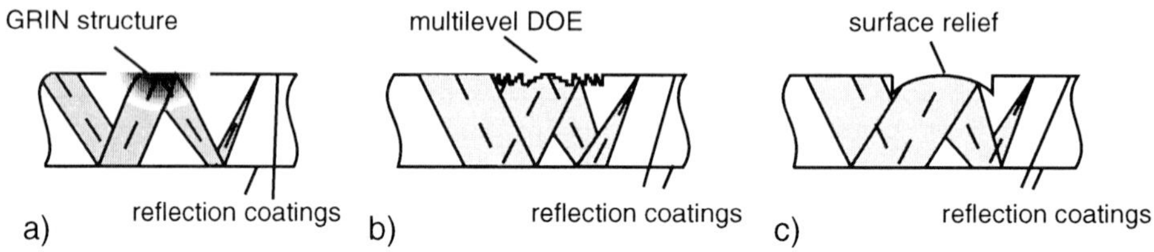

Fig. 7.7: Planar optical integration is not restricted to diffractive elements. Examples of different types of optical components which can be integrated in the planar optical scheme (Chapters 4 and 5): a) GRIN elements; b) diffractive elements; c) reflective elements.

The surfaces of the substrate including the optical components are coated with a reflective (metallic or dielectric) layer to keep the light inside the substrate. Alternatively for large propagation angles (e.g., $\alpha > 43°$ in SiO_2), total internal reflection can be used for the light reflection [16, 17, 18]. The fact that the light signal experiences only very few interfaces between different media, makes the optics insensitive to environmental influences such as dust and humidity. All components on one side of the substrate can be fabricated simultaneously using optical lithography. Replication techniques might be used to mass produce the systems thereby making the optics inexpensive (Chapter 3).

In order to exploit the advantages of planar optics fully, it is necessary to find fabrication techniques which allow the monolithic integration of all necessary components of the optical system. Focusing as well as deflecting components, i.e., lenses as well as microprisms or gratings, need to be fabricated with good quality in one single substrate. The design freedom available in diffractive optics makes it an interesting tool for planar optical integration. With diffractive optics it is possible to implement all necessary functions using a single technology, such as, e.g., reactive ion etching. This eases the monolithic systems integration. A large variety of planar optical systems based on a diffractive implementation have been demonstrated for applications in optical interconnects and sensor technology.

Planar optics is, however, not restricted to the integration of diffractive optical elements. Fig. 7.7 shows that in principle all techniques possible for component implementation can be used in planar optical systems. Problems might arise since not all of the substrates can be used for all of the fabrication techniques. We have seen previously that many processes such as RIE can be adjusted for different etching profiles and different substrates. Thus, it is possible to combine diffractive optics with lenses fabricated using the reflow process, or analog lithographic techniques and etched into the substrate through RIE or CAIBE. On the other hand, it is also possible to use substrates which are not well suited for the etching process. In this case the diffractive optics can be fabricated through thin-film deposition rather than by etching processes. This might be necessary if special substrates are to be used, e.g., if GRIN elements should be used which need specific substrates [19]. The use of holographic optical elements in a planar configuration has also been suggested [16, 20]. These examples show that a wide range of lithographic technology is available for the optimization of components for planar optical systems.

Optical elements on the top and bottom surfaces of the substrate can be aligned relative to each other with submicron precision by using a through-wafer alignment technique (Chapter 3). Due to integration on a single substrate and the fact that the light propagates inside the substrate, the mechanical and thermal stability of planar optical systems is very good. The systems are compact and require a considerably smaller volume than conventional optical systems. For example, an integrated system with dozens of optical elements has been demonstrated on a substrate with a diameter of 25 mm and a thickness of 3 mm. Several examples of planar optical systems, e.g., for optical interconnects, optical data storage, and optical sensors will be discussed throughout the chapters on microoptics applications.

7.2 Imaging systems for optical interconnects

In conclusion of this chapter on microoptical systems, we discuss optical imaging systems for interconnect applications. This provides an example of how it is possible to adjust the optical system to a specific application in order to achieve optimum results.

7.2.1 Dilute arrays

For optical interconnect applications it is necessary to connect planes of optoelectronic devices or fiber arrays optically. In Chapter 8 we shall discuss a variety of

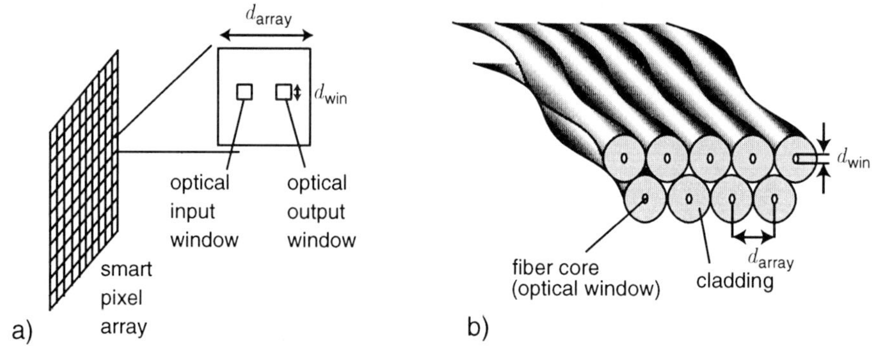

Fig. 7.8: Examples for dilute optical arrays: a) smart pixel arrays; b) 2D fiber bundels.

examples of optoelectronic components. These devices consist of optically active areas as well as driver electronics and possibly some logic functionality (e.g., in smart pixel arrays). Consequently, in arrays of such devices the optically active areas (i.e., the optical input and output (I/O) windows) are often separated by "black" areas which contain the electronics. This situation is described by the name "dilute array", illustrating the fact that the optical windows are not densely packed in the I/O planes [21]. The optical windows with possibly small extension are rather spread over a large area with possibly large distances between them (Fig. 7.8a). Typical values for the size of the optical windows are about $d_{win} \approx 1-10\ \mu m$ with a pitch of $d_{array} > 100\mu m$. This situation is similar for 2D fiber bundels. The optically interesting areas are the fiber cores ($d_{win} \approx$ 1-60 μm) which are separated for an ideally packed fiber array by the fiber cladding (d_{array}= 125 or 250 μm) (Fig. 7.8b). This "dilution" of the optical I/O planes has to be considered in order to find the optimum optical imaging systems for interconnecting two of these planes.

In order to interconnect optically two dilute arrays containing optoelectronic devices, it is necessary to image the optical output windows of the source array onto the input windows of the subsequent device array. Quality criteria for this imaging step are the efficiency of the connection as well as the crosstalk between the channels, i.e., the amount of light intensity received by the wrong (e.g., adjacent) channels. Since the optical I/O windows are not densely packed, they cover an area much larger than the sum of their own areas. Thus, the diameter of the image field is relatively large. For good interconnect quality, high resolution (i.e., small extension of the image point) is required for every optical window. In the following we want to consider three different approaches for the imaging of dilute arrays, i.e., "conventional imaging", "multichannel imaging" and "hybrid imaging". We will conclude this chapter on microoptical systems with a brief discussion of the integrated microoptical imaging systems which have been demonstrated so far.

7.2.2 Conventional imaging

In the conventional approach a single optical system is used to image the whole dilute input plane onto the output plane. Generally, a telecentric $4F$ imaging configuration is chosen for this purpose (Fig 7.9a). The distance between the lens planes is determined by the focal lengths F and can be large enough to use, e.g., a beam splitter for in/outcoupling of signals between the device planes [21]. This is necessary, for example, if electrooptical modulator arrays need to be supplied with optical readout power. The necessary resolution of the imaging system is determined by the size of the optical windows. In order to avoid energy losses, the point spread function over the whole image field has to be significantly smaller than the optical windows in the I/O planes. Due to the dilute arrangement of the optical windows, crosstalk, i.e., the coupling of one signal into the adjacent channel, will be negligible even for significant aberrations or misalignment errors. Imaging and alignment errors will only cause energy losses rather than crosstalk.

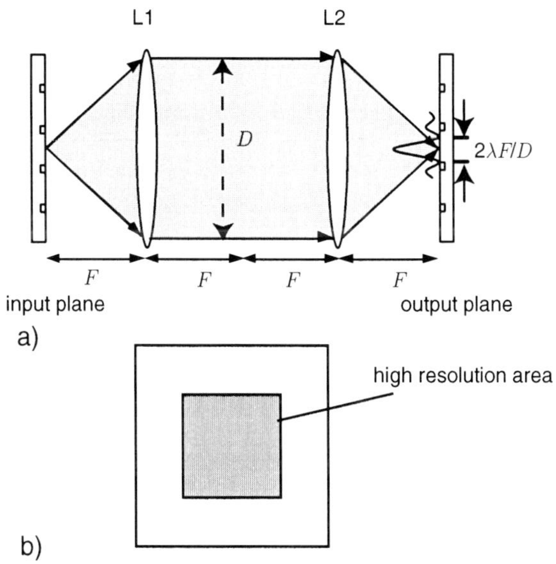

Fig. 7.9: a) Conventional optical $4F$ setup for imaging an input array onto an output array; b) illustration of the (ideally) continuous distribution of the high resolution area in the image plane.

In a conventional imaging system, high resolution is provided over a continuous area (Fig. 7.9b). The need for large image fields at high resolution significantly increases the cost for lenses in the $4F$ configuration. Lenses with large space-bandwidth product are necessary to support a sufficiently large number of channels. A big part of this SBP, however, is wasted since it is used to transmit the optically uninteresting area between the I/O windows. This wasteful system design increases the cost of $4F$ imaging systems for interconnect applications.

7.2.3 Multichannel imaging system

A more economic use of the available SBP is possible with a multichannel approach. 2D microlens arrays are used in this case to implement individual communication channels for each source/detector pair [22, 23] (Fig. 7.10). Each of the individual pairs of microlenses images only one single on-axis channel. Due to this reduced image field, the requirements on the microlenses are relatively moderate, although high resolution (small f/#) is still required. A large variety of technological approaches can be used for microlens fabrication. Implementation of the microchannel imaging system can be significantly cheaper than the conventional approach. As an additional advantage an arbitrary scheme of space-variant interconnects can be implemented using microlenses with arbitrary deflection angles.

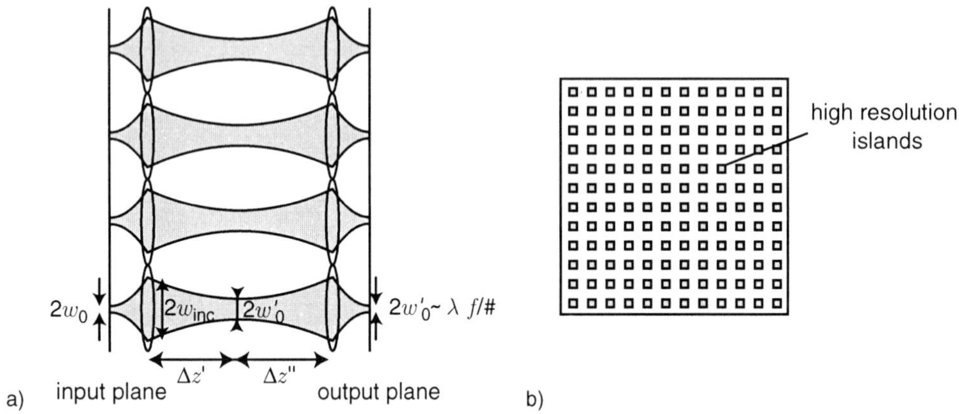

Fig. 7.10: a) Microchannel relay system for optical imaging of dilute arrays. The lines indicate the $1/e$ boundaries (isophotes) of the Gaussian beams; b) illustration of the small high resolution "islands" spread over a large area in the image plane.

The problems of microchannel imaging are related to crosstalk and a limited interconnection distance. For small lens diameters, diffraction causes a significant divergence of the optical beams [24]. In order to avoid crosstalk between the individual channels, the interconnection distance needs to be limited. For our analysis let us describe the propagating light beams as Gaussian beams characterized by a Gaussian intensity distribution $I(r, z)$ perpendicular to the propagation direction.

$$I(r, z) = I_0 e^{-2\frac{r^2}{w(z)^2}} \qquad (7.2.1)$$

$w(z)$ denotes the $1/e$ intensity beam radius and I_0 is the beam intensity on the propagation axis.

The "waist" w_0 of the Gaussian beam is defined here as the smallest radius of the beam. The beam radius is generally calculated from the location where the beam amplitude has dropped to a $1/e$ portion of the on-axis value. For a thorough discussion of the laws of Gaussian optics the reader is referred to the literature (e.g., [25, 26]). Here it is important to note that collimation of Gaussian beams is only possible for infinitely large beam diameters. This can be seen from Eq. 7.2.2 which describes the development of the beam radius upon propagation of the beam along the z-axis:

$$w(z) = w_0 \left[1 + \left(\frac{z}{z_0}\right)^2\right]^{\frac{1}{2}} \tag{7.2.2}$$

z_0 is called the Rayleigh range which also describes the depth of focus of the Gaussian beam:

$$z_0 = \frac{\pi w_0^2}{\lambda} \tag{7.2.3}$$

From Eqs 7.2.2 and 7.2.3 we see that for large values of w_0 the ratio $\frac{z}{z_0}$ approaches zero so that $w(z) = w_0$ =const, i.e., the beam is collimated. However, for decreasing w_0 the divergence of the beam becomes significant. In a microchannel system, the diameter of the microlenses is small so that we have to take into account the Gaussian beam divergence. Since beam collimation is not possible for small lens diameters, so called "beam relaying" is performed. The microlens (diameter d) is used to convert the diverging beam (diameter w_{inc}) into a slightly converging beam, focused to a waist w_0' at a distance $\Delta z'$ from the lens (Fig. 7.10a).

$$w_{\text{inc}} = w_0' \left[1 + \left(\frac{\lambda \Delta z'}{\pi w_0'}\right)^2\right]^{\frac{1}{2}} \tag{7.2.4}$$

for $\Delta z'$ results:

$$\Delta z' = \frac{\pi}{\lambda} w_0' \sqrt{w_{\text{inc}}^2 - w_0'^2} \tag{7.2.5}$$

Behind the waist (i.e., the smallest beam diameter corresponding to the geometrical focus), the beam diverges again before it is relayed onto the detector by the second microlens. At the position of the second microlens the beam diameter is determined by $a \cdot w_{\text{inc}}$, where a is a parameter [27]. The distance $\Delta z''$ between the focus and the second microlens is calculated analogously to Eq. 7.2.5. For the total distance Δz between the microlenses we obtain:

$$\Delta z = \Delta z' + \Delta z'' = \frac{\pi}{\lambda} w_0' \left(\sqrt{w_{\text{inc}}^2 - w_0'^2} + \sqrt{a^2 w_{\text{inc}}^2 - w_0'^2}\right) \tag{7.2.6}$$

Δz is maximized if:

$$w_0' = \left(\frac{a^2}{1 + a^2}\right) w_{\text{inc}} \tag{7.2.7}$$

For equal beam diameters in the first and the second microlens plane ($a = 1$) we find the maximum propagation length $\Delta z_{\max}$ for a beam waist $w_0 = \sqrt{0.5}\, w_{\text{inc}}$ halfway between the microlenses:

$$\Delta z_{max} = \frac{\pi}{\lambda} w_{\text{inc}}^2 \tag{7.2.8}$$

In order to avoid defocusing and energy losses, clipping of the beam in the lens pupils is very critical. We can calculate the total amout of energy P transmitted through a circular aperture of radius R by the integration [28]:

$$P = \int_0^R \int_0^{2\pi} I\, e^{-2\frac{\rho^2}{w^2}} \rho\, d\rho d\Phi = \pi w_0^2 I_0 \left[1 - e^{-2\frac{R^2}{w_0^2}}\right] \tag{7.2.9}$$

From Eq. 7.2.9 we immediately see that for $R \to \infty$ the maximum power transmitted through the pupil is given by:

$$P_{\max} = \pi w_0^2 I_0 \tag{7.2.10}$$

According to Eqs 7.2.9 and 7.2.10 the power loss upon transmission through the lens aperture remains below 1% as long as the ratio $\frac{R}{w}$ between the pupil aperture and the beam radius remains larger than $\frac{R}{w} > 1.52$ [24]. In spite of the high efficiency even for this clipping ratio some effects on the focusing behaviour of the Gaussian beam have been observed [29] . Here it may be sufficient, however, to consider the efficiency. If we require 99% efficiency, the diameter of the microlenses d_{lens} has to fulfil the condition:

$$d_{\text{lens}} \geq 1.52 \cdot w_{\text{inc}} \tag{7.2.11}$$

From Eqs 7.2.8 and 7.2.11 we find that the relationship between the maximum transmission distance of the microchannels and the diameter of the microlenses can be estimated from [27]:

$$\Delta z_{\max} \propto \frac{d_{\text{lens}}^2}{\lambda} \tag{7.2.12}$$

This relation illustrates the most important limitation of microchannel imaging. For small microlens diameters the interconnection distance is rather small. As a numerical example we assume a channel width $d_{\text{lens}} = 125\ \mu\text{m}$ and a communication wavelength of 850 nm. In this case Eq. 7.2.12 yields a maximum interconnection distance of $\Delta z_{\max} \approx 18.4$ mm. For highly integrated short distance communication such transmission distances can be sufficient. Very often, however, longer communication distances are either necessary or desirable. For these situations the microchannel approach has been extended to the so-called hybrid imaging configuration.

7.2.4 Hybrid imaging

For optimized and economic imaging of dilute arrays, a combination of the microchannel approach and the conventional $4F$ imaging is advantageous. To this end a $4F$ imaging setup is added to the microlens relay configuration, in order to increase the interconnection length of the system. The purpose of the additional $4F$ system is to image the pupils of the microlenses in the input array onto the corresponding pupils in the output array [21]. Thus, diffractive beam divergence between the microlens planes can be avoided and long interconnection distances become feasible. In contrast to a conventional $4F$ setup, the requirements on the resolution of the imaging lenses are reduced by the microlens arrays. For the $4F$ system in the hybrid configuration it is sufficient to resolve the microlens pupils. The job of reducing the beam divergence as well as that of tight focusing is performed by the input and output microlens arrays.

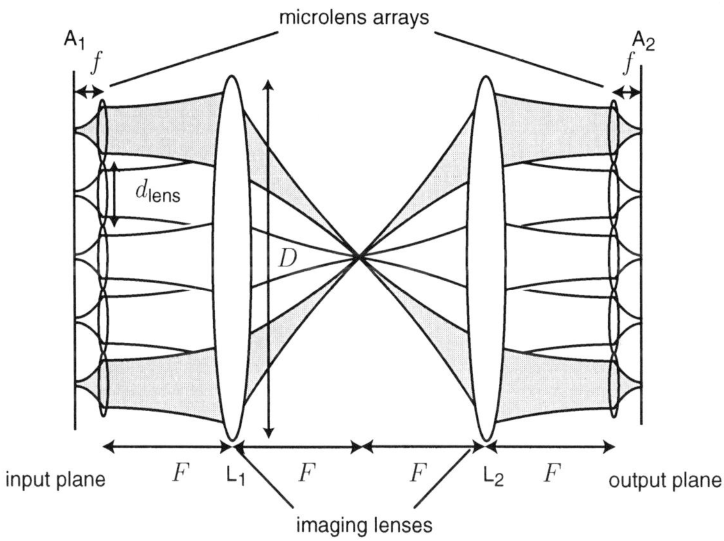

Fig. 7.11: Hybrid imaging: a conventional $4F$ imaging system is introduced after the relay microlenses, in order to increase the communication distance without introducing inter-channel crosstalk (the lines again indicate the isophotes of the Gaussian beams).

As illustrated in Fig. 7.11 the microlenses reduce the requirements of the imaging lenses by reducing the angle of divergence of the microbeams. We can estimate the advantages of the hybrid setup over the conventional $4F$ configuration by calculating the necessary diameters of the imaging lenses for both cases. The imaging optics has to be at least as big as the size of the input array which is determined by the number N and pitch $d_{\text{array}} = d_{\text{lens}}$ of the input array. In order to avoid vignetting or clipping of the beams, it is necessary to extend the lens diameter considering the beam divergence in front of the lenses. To this end we estimate the propagation angle α_{lens} behind the microlenses in the hybrid setup in the paraxial approximation from $\sin(\alpha_{\text{lens}}) = \tan(\alpha_{\text{lens}}) = \alpha_{\text{lens}} = \frac{\lambda}{d_{\text{lens}}}$. From an estimation based on simple geometrical optics rather than Gaussian beams we find the relation for the diameter of the imaging lenses L_1 and L_2 [30]:

$$D_{\text{hyb}} = N \cdot d_{\text{lens}} + 2\frac{\lambda}{d_{\text{lens}}} \cdot F \tag{7.2.13}$$

For comparison, a similar approach yields the diameter of the lenses in a conventional system. Since no quasi-collimating microlenses are introduced here, we have to consider the divergence angle of the beam as emitted by the light sources. We denote the output window of the source by d_{source}. The divergence angle is approximated by $\alpha_{\text{source}} = \frac{\lambda}{d_{\text{source}}}$. The diameter of the imaging lenses for conventional imaging is found to be:

$$D_{\text{conv}} = (N-1) \cdot d_{\text{lens}} + 2\frac{\lambda}{d_{\text{source}}} \cdot F = N \cdot d_{\text{lens}} + 2\frac{\lambda}{d_{\text{source}}} \cdot (F - f) \tag{7.2.14}$$

Here, we assumed that the numerical aperture of the microlenses is adjusted to the beam divergence of the sources: $\frac{\lambda}{d_{\text{source}}} = \frac{d_{\text{lens}}}{2f}$. Comparing Eq. 7.2.13 and Eq. 7.2.14 we find that the minimum diameter of the imaging lenses in the hybrid configuration is smaller than in the conventional setup if:

$$\frac{(F-f)}{d_{\text{source}}} > \frac{f}{d_{\text{lens}}} = (f/\#)_{\text{microlens}} \tag{7.2.15}$$

Eq. 7.2.15 shows that hybrid imaging is especially advantageous for highly dilute arrays (i.e., $d_{\text{source}} \ll d_{\text{lens}}$) and long interconnection distances (i.e., $F > f$). As a numerical example we choose $\lambda = 0.85$ μm, d = 125 μm, f = 500 μm, $F = 25$ mm and $N = 50$. The numerical aperture of the microlens is matched to the divergence angle of a light source with an output window of $d_{\text{source}} \approx 7$ μm. According to Eq. 7.2.13 and Eq. 7.2.14 we obtain minimum diameters of the imaging lenses of $D_{\text{hyb}} = 6590$ μm and $D_{\text{conv}} = 12196$ μm.

This example shows that a significant reduction of the necessary lens diameters and numerical apertures can be achieved in the hybrid configuration. Since the aberrations of the imaging system increase with increasing numerical aperture, this helps to improve the quality and feasibility of the system. An economical use of the limited SBP of the microlenses is required especially for integrated microoptical interconnection systems. This is even more important if diffractive microlenses with reduced image fields (chapter 4) are to be used for the imaging.

7.2.5 Integrated microoptical imaging systems

Several discrete imaging systems for optical interconnection applications based the use of microlenses for microchannel imaging or in hybrid imaging systems have been demonstrated [31, 23]. The size of these systems is mainly determined by the necessary alignment tools. The concept of planar integration of microoptical system has been applied to a variety of fully integrated imaging systems. A conventional $4F$ imaging system has been shown to form a good quality image of an object consisting of 8×8 densely packed pixels [32]. The image field extension in this case was about $(400\ \mu\text{m})^2$. The lenses were implemented as 4 phase level diffractive lenses with an $f/\# \approx 12$.

Using the hybrid imaging approach several integrated interconnection systems have been demonstrated. In 1992, Jahns and Acklin [33] demonstrated a monolithically integrated hybrid imaging setup in planar optics. The system implemented 1024 (32x32) parallel optical channels over an interconnection distance of $s = 18.6$ mm and with an interconnection density of 400 mm^{-2}. In this case, the hybrid configuration with microlenses and a $4F$ imaging system made implementation using diffractive optical elements possible (Fig. 7.12a). Diffractive microlens arrays (A_1, A_2) were used for collimation and coupling of the light into the planar optical substrate. A $4F$ imaging configuration also integrated in the substrate using diffractive lenses (L_1, L_2) imaged the microlens planes onto each other. A crosstalk suppression of better that 200:1 was measured over an image field of 1.6x1.6 mm^2. This illustrates the significant enhancement of the image quality compared to a conventional configuration (Fig. 7.12b).

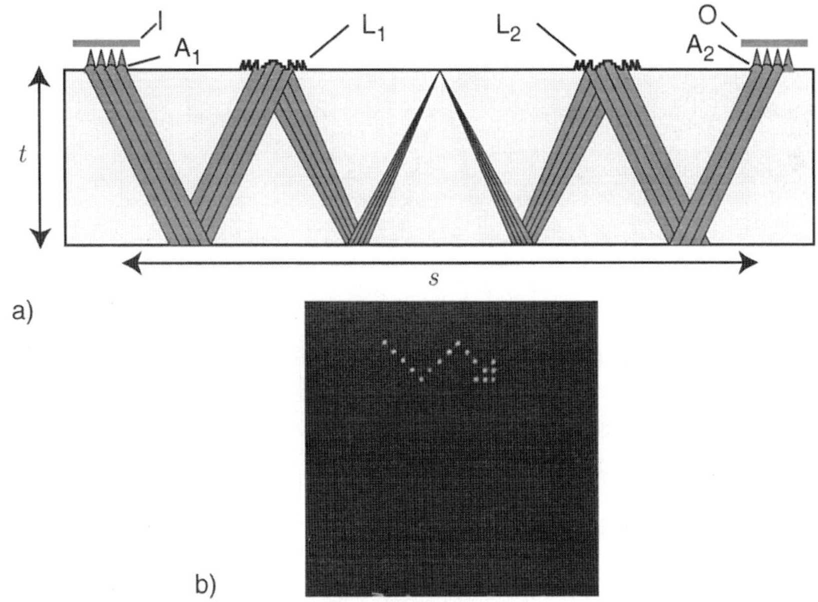

Fig. 7.12: a) Schematic of a hybrid imaging system integrated in planar optics (I, O: input and output plane; A_1, A_2: microlens arrays; L_1, L_2 imaging lenses); b) multiple exposure of the output plane, with different channels "switched on".

A similar hybrid imaging approach has also been suggested for the implementation of board-to-board interconnects in a stacked microoptical setup. Streibl et al. suggested using a stacked telecentric system for the imaging of an array of collimated laser diodes (Fig. 7.13a) [34]. Their imaging setup between the microlens arrays consisted of a three-lens telecentric system. HOEs were used for the incoupling and outcoupling of the light into a planar glass substrate so that a folded stacked optical system resulted. The three-lens telecentric imaging configuration is also referred to as light-pipe [35]. For microoptical integration, it has been shown that the imaging quality as well as the tolerances of the light-pipe imaging system are superior to more conventional setups such as the $4F$ system.

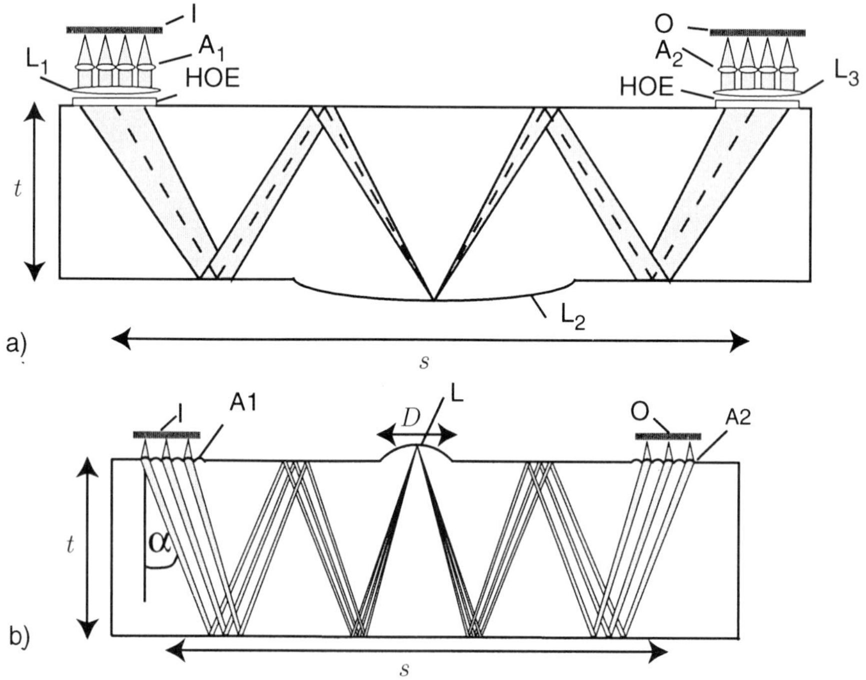

Fig. 7.13: Modified hybrid imaging setups integrated a) in stacked optics using microlens arrays (A_1, A_2, imaging lenses L_1, L_2, L_3 and holographic coupling elements (HOE); b) in planar optics using diffractive optical elements for integrating the microlenses, imaging lenses and coupling elements in one array of elements (A_1, A_2) and a diffractive imaging lens L.

A planar optically integrated hybrid imaging system based on the hybrid light-pipe configuration has been demonstrated recently (Fig. 7.13b) [36]. For better integration the functionality of the microlens arrays as well as the first lens of the light-pipe are integrated in one single array of diffractive elements. Each of the microlenses acts as a deflecting microlens which images one individual point source into the center of the field lens L. To this end the focal length as well as the deflection angle for each of the microlenses vary over the array. Individual adjustment of the microlenses is especially advantageous since the optical system is folded to be integrated in planar optics. The system integrated in a planar substrate of thickness t = 6 mm demonstrated 2500 parallel data channel over a distance of s = 8.6 mm. In the light-pipe system the pupil plane is located in the plane of the first imaging lens. In the hybrid configuration this is the plane where the microchannels are not overlapping. For optical interconnects this has the consequence that it is possible to spatially multiplex light-pipes in order to realize space-variant interconnection schemes. We will discuss more details on optical interconnection systems in Chapter 11.

7.3 Glossary

- **clipping**
of a Gaussian beam: the effect that the edges of a Gaussian beam with theoretically infinite extension are cut off upon transmission through an aperture of finite extension.

- **crosstalk**
in interconnect technology is the coupling of a transmitted signal into one or several neighbouring channels.

- **dilute array**
an array of emitters or receivers which are not densely packed. The distance between the devices is larger than their dimensions.

- **hybrid imaging system**
an imaging system which uses a combination of array optics and single lens imaging, which is well adjusted to the imaging of dilute arrays of optoelectronic devices.

- **MOEM**
acronym for micro-opto-electro-mechanical system; an approach to microoptical systems integration based on micromachining techniques mostly in silicon.

- **multichannel imaging**
optical system which images an array of input channels onto an array of output channels using separate microlenses for each channel.

- **planar optics**
technique for monolithic microoptical systems integration. The optical system is folded such that light propagation takes place inside a single planar substrate. The microoptical components are implemented as reflective elements on the surfaces of this substrate.

- **stacked optics**
integration technique for microoptical systems where several planar layers containing optical components are stacked on top of each other to form the optical system.

7.4 List of new symbols

α:	the angle by which in planar optics the optical axis is inclined to the substrate normal
d_{array}:	distance between the optical windows in a dilute array
d_{lens}:	diameter of a microlens
$d_{\text{win}}, d_{\text{source}}$:	diameter of the optical window of an optoelectronic array
D_{hyb}, D_{conv}:	diameter of the imaging lenses in a hybrid or conventional $4F$ configuration
FWHM diameter:	full-width-half-maximum; diameter of a Gaussian or sinc-like distribution measured at half the peak value.
I:	beam intensity
I/O	input/output
O(mm):	order of milimetres
O(μm):	order of micrometres
R:	radius of the aperture clipping the Gaussian beam
ρ:	integration parameter
w_{inc}:	beam diameter in the input pupil
w_0, w_0':	waist of a Gaussian beam
$w(z)$:	Gaussian beam diameter at the position z
$\Delta z'$, $\Delta z''$, Δz_{max}:	distance along the z-axis

7.5 Exercises

1. **Divergence of Gaussian beams**
   We assume that the clipping condition described by Eq. 7.2.11 is fulfilled. Calculate the diameters of a Gaussian beam (λ= 633 nm) "collimated" by (micro)lenses with diameters of d_{lens}= 50 μm, 100 μm, 250 μm, 500 μm, 1 mm, and 1 cm at a distance of z = 10 cm behind the lens. What is the percentage of light transmitted through a second lens of equal diameter in this plane?

2. **Microchannel imaging**
   Based on your knowledge of Fourier optics, derive the relation $\Delta z_{\mathrm{max}} \approx \frac{d_{\mathrm{lens}}^2}{2\lambda}$ for the maximum propagation distance between two lenses in the microchannel approach.

3. **Hybrid imaging**
   What are the advantages of hybrid imaging systems for the imaging of dilute arrays?

References

[1] W. S. Trimmer (ed.), "Micromechanics and MEMS — Classical and seminal papers to 1990", IEEE Press, New York (1997).

[2] M. Mehregany, K. J. Gabriel and W. S. N. Trimmer, "Integrated fabrication of polysilicon mechanisms", *IEEE Trans. Electron. Devices* **35** (1988), 719–723.

[3] M. E. Motamedi, "Micro-opto-electro-mechanical systems", *Opt. Eng.* **33** (1994), 3505–3516.

[4] K. S. J. Pister, M. W. Judy, S. R. Burgett and R. S. Fearing, "Microfabricated hinges", *Sens. Actuators A* **33** (1992), 249–256.

[5] M. C. Wu, "Micromachining for optical and optoelectronic systems", *IEEE Proc.* **85** (1997), 1833–1856.

[6] L. Y. Lin, J. L. Shen, S. S. Lee and M. C. Wu, "Realization of novel monolithic free-space optical disc pick-up heads by surface micromachining", *Opt. Lett.* **21** (1996), 155–157.

[7] M. E. Motamedi, A. P. Andrews, W. J. Gunning and K. Moshen, "Miniaturized micro-optical scanners", *Opt. Eng.* **33** (1994), 3616–3623.

[8] L. J. Hornbeck, "Digital light processing for high-brightness high-resolution applications", *Proc. SPIE* **3013**, Electronic Imaging, EI '97; Projection Displays III, San Jose, CA, USA (1997), 27–40 (http://www.ti.com/dlp).

[9] K. Iga, M. Oikawa, S. Misawa, J. Banno and Y. Kokubun, "Stacked planar optics: an application of the planar microlens", *Appl. Opt.* **21** (1982), 3456–3460.

[10] K.-H. Brenner, "3D-integration of digital optical systems", *OSA techn. digest* , **6**, Optical computing, Salt Lake City (1991), 25–28.

[11] W. Singer and K.-H. Brenner, "Stacked microoptical systems", in Micro-optics, H. P. Herzig (ed.), Taylor & Francis, London (1997), 199–221.

[12] J. B. Moisel K.-H. Brenner, "Demonstration of a 3D integrated refractive microsystem", *Inst. phys. conf. series* , **139**, International Conference on Optical Computing, Edinburgh (1994), 259–262.

[13] C. Passon, J. Moisel, N. McArdle, W. Eckert, K.-H. Brenner, M. Kuijk and P. Heremans, "Integration of refractive micro-optical elements with differential-pair optical-thyristor arrays", *Appl. Opt.* **35** (1996), 1205–1211.

[14] J. Jahns and A. Huang, "Planar integration of free space optical components", *Appl. Opt.* **28** (1989), 1602–1605.

[15] J. Jahns, "Planar packaging of free space optical interconnections", *Proc. of the IEEE* **82** (1994), 1623–1631.

[16] K.-H. Brenner and F. Sauer, "Diffractive reflective optical interconnects", *Appl. Opt.* **27** (1988), 4251–4254.

[17] M. Kato, Y.-T. Huang and R. K. Kostuk, "Multiplexed substrate-mode holograms", *JOSA A* **7** (1990), 1441–1447.

[18] J. M. Battiato, R. K. Kostuk and J. Yeh, "Fabrication of hybrid diffractive optics for fiber interconnects", *IEEE Phot. Techn. Lett.* **5** (1993), 563–635.

[19] H. Kurita and S. Kawai, "Quasi-toric planar microlenses for oblique incidence light beams", *Appl. Opt.* **26** (1997), 1017–1022.

[20] S. Reinhorn, S. Gorodeisky, A. A. Friesem and Y. Amitai, "Fourier transformation with a planar holographic doublet", *Opt. Lett.* **20** (1995), 495–497.

[21] A. W. Lohmann, "Image formation of dilute arrays for optical information processing", *Opt. Comm.* **86** (1991), 365–370.

[22] F. B. McCormick, F. A. P. Tooley, T. J. Cloonan, J. M. Sasian and H. S. Hinton, "Microbeam optical interconnections using microlens arrays", in Photonic Switching, H. S. Hinton and J. W. Goodmann (eds), 8 OSA Digest Series, Salt Lake City (1991), 90–96.

[23] Y. Liu, B. Robertson, D. V. Plant, H. S. Hinton and W. M. Robertson, "Design and characterization of a microchannel optical interconnect for optical backplanes", *Appl. Opt.* **36** (1997), 3127–3141.

[24] F. B. McCormick, "Free space optical interconnection techniques", in Photonics In Switching II, J. E. Midwinter (ed.), Academic Press, Boston, MA (1993).

[25] A. E. Siegman, "Lasers", Oxford University Press, Oxford (1986).

[26] B. E. A. Saleh and M. C. Teich, "Fundamentals of Photonics", John Wiley & Sons, New York (1991).

[27] F. Sauer, J. Jahns, C. R. Nijander, A. Y. Feldblum and W. P. Townsend, "Refractive-diffractive micro-optics for permutation interconnects", *Opt. Eng.* **33** (1994), 1550–1560.

[28] P. Belland and J. P. Crenn, "Changes in the characteristics of a Gaussian beam weakly diffracted by a circular aperture", *Appl. Opt.* **21** (1982), 522–527.

[29] S. A. Self, "Focusing of spherical Gaussian beams", *Appl. Opt.* **22** (1983), 658–661.

[30] J. Jahns, F. Sauer, B. Tell, K. F. Brown-Goebeler, A. Y. Feldblum, C. R. Nijander and W. P. Townsend, "Parallel optical interconnections using surface-emitting microlasers and a hybrid imaging system", *Opt. Comm.* **106** (1994), 328–337.

[31] A. Kirk, A. Goulet, H. Thienpont, N. McArdle, K.-H. Brenner, M. Kuijk, P. Heremans and I. Veretennicoff, "Compact optical imaging system for arrays of optical thyristors", *Appl. Opt.* **36** (1997), 3070–3078.

[32] J. Jahns, "Integrated-optical imaging system", *Appl.Opt.* **29** (1990), 1998.

[33] J. Jahns and B. Acklin, "Integrated planar optical imaging system with high interconnection density", *Opt. Lett.* **18** (1993), 1594–1596.

[34] N. Streibl, R. Völkel, J. Schwider, P. Habel and N. Lindlein, "Parallel optoelectronic interconnections with high packing density through a light guiding plate using grating couplers and field lenses", *Opt. Comm.* **99** (1992), 167–171.

[35] K.-H. Brenner, W. Eckert and C. Passon, "Demonstration of an optical pipeline adder and design concepts for its microintegration", *Opt. and Laser Techn.* **26** (1994), 229–237.

[36] S. Sinzinger and J. Jahns, "Integrated microoptical imaging system with high interconnection capacity fabricated in planar optics", *Appl. Opt.* **36** (1997), 4729–4735.

8 Optoelectronic devices and smart pixel arrays

Microoptical systems alone are of little use for fully functional systems without suitable optoelectronic devices and actuators being integrated with the optics. In the following sections we shall discuss some basic aspects of optoelectronic devices for integration in hybrid microsystems. We also address systems aspects leading to the concept of so-called "smart pixels". This is praticularly interesting for the application in optical interconnects. Most optoelectronic devices today are based on multilayer stacks of semiconductor materials with different properties fabricated, for example, by molecular beam epitaxy. Therefore, we shall begin by discussing these so-called "superlattices and multiple quantum wells". We then focus on specific devices such as "self-electro-optics-effect-device (SEED)" as a modulator device, and "vertical cavity surface emitting lasers (VCSELs)" as an active light source. Finally we will introduce the concept of "smart pixel arrays".

8.1 Superlattices and multiple quantum wells

The fabrication of optoelectronic devices has been revolutionized by the development and refinement of techniques for thin layer deposition, such as molecular beam epitaxy (MBE) and metal-organic chemical vapour deposition (MOCVD). With these fabrication techniques it became possible to grow thin epitaxial layers of semiconductor materials. The layer thickness can be defined with a precision of a few atomic layers. Stacks of layers with varying stochiometric material composition as well as doping concentration can be fabricated. With this technology it is possible to tailor the electronic and photonic properties of semiconductor materials. By stacking several extremely thin layers (typical layer thickness 50-100 Å) of materials with different stochiometric composition or different doping, very specific optoelectronic properties can be generated. Such material stacks are called superlattices. In superlattice structures quantum physical effects can be observed. The electronic energy bands, e.g., are subdivided into discrete minibands due to the lateral confinement of the electrons. Depending on the type and thickness of the stacked semiconductor materials, new electronic and photonic properties can be achieved which can be exploited in electronic and optoelectronic devices [1].

One generally distinguishes between two different types of superlattices: hetero-superlattices and doping-superlattices. Hetero-superlattices are stacked layers of materials with different energy band gaps. In doping-superlattices the material remains the same from layer to layer, however, the doping changes. In spite of some similarities there are also fundamental differences between these different types of superlattice.

8.1.1 Hetero-superlattices

A typical example of a hetero-superlattice is a stack composed of gallium arsenide (GaAs) and aluminum gallium arsenide ($Al_xGa_{1-x}As$). Here the parameter x describes the percentage of gallium atoms in the gallium arsenide structure which have been replaced by aluminum atoms. Due to the different energy gaps of the stacked materials, quantum wells are formed at the location of the lower band-gap material (Fig. 8.1a). Quantum wells are deformations in the electronic band structure of the semiconductor material which are formed due to bending of the bands at the layer boundaries. The multilayer stacks of alternating materials are also called multiple quantum well (MQWs) materials. Consider the case where the high band-gap material (AlGaAs) is n-doped. In this case, due to energy constraints, the free electrons will gather in the minima of the conducting band located in the low band-gap material (GaAs). The lowest energy states for holes in the valence band are the maxima of the valence band (Fig. 8.1a). Due to the thin layer structure, electrons are confined to the narrow quantum wells so that the originally continuous energy band is split into several discrete minibands. These additional energy bands have immediate impact on the absorption characteristics of the structure. Bulk semiconductor material shows a typical quasi-continuous absorption characteristic for incident radiation with a wavelength short enough to bridge the energy gap. The quantum-well material, on the other hand, exhibits sharp absorption peaks due to the miniband structure in the potential wells. [1]

A second important effect of electron confinement in the quantum wells is an increase in the stability of excitons. Excitons are weakly bound states between the free electrons and holes in the semiconductor. In bulk material the binding of the excitons is so weak that they can only be observed at extremely low temperatures (typically $T \approx 5$ K). With an electric field applied, electrons and holes are drawn into different directions so that the excitonic state is immediately destroyed. This is different in quantum well materials. Due to the lateral confinement of electrons and holes, both the probability and the stability of excitons are greatly increased. Quantum well materials thus show a strong excitonic absorption peak at energies below the band-gap energy. If an electric field is applied, electrons and holes are drawn to opposite walls of the well. This reduces the binding energy of the excitons but, for sufficient confinement, the excitonic state is not destroyed. The reduction of the exciton binding energy shows itself in a shift of the absorption peak to

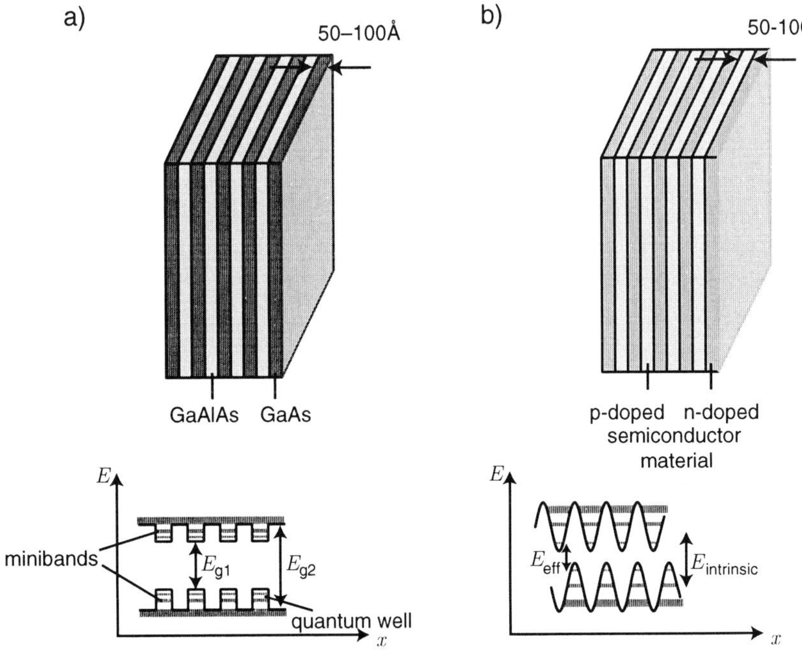

Fig. 8.1: The two basic types of superlattices and the corresponding electronic energy band structure: a) hetero-superlattice; b) doping-superlattice or nipi-structure.

lower energies, i.e., longer wavelengths. In self-electro-optic effect devices (SEEDs) excitonic absorption is used to modulate an incident beam of light [2].

8.1.2 nipi-superlattices

The second type of superlattice material consists of a stack of layers of a single material with alternating doping (Fig. 8.1b). Such a material is called doping-superlattice [3]. Often a layer of intrinsic semiconductor material (i-layer) is introduced between an n- and a p-doped layer. The structures are, therefore, also referred to as nipi-structures. Due to carrier diffusion and electron-hole recombination at the p-n junction, electrostatic potentials build up. Accumulating over the large number of junctions this potential reaches significant strength and effectively distorts the energy band. The bending of the energy bands under an applied electric field can also be observed in bulk semiconductor material. It is responsible for an extension of the tail of the absorption spectra to radiation below the band-gap energy. This is called the Franz-Keldysh effect [4].

Due to the band distortion in the nipi material, quantum wells are formed which are similar to hetero-superlattice materials. One difference, however, is that the minima in the conducting band are not located at the same lateral positions as the maxima in the valence band. Consequently, the band gap at each location remains approximately equal to the band gap of the intrinsic material. There is a certain probability for the electrons to tunnel through the potential wall between the location of the minimum in the conducting band and the maximum in the valence band. These tunneling electrons can be excited by photons with energies below the intrinsic band-gap energy ("photon-assisted tunneling"). This observation of an effective band-gap energy below the intrinsic band gap is often referred to as the quantum-confined Franz-Keldysh effect. The tunneling probability depends strongly on the height and width of the potential wall. This is influenced by the internal and external electric fields. Besides the reduced band gap, the quantum wells formed in nipi- superlattices also increase the probability of excitons in the material. Increased excitonic absorption peaks can be observed similarly to the situation in hetero-superlattices.

The potential of nipi-superlattices lies in the fact that through manipulation of the layer thicknesses as well as the doping concentration it is possible to manipulate the material to assume almost any desired optoelectronic property. The effective band gap, for example, can be given any value between zero and the band gap of the bulk material. This can be achieved for almost any semiconductor material provided there are dopants for the p- as well as the n-doping. In contrast, for hetero-superlattice structures it is necessary that the crystal lattices of the two stacked materials are well adjusted to each other. Otherwise it would not be possible for epitaxial growth of the individual layers on top of each other. Even after fabrication, the properties of a doping-superlattice structure can be influenced within a wide range, e.g., by light absorption or application of electric fields. These features are currently under investigation for the fabrication of nonlinear optoelectronic devices [5].

8.2 The SEED (self-electro-optic effect device)

8.2.1 Structure and fabrication

Self-electro-optic effect devices (SEEDs) are modulator-type devices which are based on the quantum-confined Stark effect (QCSE), i.e., the electric field induced shifting of the sharp excitonic absorption peaks in multiple quantum wells. Such electroabsorption modulators are fabricated by placing quantum well materials in the intrinsic region of a p-i-n diode. Fig. 8.2 shows the schematic layer configuration of a SEED [6].

As described above, quantum well materials exhibit sharp absorption peaks close to the band-gap energy. These absorption peaks are caused by excitons which are rather stable in the quantum confinement. If an electric field is applied perpendicular to the quantum wells, these excitonic absorption peaks become

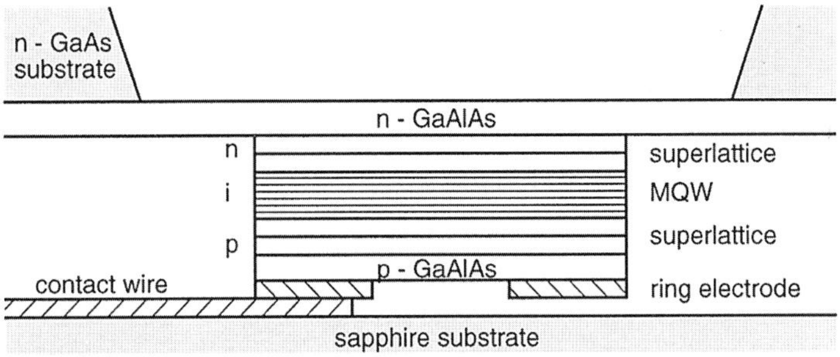

Fig. 8.2: Schematic layer structure of a SEED [6].

detuned. Fig. 8.3 shows the sharp excitonic absorption mechanism in quantum well materials. The figure further illustrates the red-shift in the absorption spectrum for a quantum well material with various voltages applied.

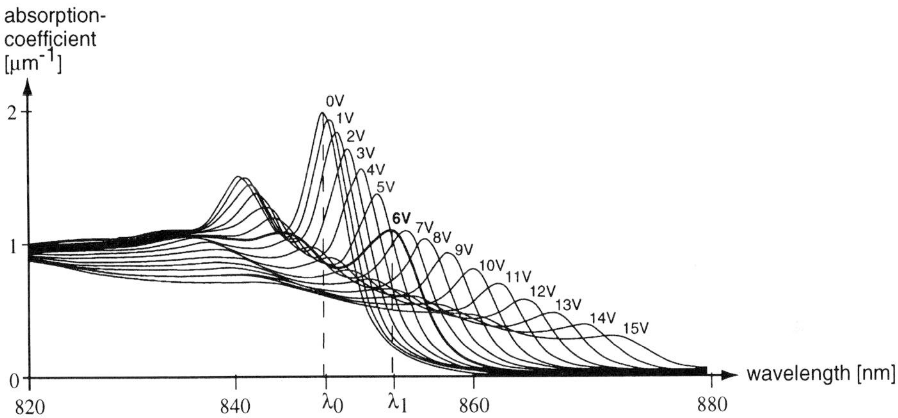

Fig. 8.3: Absorption spectrum of the SEED p-i-n diode for different voltages applied [7].

8.2.2 Energy dissipation and efficiency

By varying the electric field over the quantum well structure, significant changes in the transmission can be achieved. A layer stack with a thickness of a few micrometres is enough to change the transmission by a factor of 2 or more. This is sufficient for the fabrication of optical modulator devices.

The possibility of using the quantum wells in a reverse-biased diode configuration yields very energy-efficient devices. Since no electric current flows through the device, except for the photocurrent induced by the modulated light beam, only very little energy is dissipated during the modulation. The energy ΔE necessary to change the transmission can be calculated from the capacitance C of the device and the applied voltage V:

$$\Delta E = \frac{C\,V^2}{2} \tag{8.2.1}$$

Typical switching energy densities of several fJ/μm^2 result [8]. This is very low for optical devices and comparable to electronic circuits. Fast switching speeds below 100 picoseconds have also been demonstrated with SEEDs. The switching speed, however, depends on the power level of the optical input.

8.2.3 All-optical modulation

If a multiple quantum well p-i-n diode is driven in the reverse biased mode (Fig. 8.4), it can in principle work as an all optical device where light is used to modulate light. A light beam with suitable wavelength incident on the diode causes a photocurrent within the diode. Thus, the resistance of the diode is reduced and so is the voltage drop over the device, i.e., the electric field inside the diode. Most of the voltage drop occurs at the additionally integrated load resistor. This shifts the absorption peak and changes the absorption of the incoming beam. Depending on the wavelength, the absorption is increased or decreased resulting in positive or negative feedback, respectively. With positive feedback a bistable device results, whereas in the case of negative feedback the power absorbed by the modulator is proportional to the current which drives the diode.

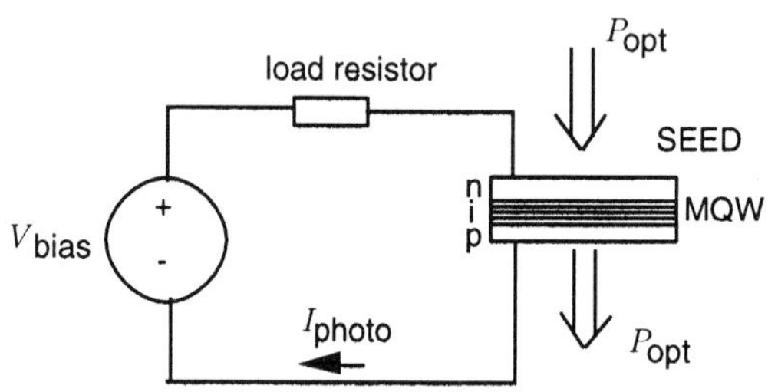

Fig. 8.4: The electrical circuit to drive the SEED in the reverse bias mode.

The next step in the evolution of SEEDs has lead to the replacement of the load resistor by a photodiode driven by an optical bias beam. The so-called D-SEEDs have advantages with respect to the integration of large device arrays as well as to their switching behaviour. The use of a photodiode rather than a load resistor reduces problems with stray capacities which occur when the size of the device is reduced. Such D-SEEDs can be fabricated in large two-dimensional arrays. Good switching energies could be demonstrated. Additionally the load resistor of the photodiode can be regulated by the optical bias beam which allows operation in a broader power range.

The problems related to the use of these D-SEEDs are typical for biased so-called "two-terminal" devices. Generally the energy of the bias beam is chosen so that it sets the device just below the threshold. Small modulations of the signal beam are then sufficient for switching the device. As a consequence of this operation mode, the biasing energy is very critical. In systems applications it is difficult to provide the very homogeneous power which is necessary over large arrays of bias beams. Additionally the bias beam also depends on the signal power level and sometimes needs to be adjusted, especially in cascaded operation. These problems have been overcome in the so-called symmetric SEED or S-SEED.

8.2.4 S-SEED

The S-SEED consists of two SEEDs which are coupled serially in an electronic circuit. Each SEED represents the load diode for the other device (Fig. 8.5a). This way the S-SEED is bistable with respect to the ratio of the two incident beams (Fig. 8.5b). Additional quantum well mirrors are integrated with the SEEDs, so that they can be used in reflection. During operation, two sets of beams are imaged onto the device. Firstly, two signal beams with different optical power levels are used to set the state of the device. A contrast ratio of 2 : 1 is generally sufficient for this purpose. Diode 1 which, e.g., receives the higher power beam, generates higher photocurrents so that most of the voltage drop occurs at diode 2. This shifts the absorption of the MQWs in the diodes so that an even higher photocurrent is generated in diode 1 which causes the positive feedback. Once the device is set, two equal intensity clock beams are used to read the state. The difference in absorption in the two diodes will change the relative power of the two beams. For the readout, higher power beams can be used, which results in an effective gain of the device. This is not an optical gain, but merely stems from the fact that higher power beams can be switched by low power signal beams. This is referred to as "time sequential gain". The requirements on the uniformity of the read-out clock signals are fairly moderate, as the device during read-out is operated in the middle of the bistable region [9].

8.2.5 Performance of S-SEEDs

The switching time in SEEDs, specifically in S-SEEDs, is determined by several factors:

- the response time of the quantum well material to the applied electric field. This time constant is a characteristic of the semiconductor material and generally very fast (e.g., several 100 fs).

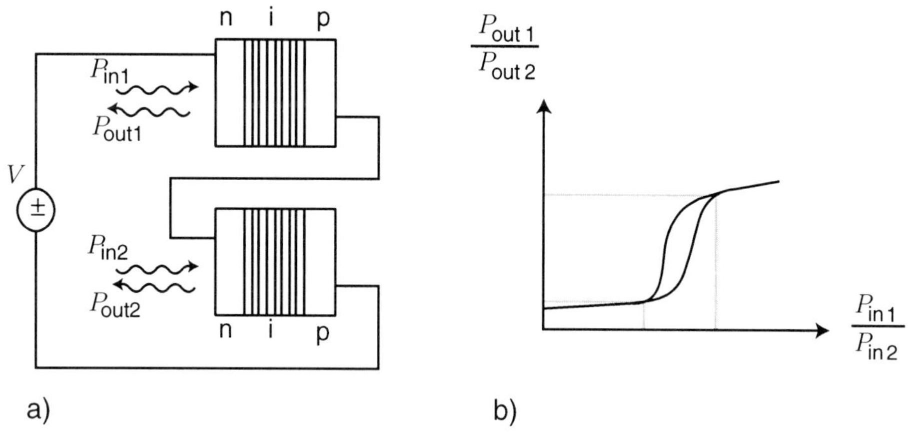

Fig. 8.5: a) The symmetric SEED (S-SEED) as a combination of two MQW modulators; b) bistability of the S-SEED in the ratio of the power of the input beams.

- the escape time of electrons and holes from the quantum wells which influences the response time of the diodes working as photodetectors. This time constant varies roughly in the range 10 ps–1 ns.
- the time it takes the photocurrent to charge and discharge the diode and parasitic capacitances. This constant is proportional to the incident optical power, the area of the device, the change in the applied electric field and inversely proportional to the responsivity of the device.
- the RC time constant of metal leads, contacts and stray capacitances. This RC constant assumes values of the order of 10 ps.

The optical switching energy is defined as the amount of optical energy, incident on one of the windows of the S-SEED, which is necessary to cause a change of the state of the device. This switching energy ΔE is approximately equal to the product of the switching time and the optical input power. Typical switching powers of ≈ 1 pJ can be achieved for devices with a window size of 5 μm $\times$ 10 μm operating at 6 V bias voltage. For a device which requires 1 pJ of switching energy an optical input beam of 1 μW would result in switching times of 1 μs. A 1 mW beam accelerates the switching to 1 ns .

Further developments have led to the integration of SEEDs with higher functionality, such a L-SEEDs (logic SEEDs) or FET-SEEDs (SEEDs integrated with field effect transistors (FETs) [10]). L-SEEDs combining several SEEDs are used to perform arbitrary logic functions. They generally consist of a set of input SEEDs and two output SEEDs which switch the incident clock signals. The topology of the interconnection between the input SEEDs results from the logic function to be performed. The L-SEED is an example of a smart pixel [9].

The FET-SEED attempts to combine electronic transistors with SEED technology. This can be achieved by developing transistors on MQW substrates (e.g., GaAs). Another approach is the use of hybrid integration techniques. Flip-chip bonding has been used to integrate CMOS logic circuitry with GaAs MQW modulators to fabricate smart pixels.

8.3 Vertical cavity surface emitting lasers

Semiconductor lasers are interesting light sources for a variety of applications, such as optical sensing, data storage and communication [11, 12, 13, 14]. Especially for integrated optics and microoptical applications the compact fabrication using lithographic techniques is an important feature. Most of the laser diodes currently used are of the edge-emitting type. The light is generated in a cavity of rectangular shape oriented parallel to the substrate surface. The cavity of such edge-emitting laser diodes has a width of O(10 μm) and a thickness of O(1 μm). This rectangular shape of the waveguiding active region of edge-emitting laser diodes results in non-circular symmetric modes. The emitted beam is astigmatic with different divergence angles in the x and y direction. For many applications this astigmatism has to be compensated for by suitable collimation and beam-shaping optics (cf. Chapter 10). Vertical cavity surface emitting lasers (VCSELs) use symmetric quantum well cavities oriented perpendicular to the substrate [15, 16]. This results in a superior beam quality, i.e., in circular symmetrical non-astigmatic beams. Additional advantages of VCSELs can be found in the packaging behaviour which is particularly interesting for applications in microoptics. For integration in 3D microoptical systems, light sources which emit vertically to a substrate surface are desirable. Conventional edge-emitting lasers can be converted into surface-emitting lasers by integration with 45° mirrors or grating couplers (Fig. 8.6a). Nevertheless, the size of the cavity parallel to the substrate surface limits the achievable packaging density. The structure of VCSELs, on the other hand, is optimized for the lithographic fabrication of densely packed 2D arrays. Thus, they are ideal light sources for 3D microoptics, where 2D data planes are processed in parallel (Fig. 8.6 b) [17, 18].

8.3.1 Structure and fabrication

In VCSELs the high carrier confinement in multiple quantum wells is necessary to excite laser activity in the material. Fig. 8.7 shows the schematic quantum layer structure of a VCSEL. In a typical design the active area consists of several (1–20) quantum wells consisting of a layer sequence of intrinsic GaAs (thickness $\approx$ 100 Å) and interlaced layers of AlGaAs (thickness $\approx$ 70 Å). This combination of active area and spacer layers forms the cavity of a Fabry-Perot resonator. The mirrors are fabricated as multilayer stacks of alternating low and high refractive index material. Each layer is grown epitaxially, e.g., by MBE or MOCVD. The refractive

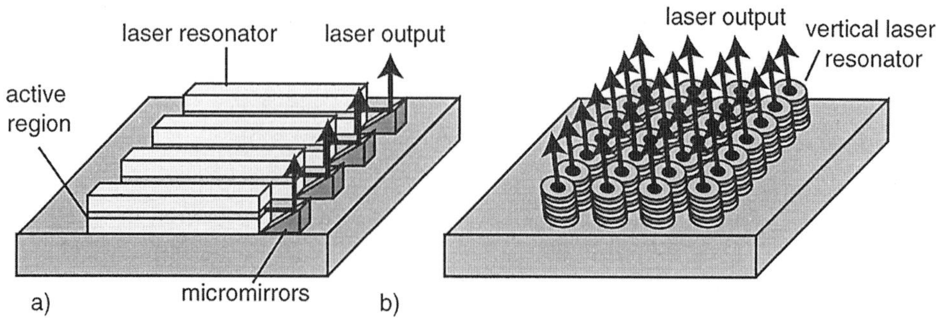

Fig. 8.6: Illustration of the array integration of a) edge-emitting laser diodes (reflected to surface normal emission) and b) vertical cavity surface emitting laser diodes.

index is controlled through the Al content in the AlGaAs lattice.

The mirror layers below and on top of the quantum wells are n-doped (e.g., Be-doping) and p-doped (e.g., Si-doping), respectively. Thus, the overall structure corresponds to that of a p-i-n laser diode. The active area of such a VCSEL is very small even compared to the wavelength λ_m in the material ($\lambda_m = \frac{\lambda_0}{n}$). Reasonable gain in these micro-resonators can only be achieved if the reflectivity of the mirrors is very high (> 0.99). The structure has to be fabricated so that the multiple quantum wells, which form the active area, are located at the maximum of the standing wave between the mirrors. An additional carrier confinement can be achieved by a graded index structure (e.g., through a gradually increasing Al content) in the spacer region. However, this results in reduced optical confinement.

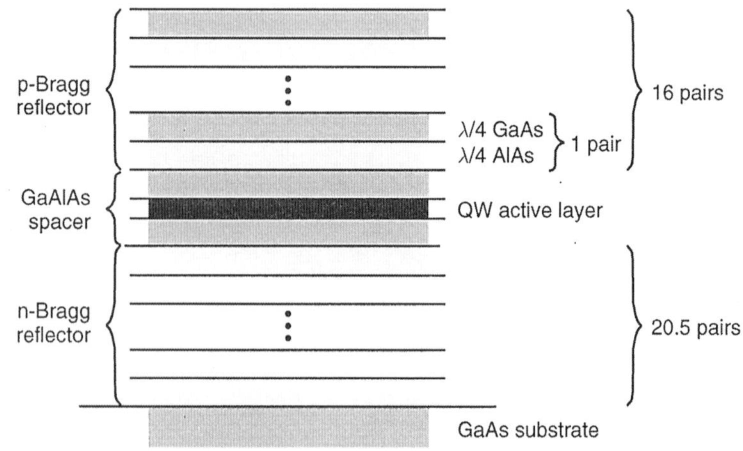

Fig. 8.7: Schematic layer structure of a AlGaAs/GaAs VCSEL [17].

The lateral confinement for the laser can be provided by index or gain guidance (Fig. 8.8). Index-guided structures are fabricated by etching cylindrical laser resonators. Through the index steps at the edges of the resonators they provide optical as well as electrical confinement to the active region. The deep etching process makes index-guided VCSELs difficult to fabricate especially in large arrays. Additionally, these laser "mesas" are sensitive to environmental influences. Gain-guided structures are fabricated through electrical isolation of the layer structure from the surroundings [19]. This can be achieved in high yield processes by ion (e.g., proton- or oxygen-) implantation. The gain-guided VCSELs can be fabricated as truly planar structures. The beam guidance, however, is not quite as high as in the case of index guidance. In general, this causes the threshold currents to be higher for gain-guided VCSELs.

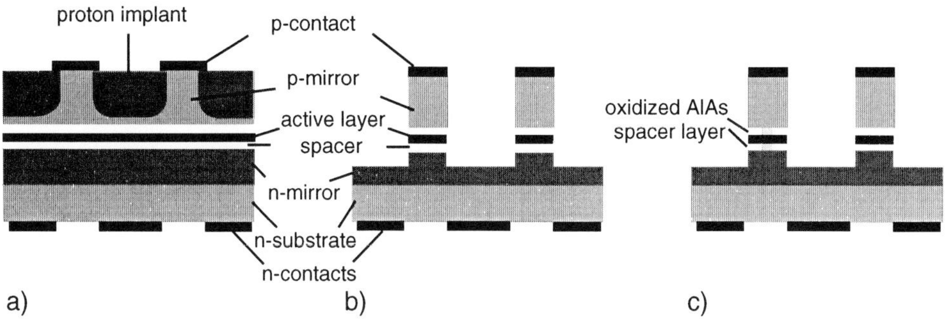

Fig. 8.8: Mechanisms for optical and electrical confinement in VCSELs: a) gain-guided resonator; b) index-guided resonator; c) index-guided resonator with oxide confinement.

On the way to reducing the lasing thresholds of VCSELs, during recent years the improvement of the lithographic etching techniques resulted in a significant increase in the yield for the etching of index-guided VCSEL resonators. Additional reduction of the threshold can be achieved with a technique called "oxide confinement" (Fig. 8.8c). In order to achieve better electrical confinement, the laser resonators are etched to just above the active area of the laser. The first layer of the top mirror consists of AlAs or GaAlAs with an Al content significantly higher than in the other layers. This AlAs layer can be selectively oxidized at elevated temperatures (T> 400°C) to Al_xO_y which provides good electrical isolation. This process is very sensitive to the Al content in the layer. The oxidation undercuts laterally under the Bragg mirror in a diffusion process [20, 21, 22]. The oxide confinement technique allows one to reduce the effective width of the laser resonator significantly by confining the current flow to a diameter much smaller than the laser resonator. As an alternative to the oxidation process it is possible to generate an isolating air gap by selectively etching the AlAs layer. However, comparisons show that inherent optical losses are lower in oxide-confined structures [23]. Optical scattering at the larger index steps of the air gap confinement is mainly held responsible for the increased losses.

8.3.2 Mirrors and resonator

The multilayer mirror structures with reflectivities of $\varrho \geq 0.99$ are critical parts of a VCSEL. The lattice structure of the GaAs substrate has to be conserved in the mirrors so that the GaAs quantum wells can be grown epitaxially on top. To achieve high reflectivities, layers of AlGaAs are grown with alternating high and low refractive index. Each of the layers is designed to have an optical thickness of $t = \frac{\lambda}{4}$ for the laser wavelength. Upon reflection at a layer with higher refractive index, a phase shift of $\delta\phi = \frac{\lambda}{2}$ occurs. For the $\frac{\lambda}{4}$ layer stack all reflected partial waves interfere constructively, independent of the reflecting surface. The reflectivity ϱ_{stack} of the stacked layer system is determined by the refractive indices n_{H}, n_{L} of the two layers and the number N of periods in the layer stack [24]:

$$\varrho_{\text{stack}} = \left(\frac{1 - \left(\frac{n_{\text{H}}}{n_{\text{L}}}\right)^{2N}}{1 + \left(\frac{n_{\text{H}}}{n_{\text{L}}}\right)^{2N}} \right)^2 \tag{8.3.1}$$

With this technique very high reflectivities can be achieved with a series of layers of moderate reflectivities. The refractive index of the AlGaAs system is controlled by the Al content of the layer according to Eq. 8.3.2:

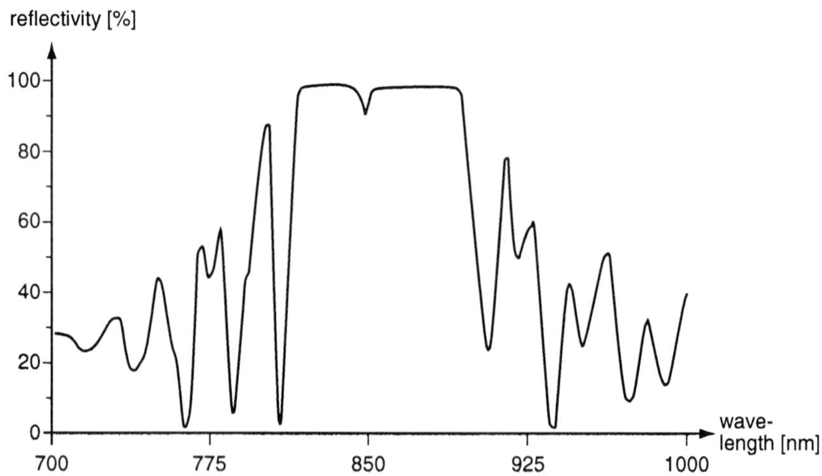

Fig. 8.9: Typical reflectivity of a Fabry-Perot resonator with multilayer AlGaAs/GaAs mirror stacks.

$$\text{at } \lambda = 1.239\ \mu\text{m}: \quad n_{\text{Al}_x\text{Ga}_{1-x}\text{As}} = n_{\text{GaAs}} - 0.54 \cdot x = 3.45 - 0.54 \cdot x \tag{8.3.2}$$

With this material combination a significant index range can be covered so that a relatively small number of layers can yield high reflectivity mirrors. For example, with a moderate number of about 25 pairs of alternating GaAs and AlGaAs layers, resulting in 50 reflections with reflectivities of $\varrho_{\text{layer}} \approx 0.085$ it is possible to achieve overall reflectivities of $\varrho_{\text{stack}} \approx 99$ %. Fig. 8.9 shows the spectral reflectivity of a Fabry-Perot resonator with such multilayer mirrors. The transmission peak of the resonator can be observed at the center of the wide high reflectivity band of the mirrors between $\lambda \approx 800$–890 nm, [19].

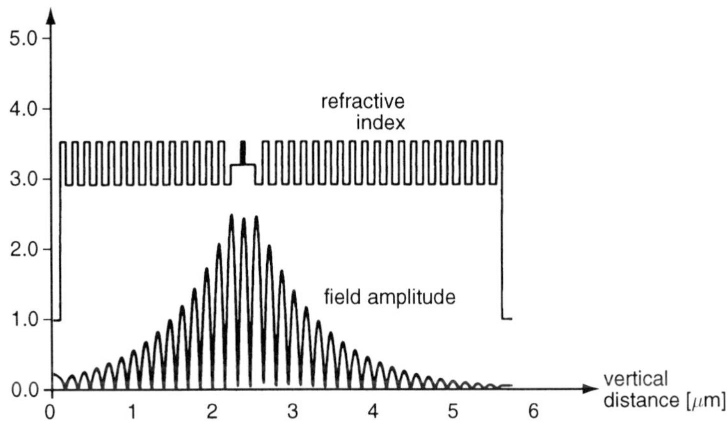

Fig. 8.10: Optical intensity (electric field distribution) and the refractive index distribution within the Fabry-Perot resonator.

The distribution of the refractive index and the corresponding electric field distribution within the layer stack is shown in Fig. 8.10. As can be observed the E-field distribution is concentrated very well at the center of the resonator. In this example only one quantum well represents the active area. The standing wave within the resonator is additionally modulated by the multilayer structure of the refractive index. The effective cavity length of the resonator can be estimated from the $1/e$-width of the field distribution. A typical value is $L_{\text{eff}} = 1.83\ \mu$m. With an average refractive index $n = 3.3$ for the GaAs multilayer stack, the free spectral range λ_{fsr} of this cavity is estimated to be [19]:

$$\nu_{\text{fsr}} = \frac{c}{2nL_{\text{eff}}} \quad \rightarrow \quad \lambda_{\text{fsr}} = \frac{\overline{\lambda}^2}{2nL_{\text{eff}}} \tag{8.3.3}$$

λ_{fsr} denotes the distance between two resonant wavelengths, i.e., between two transmission wavelengths of the resonator. Due to the small effective length L_{eff} of the resonator, a large free spectral range of $\lambda_{\text{fsr}} \approx 60$ nm results. The high reflectivity bandwidth of the mirrors $\Delta\lambda_{\mathrm{mirror}}$ can be estimated from Fig. 8.9 to be less than $\Delta\lambda_{\mathrm{mirror}} \leq 100$ nm, so that a single centered longitudinal laser mode is emitted by such a laser resonator.

The resonant modes of the multilayer laser cavity (Fig. 8.7) are extremely sensitive to the spacer thickness L. For the quantum-well resonator this dependence can be described by:

$$\frac{\Delta\lambda}{\lambda} \approx \frac{\Delta L}{L} \tag{8.3.4}$$

The slightest error during fabrication detunes the resonant wavelength. For example, an error in the layer thickness of a single atomic layer $\Delta L \approx 1$ Å, causes wavelength shifts in the resonant frequency of the cavity of about $\Delta\lambda \approx 1$ nm. The ideal configuration, i.e., the highest efficiency for the laser action results if the resonance frequency of the cavity corresponds to the band-gap energy of the quantum-well material:

$$\begin{array}{llcl} \text{GaAs:} & \Delta E = 1.424 \text{ eV} & \rightarrow & \lambda = 874.9 \text{ nm} \\ \text{Al}_x\text{Ga}_{1-x}\text{As:} & \Delta E = 1.424 + 1.247x \text{ eV} & \rightarrow & \lambda = 850 \text{ for } x = 0.0289 \text{ nm} \end{array} \tag{8.3.5}$$

The possibility of detuning the laser resonator by a small variation of the layer thickness has also been applied to fabricate arrays of VCSELs with continuously varying wavelengths. A total tuning range of $\Delta\lambda = 43$ nm has been achieved [17].

8.3.3 I-V characteristics and efficiency

Fig. 8.11 shows typical voltage-current (V-I) and optical output power-current (P-I) curves for MQW-VCSELs operating in continuous mode. The V-I-curve shows the characteristic dependence for a laser diode. Threshold currents (I_{th}) well below 1 mA have been demonstrated. The threshold voltage (V_{th}) is determined by the series resistance of the mirror stacks. High refractive index differences between the mirror layers cause high potential walls with high resistances. Bragg mirrors consisting of a larger number of layers with lower refractive index steps have been introduced in VCSELs and lead to lower threshold voltages. As another means to reduce the series resistance of the Bragg layers without reducing the effective index steps, δ-doping has been applied. Thin ($t \approx 99$ Å) layers of p- or n-doped AlGaAs are grown between the low and high index layers. This introduces additional carriers and causes tunneling currents which reduce the resistances [25]. In order to completely avoid the series resistance of the mirror, intracavity contacts have been suggested [26]. In this case only single p- and n-doped layers immediately above or below the active region are contacted. The mirror stacks at the same time remain undoped and need not be contacted. Typical threshold voltages are as low as $V_{th} = 1$–5 V.

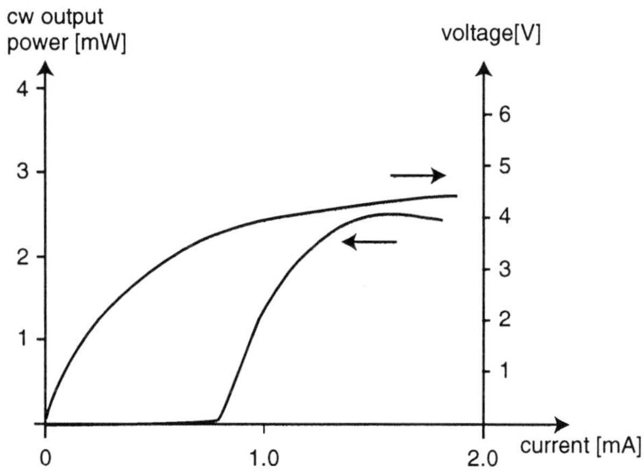

Fig. 8.11: Voltage vs. current and optical output power vs. current distribution of a typical VCSEL at room temperature.

Peak optical output powers P_{peak} of several mW can be achieved with cavity diameters smaller than 10 μm. The drop of output power for higher currents is due to intense local heating. This is one of the most important problems with the use of densely packed VCSEL arrays for interconnection applications, for example. Current packaging technology allows one the fabrication of large arrays of densely packed VCSELs which show high heat dissipation of several W/cm^2 or even in the kW/cm^2 range. Maximizing the efficiencies in order to reduce heat dissipation is, therefore, one of the major research goals with regard to VCSELs.

8.3.4 Spectral characteristics and thermal effects

As discussed above, in VCSELs generally a single longitudinal mode is excited due to the small cavity length and the spectral characteristic of the mirror reflectivities. For electrical currents close to the threshold only one transverse mode, the TEM$_{00}$ occurs. For higher currents, however, higher order transverse modes are excited due to saturation effects in the TEM$_{00}$ mode. These higher order transverse modes can be seen in the spectrum (Fig. 8.12) as secondary output peaks shifted to shorter wavelengths (higher energies). Fig. 8.12 at the same time exhibits a shift of the whole spectrum towards longer wavelengths for higher currents. This is due to heating of the VCSEL.

Other temperature-dependent effects can be seen from Fig. 8.13, where the cw output power vs. drive current at different temperatures is shown. Apart from a drop in the peak output power, an increase in the threshold current can be observed

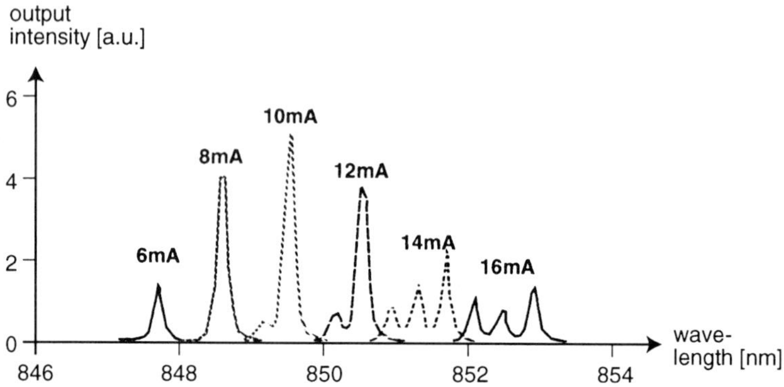

Fig. 8.12: Spectra of the optical output of a VCSEL in cw mode for different drive currents, illustrating the thermal shift and higher order transverse modes.

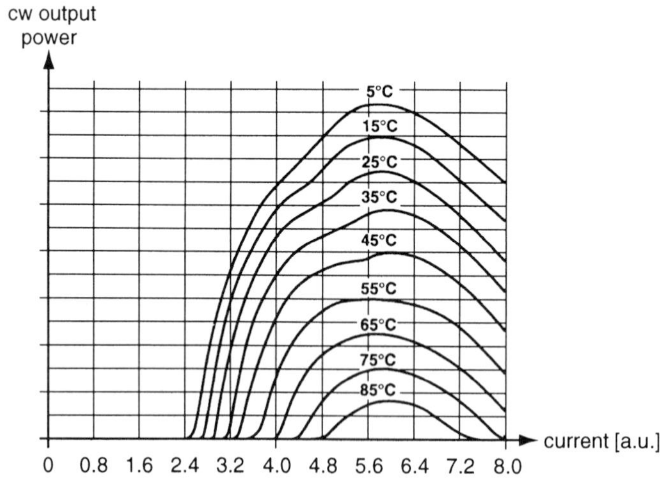

Fig. 8.13: Temperature depencence of the optical output vs. current curves.

for increasing temperatures.

8.3.5 Other material combinations

In addition to the GaAs/AlGaAs material combination, another interesting material for the fabrication of VCSELs is $In_xGa_{1-x}As_yP_{1-y}$ which can be lattice matched to InP to form quantum wells if the composition indices fulfil the equation [27]:

$$y = 2.16(1 - x) \tag{8.3.6}$$

This material is especially interesting because it provides band-gap energies between $\Delta E = 0.75$–1.34 eV, corresponding to the wavelength range $\lambda = 1.65--0.925$ μm. This wavelength regime is interesting for optical communication applications because of the dispersion minimum of optical fibers which is located at a wavelength of 1.3 μm. There are, however, some problems related to InGaAsP, which are the reason why it is difficult to achieve cw laser operation at room temperature. Non-light emitting recombination effects (Auger recombination) and intra-valence-band absorption cause the threshold carrier density to be very high and the lasing operation to be very inefficient. Secondly, it is very difficult to fabricate high quality multilayer mirrors. Because of the low index difference between the materials, many layer pairs are necessary to achieve reasonable reflectivity. Additionally the overall thickness of the structure is increased due to the longer wavelength.

Strained systems of GaInAs/GaAs have been demonstrated for VCSELs emitting at a wavelength of $\lambda = 0.98$ μm. Threshold currents I_{th} below 1 mA corresponding to threshold current densities J_{th} of about 800 A/cm^2 have been demonstrated for this material combination.

The SEEDs and VCSELs discussed in the previous sections are only two of the most important and widely used types of optoelectronic devices. Research efforts are undertaken on a much larger variety of devices. Some types, for example optical thyristors based on pnpn-superlattices, so-called V-step devices or nipi optical devices are also based on the physics of multiple quantum wells [28]. Others, e.g., liquid crystal light modulators, try to exploit different effects. A more thorough discussion of the device technology is, however, beyond the scope of this book. We shall, therefore, conclude our discussion of optoelectronic devices with some remarks about the smart pixel concept.

8.4 Smart pixel arrays (SPAs)

In the remainder of this book we shall focus on applications for microoptical systems where they generally need to be integrated with appropriate optoelectronic devices (i.e., light sources, detectors etc.). One of the most widely discussed applications of free-space microoptics is in the implementation of optical interconnects for data communication. For optical interconnect applications, the concept of the so-called "smart pixel" has been developed in particular [29]. The goal is to implement a large number of densely packed parallel data channels. This is achieved with an optical imaging system which projects an image of a 2D source array onto a 2D detector array. For systems which integrate some functionality, such as routing systems or photonic switches, the optically transmitted data channels need to be controlled individually and interconnected logically by switching and logic devices. In spite of considerable research progress in the field of nonlinear optics, pure optical switches will not be available for this purpose in the near future. Currently this control function is performed electronically through optoelectronic devices or device arrays. To this end, the detectors are integrated with electronic logic circuits of varying complexity. In order to be able to cascade the device planes, they also contain optical outputs (i.e., sources or optically read modulators). They can be used to transmit the signal optically to the next device plane. Since the individual devices combine optical input/output (I/O) with some functionality in the form of electronic logic circuits they are often referred to as "smart pixels". Fig. 8.14 shows a schematic of a smart pixel array (SPA). The figure shows optical I/O windows surrounded by electronic driver circuits and logic gates. The complexity of the logic operation performed in each pixel is often called the "smartness" of the pixel. With increasing functionality, the area of the electronic circuitry per pixel increases. In this case the distance between the optical windows needs to increase in order to provide enough area for the electronics. The grid of optical windows becomes more dilute. Thus. the term "granularity" is often used synonymously to refer to the smartness of the pixels. There is an ongoing discussion as to how smart the pixels of optoelectronic interconnection systems need to be in order to yield the best performance.

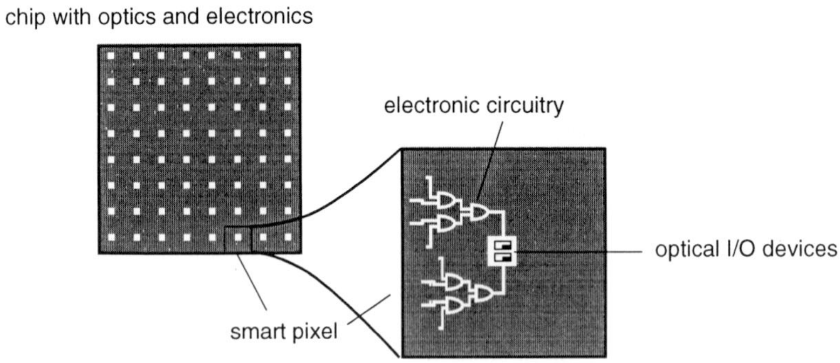

Fig. 8.14: Schematic of a smart pixel array consisting of individual devices with optical I/O windows and local electronic circuitry ("smartness").

An optoelectronic system as described above, which uses free-space optics to interconnect SPAs, can be understood as an electronic system in which the electronic interconnects are partially replaced by optical interconnects. The extent to which electrical wires are replaced by optical data channels directly corresponds to the functionality of the electronic circuitry of the SPAs. In this sense, the discussion about the ideal smartness or granularity of the SPAs is equivalent to the question of what interconnection distance applies for optical interconnects to prove advantageous over electronic interconnects. Obviously this discussion is greatly influenced by future technological developments. Optoelectronic device technology as well as integrated free-space optics will make further progress and trigger the breakthrough of optical interconnects for constantly decreasing communication distances.

The main problem with the fabrication of smart pixels has its origin in the necessary combination of optoelectronic and electronic functionality. Nowadays most electronic devices and logic circuits are based on CMOS technology in silicon substrates. Efficient light detectors in silicon are also feasible. On the other hand, the fabrication of efficient light sources in silicon-based systems has not yet reached a mature state. The physical reason for these problems with silicon-based light sources can be found in the structure of the electronic bands in silicon. As opposed to GaAs semiconductors, silicon is a so-called "indirect" semiconductor material. This means that the electron transition at the lowest band-gap energy cannot take place without an additional change of the electron momentum. This significantly reduces the transition probability and thus the efficiency of a corresponding light source. In contrast III-V semiconductor materials like GaAs are "direct" semiconductors where the radiating transition can be excited much more efficiently. It is, therefore, well suited for the fabrication of optoelectronics as described above. However, in the field of electronic logic circuits, GaAs based systems are lagging behind silicon CMOS technology with respect to functionality, efficiency and integration. Furthermore, compared to silicon, GaAs is a very expensive material composition. At the current stage of development, several approaches to smart pixel fabrication are possible [30]:

- development of detectors and logic circuits based on III-V semiconductor technology which can be integrated monolithically with the demonstrated optoelectronic sources [10, 31];
- development of detectors and emitters based on indirect band-gap materials such as Si or Ge [32];
- epitaxial growth of GaAs or similar III-V semiconductor materials on Si [33];
- hybrid bonding of optoelectronic sources and detectors based on III-V semiconductors, onto silicon CMOS circuits [34, 35].

All these technological approaches are currently under research. Which of these approaches will prevail, can only be decided by future developments in the field.

8.5 Glossary

- **direct and indirect semiconductors**
  In direct semiconductors the minimum of the electronic conduction band coincides with the maximum in the valence band. Electronic transitions at the band-gap energy can occur without change of electron momentum. Direct semiconductors (e.g., GaAs) can be used efficiently in light sources and detectors. In indirect semiconductors (e.g., Si, Ge) the change in the electronic energy has to be significantly higher than the band-gap energy or the electron needs to change its momentum upon transition. This causes difficulties for the fabrication of efficient light sources. Indirect semiconductors are, however, equally well suited for detectors as direct semiconductors.

- **exciton**
  Excitons are binding states between electrons and holes in a semiconductor. Generally in bulk semiconductor materials the excitonic binding is very weak and destroyed due to the high mobility of electrons and holes. If the mobility of the carriers is reduced, e.g., at low temperatures or in quantum-well materials, excitons can be observed.

- **MQW**
  multiple quantum wells: wells in the electronic energy bands in semiconductor materials which result from the stacking of multiple layers of materials with different band gaps. In these wells the electrons and holes are confined laterally leading to quantum effects.

- **SEED**
  self-electro-optic effect device: optoelectronic device which utilizes an electric field induced shift of the excitonic absorption peak in multiple quantum wells for switching, induced by incident optical radiation.

- **smart pixel**
  A smart pixel consists of an optical light source and a detector integrated with some electronic circuitry. The optically transmitted signals are processed by the local electronics in each pixel.

- **superlattice**
  stack of thin (thickness O(nm)) epitaxial layers of semiconductor materials which differ in doping or material composition. Due to the different band-gap energies the electronic bands are modified to form multiple quantum wells.

- **VCSEL**
  vertical cavity surface emitting laser: laser diode with the resonator structure oriented perpendicular to the substrate. Electron confinement in multiple quantum wells and high reflectivity multiple layer mirrors result in good quantum efficiencies.

8.6 List of new symbols

$Al_xGa_{1-x}As$	aluminum gallium arsenide, where the aluminum content is determined by x and the remaining lattice sites are occupied by gallium
C	capacitance of a SEED
E_g, E_{eff}, $E_{intrinsic}$	band-gap energy of a material or a multiple quantum well; effective and intrinsic band-gap energy
E_{diss}	energy dissipated, e.g., by a VCSEL diode
ΔE	switching energy of a SEED modulator
$\delta\Phi$	phase shift occurring upon reflection at a higher refractive index medium
I_{photo}, I_{th}	photo- and threshold current of an optoelectronic device
J_{th}	threshold current density
L	spacer thickness of a MQW laser resonator corresponding to the length of the resonator cavity
L_{eff}	effective length of a Fabry-Perot cavity
λ_{fsr}	free spectral range of a laser cavity
N	number of periods in a layer stack
n_H, n_L	high and low refractive indices of the materials in a layer stack
ν_{fsr}	free spectral frequency range of a laser cavity
ϱ_{stack}	reflectivity of the multilayer stack
V, V_{bias}	voltage applied to a SEED

8.7 Exercises

1. **Heat dissipation in an array of VCSELs**
   We assume a 2D array of 16x16 VCSELs with a power conversion efficiency of 15%. The VCSELs are arranged on a 250 μm grid and provide an output optical power of 1 mW each. Calculate the amout of heat dissipated per VCSEL. How high is the total heat dissipated by the array? Using air cooling it is possible to dissipate $\approx 0.1 \frac{\mathrm{W}}{\mathrm{cm}^2}$. Is this enough to keep the array at a constant temperature?

2. **Modulator vs active devices**
   SEEDs are modulator devices. What is a modulator device and how does the use of modulators rather than light emitting devices influence the optical system?

3. **Doping and hetero-superlattices**
   What is the difference between doping and hetero-superlattices?

4. **VCSELs vs edge-emitting laser diodes**
   What are the advantages of VCSELs as compared to edge-emitting laser diodes for applications in free-space microoptics?

References

[1] H. Ibach and H. Lüth, "Festkörperphysik", Springer-Verlag, Berlin (1988).

[2] D. A. B. Miller, "Quantum wells for optical information processing", *Opt. Eng.* **26** (1987), 368–372.

[3] G. Döhler, "Solid state superlattices", *Scientific American* **49** (1983), 118–126.

[4] D. A. B. Miller, D. S. Chemla and S. Schmitt-Rink, "Relation between electroabsorption in bulk semiconductors and in quantum wells: the quantum-confined Franz-Keldysh effect", *Physical Review B* **33** (1986), 6976–6982.

[5] G. H. Döhler, "Doping-Superlattices — Historical Overview", in Properties of III-V superlattices and quantum wells, P. Bhattacharaya (ed.), Data review series 15 EMIS, UK (1996), 11–25.

[6] D. A. B. Miller, D. S. Chemla, T. C. Damen, T. H. Wood, C. A. Burrus, A. C. Gossard and W. Wiegmann, "The quantum well self-electro-optic effect device: optoelectronic bistability and oscillation, and self-linearized modulation", *IEEE J. of Quant. Electr.* **QE-21** (1985), 1462–1476.

[7] R. A. Morgan, "Improvements in self-electro-optic effect devices: towards system implementation", *Proc. SPIE* **1562**, Devices For Optical Processing, San Diego (1991), 213–226.

[8] D. A. B. Miller, "Optics for low-energy communication inside digital processors: quantum detectors, sources and modulators as efficient impedance converters", *Opt. Lett.* **14** (1989), 146–148.

[9] A. L. Lentine and D. A. B. Miller, "Evolution of the SEED technology: bistable logic gates to optoelectronic smart pixels", *IEEE J. of Quant. Electr.* **29** (1993), 655–667.

[10] D. A. B. Miller, M. D. Feuer, T. Y. Chang, S. C. Shunk, J. E. Henry, D. J. Burrows and D. S. Chemla, "Field-effect transistor self-electro-optic effect device: integrated photodiode quantum well modulator", *IEEE Phot. Techn. Lett.* **1** (1989), 62–64.

[11] K. Iga, "Fundamentals of Laser Optics", Plenum Press, New York (1994).

[12] K.-J. Ebeling, "Integrierte Optoelektronik", Springer Verlag, Berlin (1992).

[13] K.-J. Ebeling, "Physics of semiconductor lasers", in Quantum Optics of confined systems, M. Ducloy and D. Bloch (eds), Kluwer Academic Publisher, Dordrecht, NL (1995), 283–308.

[14] L. A. Coldren and S. W. Corzine, "Diode Lasers and Photonic Integrated Circuits", John Wiley & Sons, New York (1995).

[15] K. Iga, F. Koyama and S. Kinoshita, "Surface emitting semiconductor lasers", *IEEE J. Quant. Electron.* **24** (1988), 1845–1855.

[16] J. L. Jewell, J. P. Harbison, A. Scherer, Y. H. Lee and L. T. Florez, "Vertical-cavity surface-emitting lasers: design, growth, fabrication, characterisation", *IEEE J. Quant Electron.* **27** (1991), 1332–1346.

[17] C. J. Chang-Hasnain, "Vertical cavity surface emitting laser arrays", in Diode Laser Arrays, D. Botez and D. R. Scifres (eds), Cambridge University Press, Cambridge (1994), 368–413.

[18] D. L. Huffaker, L. A. Graham and D. G. Deppe, "Fabrication of high packaging density vertical cavity surface emitting laser arrays using selective oxidation", *IEEE Photon. Techn. Lett.* **8** (1996), 596–598.

[19] R. A. Morgan, L. M. F. Chirovsky, M. W. Focht, G. Guth, M. T. Asom, R. E. Leibenguth, K. C. Robinson, Y. H. Lee and J. L. Jewell, "Progress in planarized vertical cavity surface emitting laser devices and arrays", *Proc. SPIE* **1562**, Devices For Optical Processing, San Diego (1991), 149–158.

[20] D. L. Huffaker, D. G. Deppe, K. Kumar and T. J. Rogers, "Native-oxide defined ring contact for low threshold vertical cavity lasers", *Appl. Phys. Lett.* **65** (1994), 97–99.

[21] B. J. Thibeault, E. R. Hegblom, P. D. Floyd, R. Naone, Y. Akulova and L. A. Coldren, "Reduced optical scattering loss in vertical cavity lasers using a thin (300 Å) oxide aperture", *IEEE Photon. Technol. Lett.* **8** (1996), 593–595.

[22] M. Grabherr, R. Jäger, R. Michalzik, B. Weigl, G. Reiner and K. J. Ebeling, "Efficient single mode oxide confined GaAs VCSELs emitting in the 850 nm wavelength regime", *IEEE Photon. Techn. Lett.* **9** (1997), 1304–1306.

[23] P. D. Floyd, B. J. Thinbeault, E. R. Hegblom, J. Ko, L. A. Coldren and J. L. Merz, "Comparison of optical losses in dielectric-apertures vertical cavity lasers", *IEEE Photon. Technol. Lett.* **8** (1996), 590–592.

[24] H. A. McLeod, "Thin-film optical filters", Adam Hilger Ltd., Bristol (1986).

[25] E. F. Schubert, "Doping of III-V semiconductors", Cambridge University Press, Cambridge, UK (1993).

[26] J. W. Scott, B. J. Thinbeault, D. B. Young, L. A. Coldren and F. H. Peters, "High efficiency submilliamp vertical cavity lasers with intracavity contacts", *IEEE Phot. Technol. Lett.* **6** (1994), 678–680.

[27] J. J. Coleman, "Strained layer quantum well heterostructure laser arrays", in Diode laser arrays, D. Bitez and D. R. Scifres (eds), Cambridge University Press, Cambridge (1994), 336–367.

[28] P. Battacharaya, "Semiconductor optoelectronic devices", Prentice Hall, Englewood Cliffs, NJ, USA (1994).

[29] H. S. Hinton, "Architectural Considerations for Photonic Switching Networks", *IEEE J. on Selected Areas Commun.* **6** (1988), 1209–1226.

[30] F. B. McCormick, "Free Space Optical Interconnection Techniques", in Photonics In Switching II, J. E. Midwinter (ed.), Academic Press, Boston, MA (1993).

[31] O. Sjölund, D. A. Louderbeck, E. R. Hegblom, J. Ko and L. A. Coldren, "Individually optimized bottom-emitting vertical cavity lasers and bottom-illuminated resonant photodetectors sharing the same epitaxial structure", *EOS Top. Meet. Digest* , **3490**, Optics in Computing, Brugge, BE (1998), 6–9.

[32] S. L. Zhang, Y. Chen, L. Jia, J. J. Li, F. M. Huang, T. Zhu, X. Wang, Z. F. Liu, S. M. Cai, S. P. Wong and I. H. Wilson, "Multiple Source Quantum Well Model of Porous Silicon Light Emission", *J. Electrochem. Soc.* **143** (1996), 1394–1398.

[33] K. W. Goossen, G. D. Boyd, J. E. Cunningham, W. Y. Jan, D. A. B. Miller, D. S. Chemla and R. M. Lum, "GaAs-AlGaAs multiquantum well reflection modulators grown on GaAs and Silicon Substrates", *IEEE Photon. Techn. Lett.* **1** (1989), 304–306.

[34] K. W. Goossen, A. L. Lentine, J. A. Walker, L. A. D'Asaro, S. P. Hui, B. Tseng, R. Leibenguth, D. Kossives, D. Dahringer, L. M. F. Chirovsky and D. A. B. Miller, "Demonstration of a dense, high-speed optoelectronic technology integrated with silicon CMOS via flip-chip bonding and substrate removal", *OSA Techn. Digest* ,**10**, Optical Computing, Salt Lake City, UT, USA (1995), 142–144.

[35] T. J. Drabik, "Optoelectronic integrated systems based on free-space interconnects with an arbitrary degree of space variance", *IEEE Proc.* **82** (1994), 1595–1622.

9 Array illuminators

Free-space optical systems are used in a number of applications. A particular case which has been investigated a great deal is for use as an interconnection technology in VLSI computing [1]. In that case, it is necessary to construct compact and often unusual optical arrangements. The high degree of integration in electronics makes the use of miniaturized optics mandatory and has motivated in part the ongoing research on the integration of optical systems, using both waveguide and free-space optics (Chapters 6 and 7). The interest in free-space optics for interconnects is based largely on its potential to serve as a 3D interconnection technology. This means that, as opposed to electrical and waveguide optical interconnections, free-space propagation enables one to transmit data via the third dimension. In addition, free-space optics provides a large space-bandwidth product. This means that, it allows one to implement a large number of data channels (several thousand or more), which is referred to as the "parallelism" of optics. As input/output devices to the electronics, arrays of smart pixels are of interest (see previous chapter). The specific task of (micro)optical imaging between smart pixel arrays has been considered already in Chapter 7. Another topic related to the issue of optical interconnections is "array illumination." An array illuminator is an optical (sub-)system that splits an incoming wave into a 1D or 2D array of equal intensity beams. Array illumination can, therefore, be viewed as a specific aspect of the more general problem of multiple beam-splitting. Multiple beam-splitting is not restricted to optical interconnections, but has applications in laser technology (for example, for the coupling of many laser resonators) and for illumination purposes (for example, in sensors).

There is an ongoing discussion about the best ways to generate a uniform array of optical beams. The use of active 2D arrays of individually addressed laser diodes, e.g., VCSELs, gives the largest flexibility (Chapter 8). In some cases, however, it is advantageous to use passive optical components which split a single high power beam into an array of equal intensity beamlets. Such a passive array illuminating system has advantages, e.g., with respect to cost and wavelength uniformity of the individual beamlets [2]. For VCSEL arrays, the problem of local heating due to the power dissipated needs to be addressed with specific cooling concepts. If all the power is dissipated at one central source, cooling can be significantly easier.

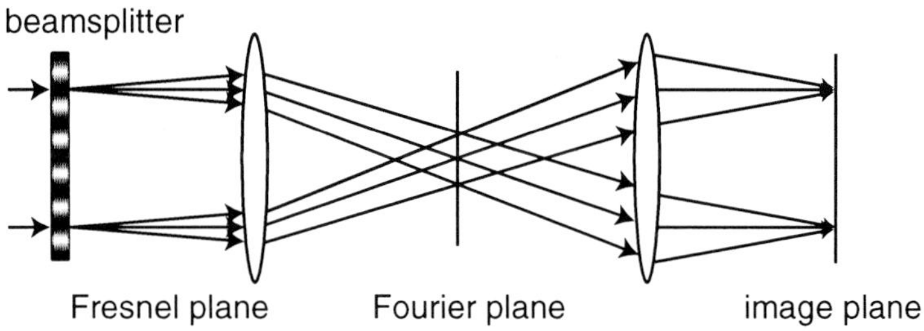

Fig. 9.1: Classification of array illuminators according to the location of the generated beam or spot array.

In some situations it can also be advantageous to combine passive beam-splitting elements with source arrays containing a small number of sources [3].

In order to categorize the large variety of approaches for array illumination, it is possible to use the location of the generated array of beamlets relative to the passive beam-splitting component. Thus, we find components or systems which can be used to illuminate device arrays located in the Fresnel-, Fourier-, or the image plane (Fig. 9.1). The sections of this chapter reflect this categorization.

However, before starting our discussion of the various kinds of beam-splitting elements we want to determine the critical parameters of passive beam-splitting components in array illuminating systems (Fig. 9.2).

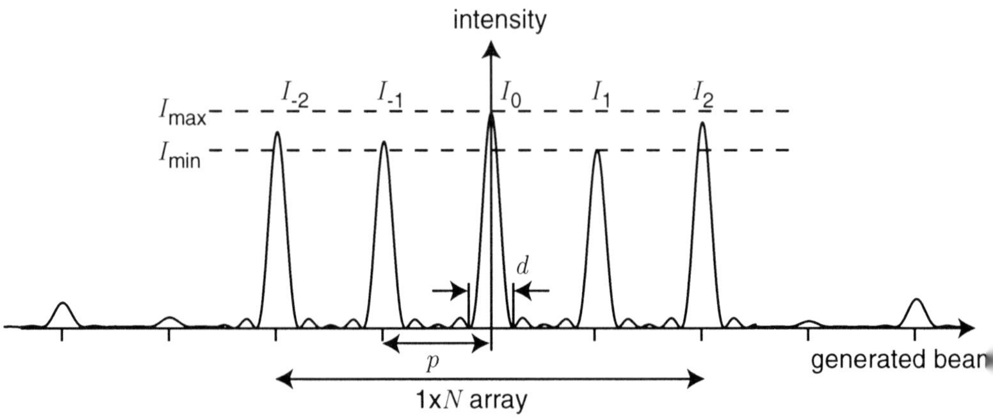

Fig. 9.2: Output plane of an array illuminator.

The most important features of array illuminators (AILs) are:

- *high efficiency* η*:* For obvious reasons it is desirable that as little light as possible is lost during the splitting process. The efficiency of an array illuminator is calculated from (Fig. 9.2):

$$\eta = \frac{\sum\limits_{\text{array}} I_n}{\sum\limits_{\text{all}} I_n} \tag{9.0.1}$$

- *uniformity*: It is important that all beamlets contain equal amounts of light intensity. The uniformity error ΔI_{array} is defined as:

$$\Delta I_{\text{array}} = \frac{I_{\max} - I_{\min}}{I_{\text{av}}} \quad \text{with} \quad I_{\text{av}} = \frac{\sum\limits_{\text{array}} I_n}{N} \tag{9.0.2}$$

- *splitting ratio*: The number N of beams into which the incident wavefront is split. The splitting ratio should correspond to the number of devices to be addressed.

- *compression ratio* $\mathcal{C}$: The ratio between the area covered by the beamlets and the dark areas between the beams. The compression ratio is given by the ratio of the device pitch and the size of the optical window:

$$\mathcal{C} = \frac{d}{p} \quad \text{or in 2D:} \quad \mathcal{C} = \frac{d^2}{p^2} \tag{9.0.3}$$

- *contrast* κ: provided for the optical beam intensities:

$$\kappa = \frac{\overline{I}_{\text{array}} - \overline{I}_{\text{noise}}}{\overline{I}_{\text{array}} + \overline{I}_{\text{noise}}} \tag{9.0.4}$$

9.1 Image plane array illumination

The simplest way to generate an array of equally bright spots is to use an absorbing mask with an array of holes which is illuminated homogeneously. This mask can be imaged onto the device array. Of course, due to the absorption in most of the mask area this is a very inefficient approach to array illumination. For a more efficient AIL it is desirable to use non-absorbing phase-only structures which need to be converted into the intensity distribution. This is possible using, e.g., "phase-contrast imaging" or "aperture division".

9.1.1 Phase-contrast array illumination

Phase-contrast imaging is a very widespread method for the imaging of weak phase objects. Conversion of the phase distribution into an amplitude distribution is achieved by spatial filtering [4]. To perform phase-contrast imaging the spatial filter consists of a phase platelet shifting the phase of the zeroth order (Fig. 9.3).

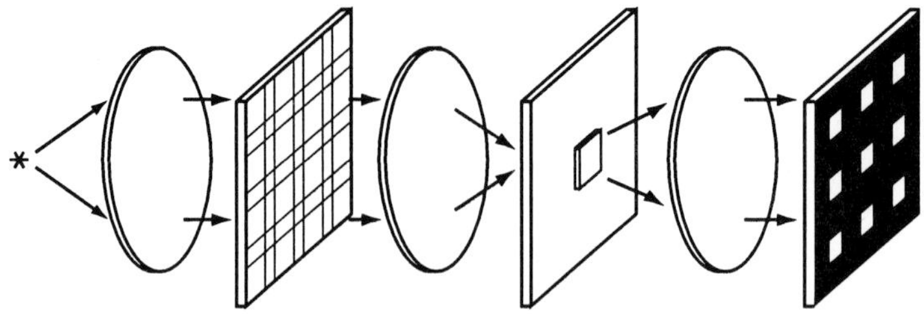

Fig. 9.3: Array illumination by phase-contrast imaging of a phase grating: the basic configuration.

If we assume a phase object with relatively low phase modulation $\Phi(x,y)$, the object transmittance is written as [5]:

$$t(x,y) = e^{i\Phi(x,y)} \tag{9.1.1}$$

This transmittance consists of the strongly dominating bias term and the relatively weak modulated term. Thus, $t(x,y)$ can be approximated as:

$$t(x,y) \approx 1 + i\Phi(x,y) \tag{9.1.2}$$

If a conventional image of such an object is recorded, the intensity $I(x,y)$ is distributed quasi uniformly over the image plane:

$$I(x,y) = I = |t(x,y)|^2 \approx |1 + i\Phi(x,y)|^2 = 1 - \Phi^2(x,y) \approx 1 \tag{9.1.3}$$

However, through spatial filtering in phase-contrast imaging, the zeroth order (i.e., the dc-bias) in the Fourier plane is shifted in phase (e.g., by $\frac{\pi}{2}$) so that the recorded image is expressed as:

$$I_{\text{pc}} = |e^{i\frac{pi}{2}} + i\Phi(x,y)|^2 = |i(1 + \Phi(x,y)|^2 \approx 1 + 2\Phi(x,y) \tag{9.1.4}$$

Thus, with the help of the phase shifting plate in the Fourier plane, we convert the phase object with negligible optical contrast into an object with contrast which is linearly dependent on the object phase. Phase contrast imaging can be illustrated by drawing the vectors of the phase object in the complex plane (Fig. 9.4). The complex amplitudes of the phase object can be represented by two vectors in a

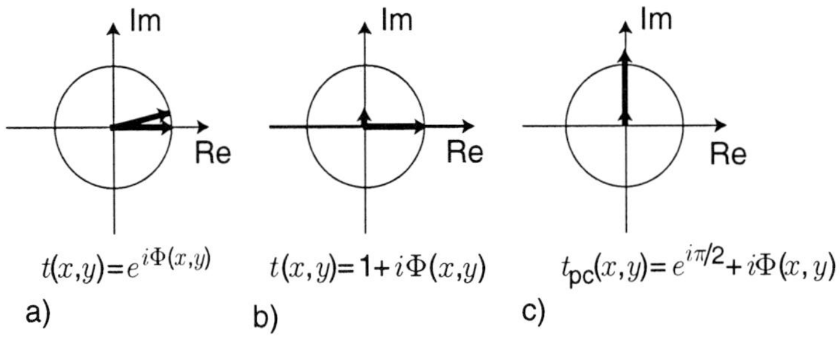

Fig. 9.4: Principle of phase-contrast imaging: a) graphical representation of a weak phase object; b) approximation using Eq. 9.1.2; c) after spatial filtering with a phase filter (phase shift of the dc-component).

unit circle. Fig. 9.4a shows the weak phase object and Fig. 9.4b the approximation according to Eq. 9.1.2. Calculation of the different intensity values for weak phase modulation hardly yields any contrast for small phase variations. However, if the zeroth order (i.e., the mean value of the complex amplitudes) is shifted by $\Phi_0 = \frac{\pi}{2}$, the situation in the complex plane looks like Fig. 9.4c. Calculating the intensity value for the different phases, we find that the contrast in the image is now proportional to the phase in the object (Eq. 9.1.4).

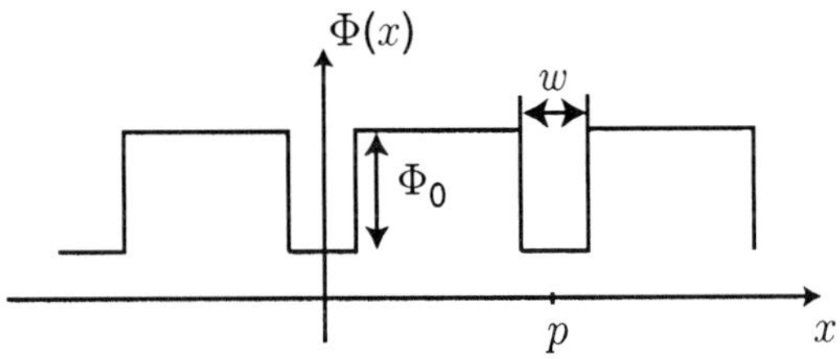

Fig. 9.5: Phase grating used for the phase-contrast array illumination.

For the implementation of array illuminators this phase-contrast method can be generalized to binary phase objects [6]. A suitable phase element for this purpose is, e.g., a phase grating with duty cycle $\mathcal{D} = w/p$ (w: width of the phase shifting region; p: period of the grating) and phase depth Φ_0 (Fig. 9.5). Such phase gratings can be converted into amplitude elements using the phase-contrast method. For illustration we again observe the two different states of the light behind the grating in the complex plane. Fig. 9.6a shows the two states of the complex amplitude of the light wave $U(x)$ behind the binary phase grating. In order to demonstrate the phase-contrast effect it is useful to separate the complex wavefront into a constant dc-part $\bar{U}$ (corresponding to the zeroth diffraction order) and a variational part

$\Delta U(x)$:

$$\begin{aligned} U(x) &= e^{i\Phi(x)} \quad ; \quad \Phi(x) = \begin{cases} 0 \\ \Phi_0 \end{cases} \\ \bar{U} &= 1 \cdot \mathcal{D} + (1-\mathcal{D}) e^{i\Phi_0} \\ \Delta U(x) &= \begin{cases} U_1 = 1 - \bar{U} \\ U_2 = e^{i\Phi_0} - \bar{U} \end{cases} \end{aligned} \tag{9.1.5}$$

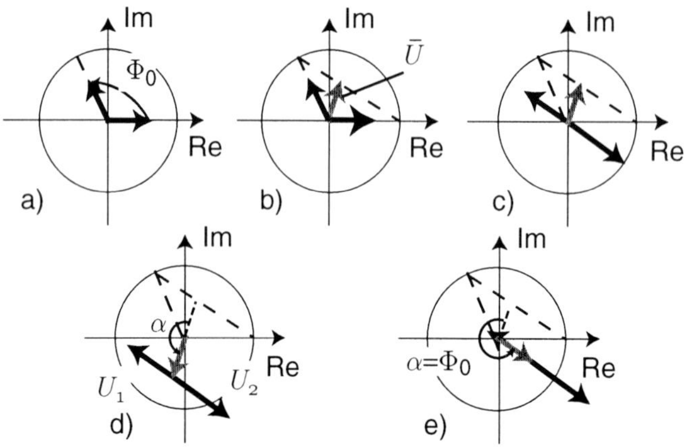

Fig. 9.6: Phase-contrast array illumination. The conversion of a phase grating into an amplitude grating with maximum contrast (see text for explanation).

As illustrated in Fig. 9.6b the vector representing the dc-term is located on the straight line between 1 and $e^{i\Phi_0}$. The two states U_1 and U_2 of the variation $\Delta U(x)$ are shown in Fig. 9.6c. They can be calculated as:

$$\begin{aligned} U_1 &= (1-\mathcal{D})(1-e^{i\Phi_0}) \\ U_2 &= -\mathcal{D}(1-e^{i\Phi_0}) \end{aligned} \tag{9.1.6}$$

Since the duty cycle of the grating is always positive and smaller than unity, the arguments of U_1 and U_2 always differ by π. For phase-contrast imaging a phase shift is applied to the zeroth diffraction order of the grating in the Fourier plane. The zeroth order corresponds to the dc-term $\bar{U}$. Consequently, the effect of the phase shift in the filter plane is illustrated in Fig. 9.6 as a rotation of the vector representing $\bar{U}$ around the origin [6]. Depending on the amount of phase shift applied to the zeroth order, different contrast values between the two states of the wavefront can be achieved. For a complete transfer of energy from the area $\frac{w}{2} < x < p - \frac{w}{2}$ to the area $\frac{w}{2} \leq x \leq -\frac{w}{2}$, the light in the former area is to be eliminated completely. This is only possible if U_2 is equal in modulus with $\bar{U}$ and differs in phase by π. These considerations lead to the relations:

$$\begin{aligned} |\bar{U}| &= \sqrt{\mathcal{D}^2 + 2\mathcal{D}(1-\mathcal{D})\cos(\Phi_0) + (1-\mathcal{D})^2} \\ &= |U_2| = |e^{i\Phi_0} - \bar{U}| = 2\mathcal{D}\sin\left(\frac{\Phi_0}{2}\right) \end{aligned} \qquad (9.1.7)$$

$$\arg[\bar{U}] = \arctan\left(\frac{(1-\mathcal{D})\sin(\Phi_0)}{\mathcal{D} + (1-\mathcal{D})\cos(\Phi_0)}\right) + \alpha = \arg[U_2] + \pi \qquad (9.1.8)$$

where α is the phase angle introduced by the spatial filter. From Eq. 9.1.7 (using the relation: $\sin(\frac{\Phi}{2}) = \sqrt{\frac{1-\cos(\Phi)}{2}}$) we find a relationship between the duty cycle and the phase depth (Φ_0) of the grating:

$$\cos(\Phi_0) = 1 - \frac{1}{2\mathcal{D}} \qquad (9.1.9)$$

Since the cosine function is limited to the range [-1;1] we conclude that maximum contrast is only possible for $\frac{1}{4} \leq \mathcal{D} \leq 1$.

In order to calculate the necessary phase shift in the filter plane we solve Eq. 9.1.8 and obtain:

$$\alpha = -\Phi_0 \qquad (9.1.10)$$

The conditions in Eqs 9.1.9 and 9.1.10 are derived for the case of 1D phase gratings. A generalization to the 2D case can be achieved by replacing $\mathcal{D}$ by a 2D duty cycle $\mathcal{D}^2$, corresponding to the ratio of phase shifted area to non phase shifted area in one period of a 2D phase grating. In this case, the condition in Eq. 9.1.9 leads to a minimum duty cycle $\mathcal{D}^2 \geq \frac{1}{4}$. This condition applies if a phase shifting platelet is used in the filter plane which only shifts the (0,0) order of the 2D grating (Fig. 9.7a). Smaller duty cycles can be used in phase-contrast systems, if the grating patterns are separable. In this case it is possible to shift the zeroth orders in the two perpendicular directions separately. The complete phase filter consists of an overlay of two phase strips [15]. This results in a more complex three phase level spatial filter (Fig. 9.7b). The condition for the duty cycle of the separable grating is:

$$\mathcal{D}^2 \geq \frac{1}{16} \qquad (9.1.11)$$

Since for phase-contrast imaging the size of the illuminated area is determined by the extension of the phase shifted area in the grating period, the ratio $\frac{1}{\mathcal{D}^2}$ can be identified as the compression ratio of the array illuminator. In the previous discussion we have seen that the compression ratio of phase contrast array illuminators is limited to 16 if the light intensity in the non-illuminated area is suppressed theoretically to zero. By alleviating this condition, i.e., accepting a reduction of

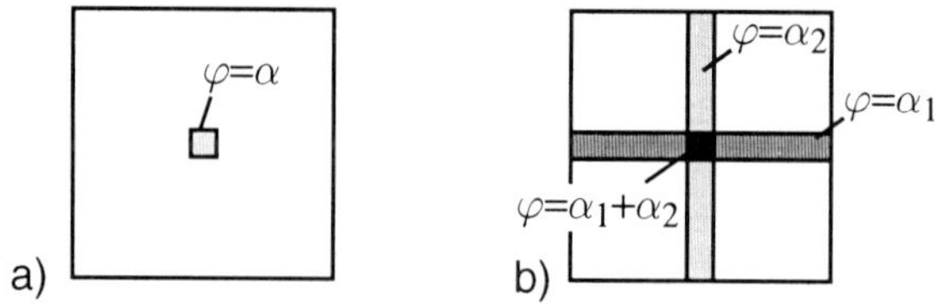

Fig. 9.7: Illustration of the phase filters for 2D phase-contrast array illuminators. a) Phase shifting of only the (0,0) order allows compression ratios of about 4; b) a three or four phase level filter allows compression ratios up to 16.

the efficiency of the array illuminator, the phase-contrast method has been demonstrated using gratings with duty cycles of $\mathcal{D}^2 = \frac{1}{81}$ corresponding to compression ratios of 81.

9.1.2 Multiple beam-splitting through aperture division

A variety of multiple beam-splitters are based on aperture division. Most of them use microlens arrays fabricated as HOEs, thin diffractive elements, or refractive microlenses [8, 9, 10]. Such microlens arrays may be used simply for focusing an incident plane wave into an array of focus spots. Alternatively, micro-telescope arrays of different geometries can be fabricated. Arrays of microprisms have also been suggested for the beam compression through anamorphotic imaging in a Brewster telescope arrangement [11]. Fig. 9.8 shows a variety of different configurations for performing multiple beam-splitting through aperture division.

Common to all image plane beam-splitters mentioned so far is that the uniformity of the intensities of the resulting beams is influenced by the incident beam. A Gaussian intensity profile of the quasi collimated incident beam can often be assumed. The intensity of the individual beams can be determined by integration of the light amplitude over the specific aperture area which is covered by the array components. According to such a calculation it is also possible to adjust the diameter of the array components in order to obtain equal energy content in all beamlets. Alternatively, beam-shaping components which generate a quasi-rectangular (i.e., a so-called super-Gaussian profile) can be used to adjust the shape of the incident beam (see following chapter).

9.1.3 Multiple beam-splitting through waveguide coupling

Another kind of multiple beam-splitter has been suggested by T. Kubota and M. Takeda [12]. They used a sequence of coupling gratings in a waveguide to convert portions of the guided mode into radiating modes sequentially. This results in an array of beams which are released from the waveguide as free-space beamlets.

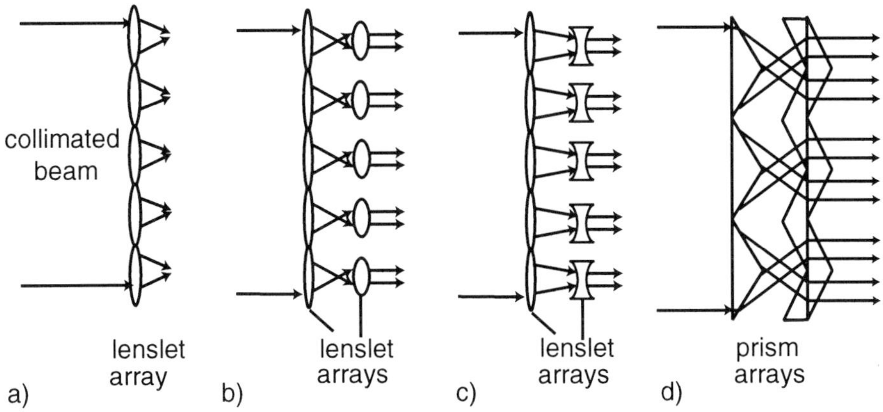

Fig. 9.8: Various configurations for multiple beam-splitting through aperture division: a) focusing through a lenslet array; b) Kepler telescope array using two arrays of positive lenslets; c) Galilean telescope using arrays of positive and negative lenses; d) anamorphotic Brewster telescopes using prism arrays.

The beam separation is determined by the thickness of the waveguide structure as well as the propagation angle of the guided mode. In order to achieve good beam uniformity it is necessary to adjust the coupling efficiency of the grating couplers to the decreasing beam intensity. An increasing relative coupling efficiency is necessary in order to achieve similar absolute intensity values in the radiated beams. 2D arrays of beamlets have been generated by two waveguide layers with coupling gratings oriented perpendicular to each other. Alternatively, a 1D waveguide with a 2D array of coupling grating patches can be used.

9.2 Fresnel plane array illuminators

For array illumination in the Fresnel plane it is possible to exploit self-imaging properties of a periodic wavefield. The British scientist H. F. Talbot observed that laterally periodic wavefields also exhibit periodicity in the z-direction of propagation [13, 14]. This results in the so-called Talbot self-imaging of periodic structures. When, e.g., a periodic grating (period p) located at $z = 0$ is illuminated by a plane wave (wavelength λ), an "image" of this grating can be observed at a distance $z_T = \frac{2p^2}{\lambda}$ without any lenses. Here z_T is called the Talbot distance. It is important to note that Talbot "self-imaging" is different from the image formation through lenses. The self-imaging of the periodic wavefield occurs in the planes where the different diffraction orders behind the grating interfere with phase differences of multiples of 2π. This happens periodically at distances which are integer multiples of the Talbot distance along the optic axis. However, for large distances from the grating, a decay of the structures at the edge of the grating can be observed due to the so-called "walk-off" effect. More and more of the diffracted beams at the edges leave the image field and are lost for formation of the self-image.

According to Lohmann [15], the Talbot effect can be illustrated by a phasor description of the wavefield during the propagation. As described above in the context of phase-contrast imaging, a complex wavefield can be described by the vectors representing the individual components in the complex plane. The propagation of this wavefield along the z-axis effectively results in a rotation of the mean value of the components around the origin (Fig. 9.9b). The orientation of the individual components relative to the mean value remains constant during propagation. Propagation of one Talbot distance corresponds to a full 2π circle around the origin, thus reconstructing the original grating.

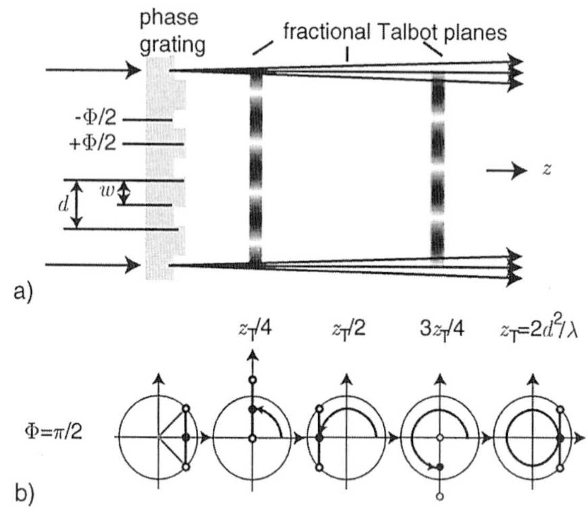

Fig. 9.9: Principle of Talbot array illumination: a) a phase Ronchi grating is converted into amplitude gratings at fractional Talbot distances; b) phasor representation of the individual components of the wavefield.

Fig. 9.9 illustrates the principle of Talbot array illumination. A phase grating illuminated by a plane wave is converted into a periodic amplitude distribution at fractions of the Talbot distance. In the phasor representation (Fig. 9.9b), the wavefield is divided into the mean value and the individual components. By a rotation of the mean value around the origin during the propagation, the grating is converted periodically into an amplitude and a phase grating. In order to illuminate an array of optoelectronic devices efficiently, the array is positioned in a plane with amplitude distribution. Since no absorption occurs, the energy in this plane is compressed to the illuminated areas. This corresponds precisely to the desired performance of an array illuminator. Higher compression ratios are possible using gratings with different duty cycle. In this case, for best performance the phase depth of the grating has to be adjusted as well. The described graphical approach yields two solutions for the Talbot AIL with optimum contrast. A Ronchi phase grating ($\mathcal{D} = \frac{1}{2}$) with a phase depth of $\Phi_0 = \frac{\pi}{2}$ is converted into an amplitude grating at a distance $z = \frac{z_T}{4}$ (see Fig 9.9). A second solution can be found for a phase grating with duty cycle $\mathcal{D} = \frac{1}{3}$ with phase depth $\Phi_0 = \frac{2\pi}{3}$. This grating is converted into an amplitude structure after a propagation of $z = \frac{z_T}{3}$ [16, 17].

For a mathematical description of Talbot self-imaging we describe the periodic object $g(x)$ by the Fourier series:

$$g(x) = \sum_m G_m e^{2\pi i m \frac{x}{p}} \tag{9.2.1}$$

When illuminated with a plane wave, the complex amplitude distribution immediately behind the grating is described as:

$$u(x, z = 0) = g(x) = \sum_m G_m e^{2\pi i m \frac{x}{p}} \tag{9.2.2}$$

During free-space propagation, the phases of the individual Fourier coefficients G_m are influenced according to:

$$u(x, z) = \sum_m G_m e^{ikz\sqrt{1-\left(\frac{m\lambda}{p}\right)^2}} e^{2\pi i m \frac{x}{p}} \tag{9.2.3}$$

In the quadratic (paraxial) approximation $u(x, z)$ is given by:

$$u(x, z) \approx e^{ikz} \sum_m G_m e^{-i\pi z\lambda\left(\frac{m}{p}\right)^2} e^{2\pi i m \frac{x}{p}} = e^{ikz} \sum_m G'_m(z) e^{2\pi i m \frac{x}{p}} \tag{9.2.4}$$

Eq. 9.2.4 again describes a Fourier series with modified Fourier coefficients:

$$G'_m(z) = G_m e^{-2\pi i m^2 \frac{z}{z_T}} \tag{9.2.5}$$

with $z_T = \frac{2p^2}{\lambda}$. From Eq. 9.2.5 we find that the modified coefficients G'_m are periodic in z:

$$G'_m(z + N z_T) = G'_m(z) \tag{9.2.6}$$

For $z = 0$, Eqs 9.2.5 and 9.2.6 describe the Talbot self-imaging of periodic objects. After a propagation distance of z_T, which is generally called the Talbot distance, the grating is reconstructed. However, the periodicity of the grating is also inherent in the complex light amplitude at smaller distances. The amplitude distributions occurring at these smaller distances are called fractional Talbot images. A simple example is the image at $z = \frac{z_T}{2}$. Here, the modified Fourier coefficients are:

$$G'_m(z) = \frac{z_T}{2} = G_m e^{-\pi i m^2} = G_m(-1)^{m^2} \tag{9.2.7}$$

Transformation of the Fourier coefficients as described in Eq. 9.2.7 leads to a grating which is shifted laterally by half a period p [14]. Thus, we find that after a propagation length of $\frac{z_T}{2}$ the grating is reconstructed except for a lateral shift of half a period. Similar numerical considerations lead to configurations where a periodic phase object is converted into an amplitude distribution at certain fractions of the Talbot distance. For binary phase gratings a significant number of configurations are suitable for array illumination with optimum contrast [16, 18, 19, 20, 21]. However, only a few of these solutions are fundamentally different. The π phase grating with duty cycle $\frac{1}{2}$, e.g., generates a binary amplitude distribution at $\frac{z_T}{4}$ as well as at $\frac{3z_T}{4}$. We expect this from the considerations above. After the additional propagation by $\frac{1}{2}$ of the Talbot length the distribution at $\frac{z_T}{4}$ is self-imaged except for a lateral shift by half a period. Similar relationships can be found for the $\frac{1}{3}$ and $\frac{2}{3}$ gratings which are converted into each other during the propagation. An interesting possibility is observed with the $\frac{1}{2}$ grating which exhibits an amplitude distribution with half the period length at a distance of $\frac{z_T}{16}$ behind the grating. This demonstrates that a spacing smaller than that of the original grating period can be achieved at specific distances behind the grating.

These examples show some possible configurations where optimum contrast can be achieved in the amplitude distribution. The light amplitude is theoretically zero in the dark areas and all the light is concentrated in the bright areas. If this condition is slightly alleviated by allowing some background light, a larger variety of gratings and compression ratios can be achieved. This increased flexibility is traded for a reduction in efficiency and contrast ratio [22].

An even larger variety of spot patterns as well as compression ratios is possible using multiple phase level gratings [17, 23]. Similarly the use of periodic microlens arrays has been considered for array illumination in the Fresnel plane. Instead of using the focal points of the microlenses, the amplitude distribution generated at fractional Talbot distances is exploited [24]. At fractional Talbot distances it is also possible to achieve spot densities higher than the density of the microlenses [25].

9.3 Fourier plane array illuminators

The most widespread technique for array illumination uses specially designed diffractive optical elements (DOEs) and a Fourier setup (Fig. 9.10). The DOE is designed in such a way that the light from the collimated input beam is distributed over M diffraction orders [26, 27]. For array illuminators, an array of equally bright diffraction orders is generated. Other non-symmetrical light distributions over the diffraction orders are also feasible. In order to achieve the highest possible light efficiency the DOE is designed as a pure phase grating. A large variety of design approaches and algorithms has been developed over the years. Each of the design approaches was influenced by the available fabrication technology [28]. One of the earliest approaches was the design of binary phase gratings for multiple imaging applications suggested by Dammann in the early 1970s. At that time the lithographic technology for the fabrication of phase gratings was just about to emerge. This explains the popularity of Dammann's approach although it was originally limited to binary grating profiles. Later on other design approaches became of interest, e.g., iterative Fourier transform algorithms. In combination with suitable quantization schemes these algorithms are better suited for the design of multilevel phase gratings. In this section we will discuss some of the basic issues related to the Dammann design technique as well as the iterative Fourier transform approach.

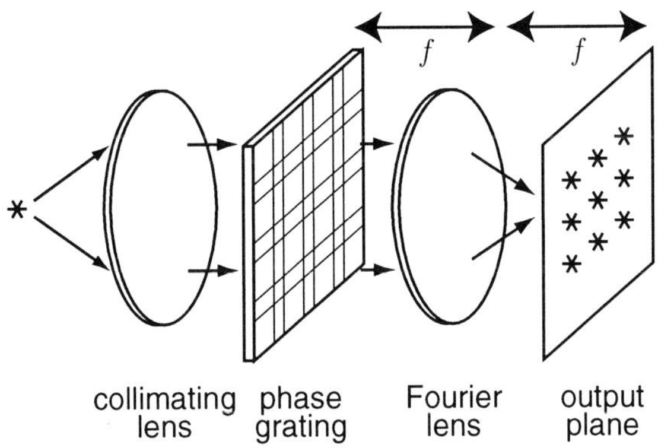

Fig. 9.10: The configuration for array illumination using a Fourier plane AIL.

9.3.1 Dammann gratings

The term Dammann grating is generally used to denote binary phase gratings which are used for multiple beam-splitting [29]. As illustrated in Fig. 9.11a one period of a 2D Dammann grating consists of rectangular patches which introduce a phase shift of π. The only free parameters are the locations of these transition points. At each of these transition points the phase changes from 0 to π and vice versa.

In the simplest case, Dammann gratings are separable in the x- and y-direction. In this case it is sufficient to consider a 1D grating (Fig. 9.12) which is periodic, $g(x) = g(x + p)$ and binary in phase. An additional assumption is the symmetry of the grating:

$$g(x) = g(-x) \tag{9.3.1}$$

In order to simplify the design procedure we concentrate on one half of the grating period. Mathematically the complex transmission function of a unit cell of a Dammann grating with N transition points can be written as [31]:

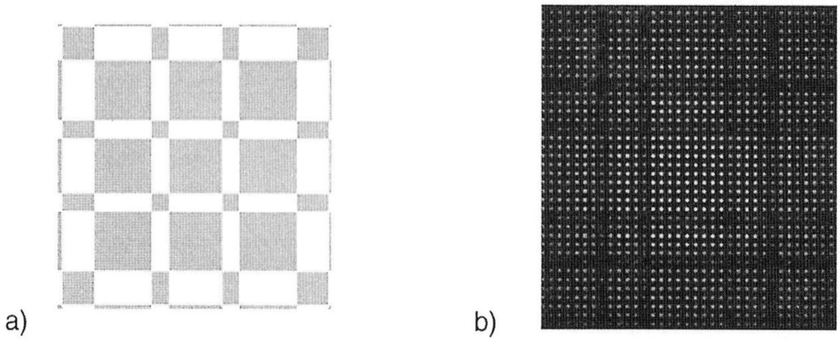

Fig. 9.11: Dammann grating: a) Schematic of one period of a 2D Dammann grating; b) simulation of the diffraction plane of the Dammann grating with a 9×9 array of equal intensity diffraction orders. The pictures were generated with the design and simulation tool DIGIOPT [30].

$$g(x) = \sum_{n=0}^{N} (-1)^n \mathrm{rect}\left[\frac{x - \frac{x_{n+1}+x_n}{2}}{x_{n+1} - x_n}\right] \qquad 0 \leq x < 0.5 \tag{9.3.2}$$

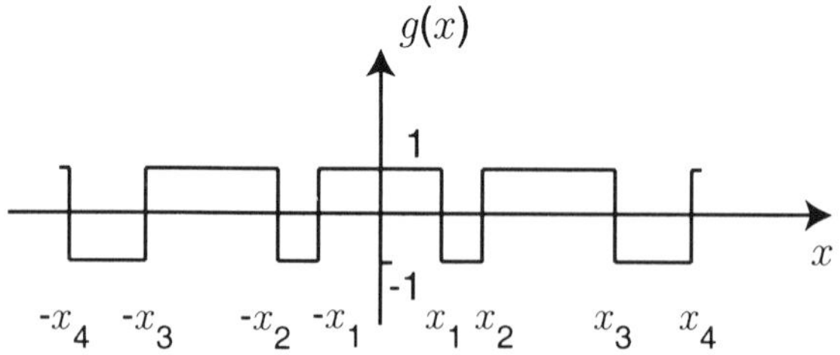

Fig. 9.12: Phase profile of one period of a Dammann grating.

The x_n $(n = 1, \ldots, N)$ in this case are the transition points of the grating. For an infinite extension of the grating, $g(x)$ can be expanded as a Fourier series:

$$g(x) = \sum_{-\infty}^{+\infty} A_m \exp(2\pi i m x) \tag{9.3.3}$$

Considering the symmetry of the grating, real Fourier coefficients A_m result:

$$A_m = 2 \int_0^{0.5} g(x) \cos(2\pi m x) dx \tag{9.3.4}$$

We thus get for the diffraction amplitudes:

$$A_0 = 4\sum_{n=1}^{N}(-1)^{n+1}x_n + (-1)^{N+1}$$
$$A_m = \frac{2}{m\pi}\sum_{n=1}^{N}(-1)^{n+1}\sin(2\pi m x_n) \quad \text{with} \quad m \neq 0 \tag{9.3.5}$$

Due to the symmetry and reality of the grating transmission function the diffraction amplitudes are real-valued ($A_m = A_m^*$) and symmetrical themselves. Furthermore, from Parseval's theorem which states the conservation of energy it follows that:

$$\sum_{-\infty}^{+\infty}|A_m|^2 = \sum_{-\infty}^{+\infty} I_m = 1 \tag{9.3.6}$$

In order to use Dammann gratings for array generation it is necessary to find a set of transition points for which the corresponding diffraction intensities become equal. The solution of the corresponding set of N equations is a complex nonlinear optimization problem for which no analytical solution exists. Various numerical algorithms (e.g., the Newton-Raphson technique, simulated annealing or the method of steepest descent) can be used to find a suitable solution [32]. Since the diffraction amplitudes are real-valued, they are allowed to have positive or negative signs. This leads to a set of 2^N solutions to the problem. The best solution can be found, e.g., as the grating profile with the highest efficiency [33]. A comprehensive comparison of some of the optimization techniques for Dammann gratings as well as multiple phase level gratings has been presented by Mait [32].

One of the crucial aspects for a fast and good convergence of the optimization process is the choice of a suitable cost function. A straightforward possibility is the standard deviation of the diffraction intensities:

$$C_I(x_1,\ldots,x_N) = \frac{1}{2N+1}\sum_{m=-N}^{N}|I_m - \bar{I}|^2 \quad \text{with} \quad \bar{I} = \frac{1}{2N+1}\sum_{m=-N}^{N} I_m \tag{9.3.7}$$

Alternatively the deviation in the diffraction amplitudes is an appropriate cost function C_A [34]:

$$C_A(x_1,\ldots,x_N) = \sum_{m=1}^{N}[A_0 - \text{sign}(A_m)\cdot A_m]^2 \tag{9.3.8}$$

Using C_A as the cost function reduces the number of local minima. Nevertheless, local optima of the cost function are the main obstacle for finding the most efficient solution. There is no direct way for finding this global optimum except by individually comparing the resulting solutions.

Fabrication issues

The most critical issue for the fabrication of Dammann gratings is related to the minimum feature size of the technological process. Just as in Fresnel zone lenses (Chapter 5), in Dammann gratings the information of the diffractive element is contained exclusively in the local distribution of the phase shifting patches. With increasing functionality of the grating performance (i.e., increasing number of optimized diffraction orders) the complexity of the spatial pattern of the grating is increased [35]. As a rule of thumb the number N of transition points per half a grating period determines the number M of diffraction orders which can be optimized:

$$M = 2N + 1 \tag{9.3.9}$$

Most likely, every additional phase transition increases the requirements on the fabrication process. Slight errors in the positioning of the transition points may cause significant errors in the diffraction pattern [31]. It is important to beware of the fact that the minimum feature size which is required for an exact fabrication of the grating cannot be identified with the size of the smallest patch in the grating. The crucial factor is rather the location of the transition points of the grating. If the minimum feature size is a common denominator of these locations, all the transition points fit into the pixelation grid. In this case, all the phase shifting patches can be fabricated ideally without positioning error. If, however, larger pixel sizes are used, then errors are automatically introduced during the fabrication. This is illustrated in Fig. 9.13 for the example of a quasi-continuous phase function. In Fig. 9.13a a function is quantized into a binary phase grating assuming infinite spatial resolution. For this example a different design algorithm has been used to calculate the continuous phase function which performs as a 1×7 beam-splitter. Nevertheless, this situation corresponds to the quantization problems in binary Dammann gratings. In Fig. 9.13b the effect of quantization with a finite resolution of one half of the smallest phase shifting patch of the designed grating is illustrated. Obviously significant positioning errors occur due to the finite resolution.

Etching errors are the second source of imperfections generated during the fabrication of Dammann gratings. As described above the information about the diffraction pattern is completely encoded in the locations of the transition points. Therefore, errors in the etching depth do not fundamentally change the diffraction pattern but mainly cause a reduction of the diffraction efficiency [31]. Compared to the ideal diffraction pattern the intensity of the zeroth order, which collects the non-diffracted light, is higher than originally designed. The same amount of light energy is lost for the higher diffraction orders. The reduction of the diffraction efficiency does not cause too many problems in many cases. The performance of the grating as array illuminator is more severely affected by the uniformity error introduced by the increased zeroth order.

The same effect as observed for errors in the etching depth is caused by a wavelength shift in the reconstructing light. The phase depth is matched to precisely

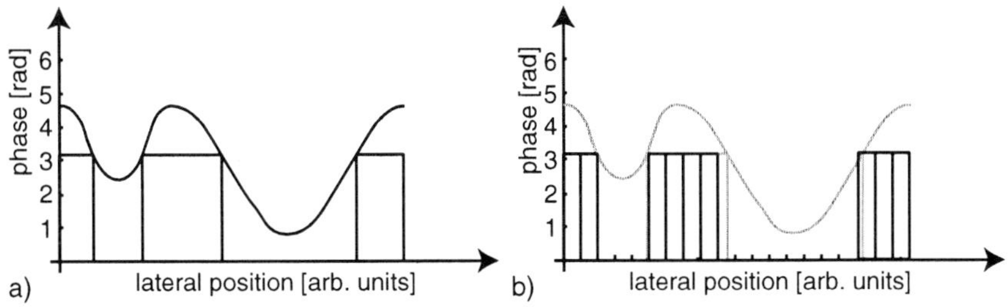

Fig. 9.13: The effect of phase quantization with finite spatial resolution: a) continuous phase function quantized into a binary phase function with infinite resolution; b) quantization with a spatial resolution of one half of the smallest absolute feature in the design.

one design wavelength. For other wavelengths the phase step is no longer identical to π, which has the same consequences as an error occurring during the etching process.

9.3.2 Modifications of Dammann's design procedure

Non-symmetrical and non-binary Dammann gratings

Several modifications of the design procedure of Dammann gratings have been suggested in order to allow more flexibility and improved performance. Killat et al. extended the design to non-symmetrical gratings and allowed phase steps which are different from π for binary implementation [36]. This is possible without a change of the basic optimization process. Their best designs show an increase in efficiency of a few percent. However, this is achieved at the expense of a more complex fabrication process. In the 2D case gratings result which are no longer binary but consist of 3 phase levels. Further adaptations of Dammann's design process are possible which lead to multilevel gratings [32, 37, 38].

Even-numbered symmetrical array generation

From Fourier optics we can find a further modification of the design considerations for Dammann gratings. Due to symmetry considerations the typical Dammann design procedure results in gratings which generate an odd number of optimized diffraction orders (Eq. 9.3.9). These orders include the zeroth order as signal order. Since the zeroth order is very sensitive to errors in the etching depth it is desirable to design beam-splitter gratings which optimize arrays of diffraction orders which do not include the zeroth order. To this end we consider the diffraction plane of a binary (π-)phase grating with a duty cycle of 0.5. Such a grating is generally called a Ronchi phase grating. By a Fourier transformation of the phase profile we can easily show that in the diffraction plane of such a Ronchi phase grating every even diffraction order disappears due to destructive interference. We now consider a Dammann grating overlayed with a Ronchi phase grating such that every second period is phase shifted by π. In the diffraction plane we get the convolution of the original spot array with the diffraction pattern of the Ronchi grating. This convolution is illustrated in Fig. 9.14 where we considered the phase Ronchi grating to consist of mainly 2 diffraction orders, namely the 1st and -1st order. The other orders can be neglected because their intensities are considerably lower. Since the Ronchi phase grating has no zeroth diffraction order, in the overall diffraction pattern every even order is extinguished. The diffraction orders of the combined element result from interference. For equal intensity in these orders, the phases of the interfering beams have to be controlled. This requires additional degrees of freedom in the grating profile (i.e., additional phase transitions in the binary phase grating). Using the same symmetry considerations as for Dammann gratings, the optimization of N diffraction orders in an even numbered array requires N transition points in a binary phase grating. The interesting feature of these gratings is that the zeroth diffraction order does not contribute to the signal orders [39]. Thus, the uniformity of the output is not affected if an error in the etching depth occurs or if the wavelength is detuned slightly from the design wavelength.

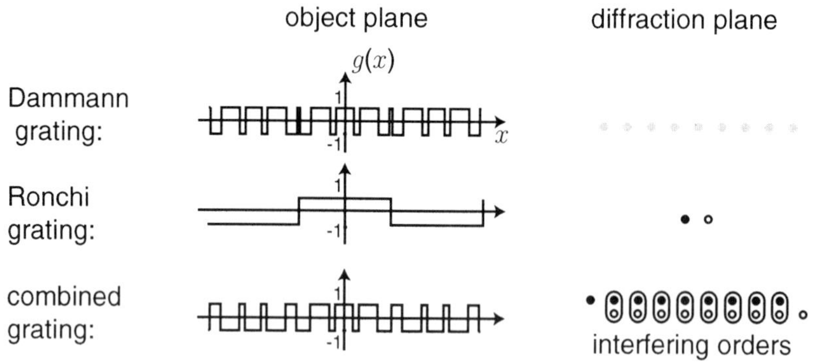

Fig. 9.14: Diffraction plane of the even-numbered array generator resulting from a convolution of the diffraction orders of the Dammann grating with the orders of the Ronchi phase grating.

Generalized array generation with multilevel phase gratings

With the development of lithographic techniques for the fabrication of microoptical elements, multilevel phase gratings became feasible and have become standard elements in the meantime. Although the design process introduced for Dammann gratings was successfully generalized, other numerical design approaches are better suited for such multi-phase level gratings. Nowadays, iterative Fourier transform algorithms are used generally to design phase functions with specific diffraction patterns. The basic design algorithm has been described already in Chapter 3 [40, 41]. This algorithm is not restricted to symmetrical spot arrays or equal intensity diffraction orders. It is also possible to generate diffraction orders with predetermined phases if a sufficient number of free design parameters is available [42].

Different from the Dammann design process, the direct numerical solution of the DOE design problem using the iterative Fourier transform does not require knowledge about the symmetry relations of Fourier optics. Fig. 9.15 shows the design of a multiple beam splitter grating generated with the iterative Fourier transform algorithm. Numerous modifications of the basic process have been suggested in order to improve the convergence and detect the optimum solution. It is beyond the scope of this chapter to discuss all the design procedures for diffractive multilevel elements. The interested reader can be referred to a number of excellent publications on this subject [27, 43, 44].

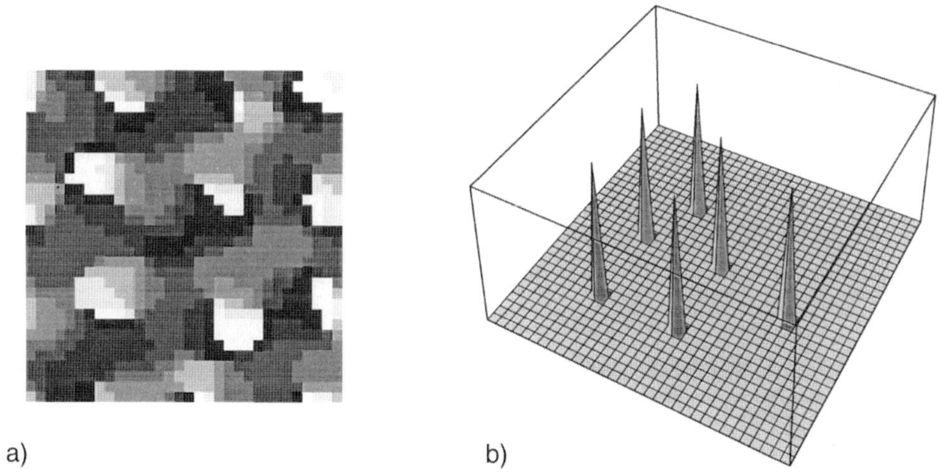

Fig. 9.15: Example of a diffractive filter designed by an iterative Fourier transform algorithm; a) one period of the 8 phase level filter (phase levels are encoded as grey-scales); b) amplitude distribution of the spot array generated in the diffraction plane of the periodic filter.

9.3.3 Lenslet arrays as Fourier plane array illuminators

We previously discussed the use of microlens arrays for array illumination by aperture division or Fresnel diffraction. These approaches suffer from a lack of uniformity in the illuminating beam. Due to the periodicity of microlens arrays it is, however, also possible to use them as diffraction gratings in Fourier plane AILs (Fig. 9.16) [37, 45].

Mathematically, we first want to consider a single diffractive microlens of the array. The transmission function of a diffractive lenslet is expressed by the Fourier

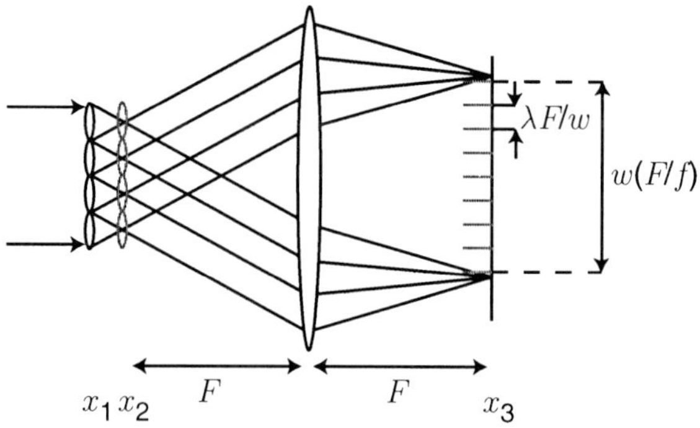

Fig. 9.16: Schematic of the Fourier AIL setup using lenslet arrays.

series:

$$u(x_1) = \text{rect}\left(\frac{x_1}{w}\right)\sum_{k=0}^{\infty} A_k e^{2\pi i k \frac{x_1^2}{R}} \tag{9.3.10}$$

In Eq. 9.3.10, w is the diameter of the lens while R denotes the radius of the first zone of the diffractive lens. In the focal plane of this lens (at $z_f = \frac{R^2}{2\lambda}$) the point spread function of the lens is formed:

$$u(x_2) \propto A_1 e^{\frac{i\pi x_2^2}{\lambda z_f}} \text{sinc}\left(\frac{x_2 w}{\lambda z_f}\right) \tag{9.3.11}$$

It is important to notice that the sinc-function resulting from the rectangular aperture of the microlens is modulated with a phase factor [37]. In the Fourier plane this phase factor forms a phase envelope function to the wide rect-function:

$$u(x_3) \propto a_1 e^{\frac{-i\pi z_f x_3^2}{\lambda F^2}} * \text{rect}\left(\frac{x_3 z_f}{wf}\right) \tag{9.3.12}$$

Due to this envelope phase factor, for an array of microlenses the uniformity of the diffraction orders is disturbed by interference effects. It is, however, possible to compensate for this phase factor by an additional array of field lenses in the plane z_f. We can extend our considerations to a microlens array and obtain for the intensity distribution in the output plane:

$$I(x_3) \propto |A_1|^2 \text{rect}\left(\frac{x_3 f}{wF}\right) \sum_{m=-M/2}^{M/2} \delta\left(x_3 - m\frac{\lambda}{w}\right) \tag{9.3.13}$$

It has been demonstrated that for the configuration described above very large spot arrays can be generated with very good uniformity. The maximum number M of diffraction orders with uniform intensity is determined by the numerical aperture NA of the lenslets [46]:

$$M = \frac{2w}{\lambda}\mathrm{NA} \tag{9.3.14}$$

9.3.4 Cascading of beam-splitter gratings

As we have seen in the previous sections, the splitting ratio of beam-splitter gratings depends on the fabrication complexity. In order to reduce the grating complexity while achieving a sufficient splitting ratio, it is possible to cascade beam-splitting gratings. In order to achieve the desired splitting ratio, we have the choice of using a large number of gratings with small splitting ratio (e.g., 1×2 splitting gratings). The advantage here is that the individual 1×2 beam-splitter can be fabricated very efficiently as high frequency grating [47]. Alternatively, only few large spot array generators can be used to generate very large arrays. The cascading of 4 Dammann gratings generating 1×9 spots each has been demonstrated. This yields a 2D array of 81×81 equal intensity diffraction orders [48].

9.4 Summary

In this chapter we have discussed the most important techniques for array illumination. The choice of which technique is to be used for a certain project depends on the circumstances, especially the available technological equipment. Nevertheless, some general statements can be made about what techniques can be used under certain circumstances.

For image plane array generators we generally find that the size of the spot array only depends on the size of the optical element and the maximum possible diameter of the input beam. In the case of phase-contrast imaging, however, we are restricted as to the variety of compression ratios. In this respect aperture dividing techniques are more flexible.

Fresnel-plane array illuminators are more flexible in the choice of compression ratios, especially if we take into account the use of multilevel phase gratings. Similar to image plane AILs it is also possible to generate extremely large arrays, depending on the size of the optical element. The most intriguing feature of the Fresnel plane AILs is the simplicity of the setup [49]. If homogeneous illumination is available, no additional lenses or other optical elements are necessary behind the AIL to generate the spot array. However, one of the problems of both image plane as well as Fresnel plane AILs stems from the dependence of the uniformity of the spot array on the illumination wave. Ideal uniformity is only possible with an ideal incident plane wave. Several techniques are available for enhancing the uniformity of the illumination, among them, e.g., beam-shaping elements. Alternatively only the central part of the incident Gaussian beam could be used at the expense of efficiency.

grating type	array size; depends on:	uniformity; depends on:	efficiency	compression ratio	beam shape
image plane array illumination					
phase-contrast grating	large; grating size	moderate illumination	< 100%	< 81	collimated
lenslet array	large; array size	moderate; illumination	< 100%	$\propto$ NA	focus
telescopelet arrays	large; array size	moderate; illumination	< 100%	moderate	focus or collimated
Fresnel plane array illumination					
Talbot grating	large; grating size	moderate; illumination	< 100%	4; 9 (binary)	focus
lenslet array	large; > array size	moderate; illumination	< 100%	$\propto$ NA	focus
Fourier plane array illumination					
Dammann grating	moderate; grating complexity	good; fabrication	< 80%	# of grating periods	focus
multilevel phase gratings	moderate; grating complexity	good; fabrication	> 90%	# of grating periods	focus
lenslet array	large; NA	good; field lenslets	< 100%	$\propto$ NA	focus

Tab. 9.1: Summary of the most important aspects of possible configurations for array illumination.

The technique most widely used for applications is the Fourier approach. This might be explained by the flexibility in the spot pattern which can be generated or by the fact that people are used to thinking in terms of Fourier optics. For Fourier-type AILs a poor uniformity in the input beam does not affect the spot uniformity. The shape of the illumination is rather reflected in the shape of the individual beamlets. The limitation of Fourier-AILs is due to the complexity of the diffraction grating. The larger the number of output spots, the larger the complexity of the diffracting grating. The uniformity in the spots is determined mainly by the fabrication precision. It is affected more severely if the complexity of the grating is very high. With highly developed lithographic fabrication, large array sizes with good uniformity have been demonstrated. Some of the basic aspects of array illumination are summarized in Table 9.1.

9.5 Glossary

- **array illuminator**
  optical system which provides an array of light beams, e.g., to illuminate an array of optoelectronic devices.

- **aperture or pupil division**
  splitting of an incident wavefront by placing several deflecting components (i.e., prisms or lenses) next to each other in the pupil of an optical system.

- **compression ratio**
  With an array illuminator, light which is continuously distributed over the input pupil is concentrated in an array of small areas in the output plane. The ratio of the size of the illuminated area to the area covered by the incident beam is called the compression ratio.

- **Gerchberg-Saxton algorithm or iterative Fourier transform algorithm (IFTA)**
  algorithm, e.g., for designing diffractive optical elements. The algorithm alternates between the complex amplitude of the optical component in the spatial domain and its Fourier transform in the frequency domain. In both domains constraints can be applied which determine how the respective plane should appear. For example, for designing a phase filter, in the spatial domain the amplitude of the element is set homogeneously to the value of unity.

- **phase-contrast imaging**
  spatial filtering technique which converts weak phase objects into an amplitude distribution in the image plane.

- **Talbot effect**
  A wavefield which is periodic in the lateral direction also exhibits periodicity in the z-direction. This results in self-imaging of periodic objects at discrete distances z_T ("Talbot distance") along the optic axis.

9.6 List of new symbols

α:	phase angle introduced by the phase filter in the phase-contrast technique
C_I, C_A:	cost functions for the design of DOEs
$\mathcal{C}$:	compression ratio of an AIL
$\mathcal{D}$:	duty cycle of a grating
M:	number of diffraction orders
N:	number of phase transition points in a Dammann grating
$t(x, y)$:	transmittance function of a phase object
$\Phi(x, y)$:	phase of a function at the coordiantes x, y
$U(x)$:	complex wavefront
$\hat{U}$:	mean value of a complex wavefront
$\Delta U(x)$:	variational part of the complex wavefront
$x_1, x_2, \ldots, x_N$:	transition points in a phase grating, where the phase changes, e.g., from 0 to π for a binary grating.
Z:	number of periods of a Fourier array illuminator
z_T:	Talbot distance for self-imaging of a periodic grating through Fresnel propagation

9.7 Exercises

1. **Compression ratios of aperture dividing array illuminators**
   What determines the compression ratios of the aperture-dividing array illuminators illustrated in Fig. 9.8?

2. **Optimization of a Dammann grating**
   Calculate the phase function of a symmetrical Dammann grating with two phase transitions x_1 and x_2. Make a drawing of the phase function of this Dammann grating. How many diffraction orders can be optimized? How many solutions result from an optimization with the cost functions defined in Eq. 9.3.7?

3. **Compression ratio of Fourier array generators**
   What determines the compression ratio of a Dammann grating?

References

[1] J. Jahns, "Free-space optical digital computing and interconnection", in Progress in Optics **XXXVIII** (1998), E. Wolf (ed.), Elsevier, Amsterdam, 419–513.

[2] N. Streibl, "Beam shaping with array generators", *Journal of Mod. Opt.* **36** (1989), 1559–1573.

[3] A. W. Lohmann and S. Sinzinger, "Improved Array Illuminators", *Appl. Opt.* **31** (1992), 5547–5552.

[4] F. Zernike, "Das Phasenkontrastverfahren bei der mikroskopischen Beobachtung", *Z. Techn. Phys.* **16** (1935), 454.

[5] J. W. Goodman, "Introduction to Fourier Optics", McGraw Hill, New York (1968).

[6] A. W. Lohmann, J. Schwider, N. Streibl and J. Thomas, "Array illuminator based on phase contrast", *Appl. Opt.* **27** (1988), 2915–2921.

[7] A. W. Lohmann, W. Lukosz, J. Schwider, N. Streibl and J. A. Thomas, "Array illuminators for the optical computer", *Proc. SPIE* **963**, Optical Computing, Toulon, France (1988), 232–239.

[8] G. Groh, "Multiple imaging by means of point holograms", *Appl. Opt.* **7** (1968), 1643–1644.

[9] A. W. Lohmann and F. Sauer, "Holographic telescope arrays", *Appl. Opt.* **27** (1988), 3003.

[10] A. C. Walker, M. R. Taghizadeh, J. G. H. Mathew, I. Redmond, R. J. Campbell, S. D. Smith, J. Dempsey and G. Lebreton, "Optically bistable thin-film interference devices and holographic techniques for experiments in digital optics", *Opt. Eng.* **27** (1988), 38–44.

[11] A. W. Lohmann, S. Sinzinger and W. Stork, "An array of Brewster telescopes", *Appl. Opt.* **28** (1989), 3835–3836.

[12] T. Kubota and M. Takeda, "Array illumination using grating couplers", *Opt. Lett.* **14** (1989), 651–652.

[13] H. F. Talbot, "Facts related to optical science", *Philosophical Magazine* **9** (1836), 401.

[14] J. T. Winthrop and C. R. Worthington, "Theory of Fresnel images: I. plane periodic objects in monochromatic light", *JOSA* **55** (1965), 373–381.

[15] A. W. Lohmann, "An array illuminator based on the Talbot effect", *Optik* **79** (1988), 41–45.

[16] A. W. Lohmann and J. Thomas, "Making an array illuminator based on the Talbot effect", *Appl. Opt.* **29** (1990), 4337–4340.

[17] J. R. Leger and G. J. Swanson, "Efficient array illuminator using binary-optics phase plates at fractional Talbot planes", *Opt. Lett.* **15** (1990), 288–290.

[18] X.-Y. Da, "Talbot effect and the array illuminators that are based on it", *Appl.Opt.* **31** (1992), 2983–2986.

[19] V. Arrizón and J. Ojeda-Castañeda, "Talbot array illuminators with binary phase gratings", *Opt. Lett.* **18** (1993), 1–3.

[20] V. Arrizón and E. López-Olazagasti, "Binary phase gratings for array generation at 1/16 of Talbot length", *JOSA A* **12** (1995), 801–804.

[21] T. J. Suleski, "Generation of Lohmann images from binary-phase Talbot array illuminators", *Appl. Opt.* **36** (1997), 4686–4691.

[22] V. Arrizón and J. G. Ibarra, "Trading visibility and opening ratio in Talbot arrays", *Opt. Comm.* **112** (1994), 271–277.

[23] V. Arrizón and J. Ojeda-Castañeda, "Multilevel phase gratings for array illuminators", *Appl. Opt.* **33** (1994), 5925–5931.

[24] E. Bonet, P. Andres, J. C. Barreiro and A. Pons, "Self-imaging of a periodic microlens array: versatile array illuminator realization", *Opt. Comm.* **106** (1994), 39.

[25] B. Besold and N. Lindlein, "Fractional Talbot effect for periodic microlens arrays", *Opt. Eng.* **36** (1997), 1099–1105.

[26] W.-H. Lee, "High efficiency multiple beam gratings", *Appl. Opt.* **18** (1979), 2152–2158.

[27] J. N. Mait, "Fourier array generators", in Micro-optics: elements, systems, and applications, H.-P. Herzig (ed.), Taylor & Francis, London, UK (1997), 293–323.

[28] H.-P. Herzig, D. Prongue and R. Dändliker, "Design and fabrication of highly efficient fan-out elements", *Jap. J. Appl. Phys.* **29** (1990), L1307–L1309.

[29] H. Dammann and K. Görtler, "High-efficiency in-line multiple imaging by means of multiple phase holograms", *Opt. Comm.* **3** (1971), 312–315.

[30] H. Aagedahl, T. Beth, H. Schwarzer and S. Teiwes, "Design of paraxial diffractive elements with the CAD system DigiOpt", *Proc. SPIE* **2404**, Diffractive and holographic optics technology, San Jose, CA, USA (1995), 50–58.

[31] J. Jahns, M. M. Downs, N. Streibl, M. E. Prise and S. J. Walker, "Dammann gratings for laser beam-shaping", *Opt. Eng.* **28** (1989), 1267–1275.

[32] J. N. Mait, "Design of binary-phase and multiphase Fourier gratings for array generation", *JOSA A* **7** (1990), 1514–1528.

[33] U. Krackhardt, J. N. Mait and N. Streibl, "Upper bound of the diffraction efficiency of phase-only fanout elements", *Appl. Opt.* **31** (1992), 27–37.

[34] U. Krackhardt and N. Streibl, "Design of Dammann-gratings for array generation", *Opt. Comm.* **74** (1989), 31–36.

[35] H. Dammann and E. Klotz, "Coherent optical generation and inspection of two-dimensional periodic structures", *Opt. Act.* **24** (1977), 505–5015.

[36] U. Killat, G. Rabe and W. Rave, "Binary Phase gratings for star couplers with high splitting ratio", *Fib. Int. Opt.* **4** (1982), 159–1567.

[37] J. Jahns, N. Streibl and S. J. Walker, "Multilevel phase structures for array generation", *Proc. SPIE* **1052**, Holographic Optics: Optically and Computer-Generated, Los Angeles (1989), 198–203.

[38] S. J. Walker and J. Jahns, "Array generation with multilevel phase gratings", *JOSA A* **7** (1990), 1509–1513.

[39] R. L. Morrison, "Symmetries that simplify the design of spot array phase gratings", *JOSA A* **9** (1992), 464–471.

[40] R. W. Gerchberg and W. O. Saxton, "A practical algorithm for the determination of phase from image and diffraction plane figures", *Optik* **35** (1972), 237–246.

[41] J. R. Fienup, "Phase retrieval algorithms: a comparison", *Appl. Opt.* **21** (1982), 2758–2769.

[42] F. Wyrowski, "Design theory of diffraction elements in the paraxial domain", *JOSA A* **10** (1993), 1553–1561.

[43] H. Aagedal, F. Wyrowski and M. Schmid, "Paraxial beam splitting and shaping", in Diffractive optics for industrial and commercial applications, J. Turunen and F. Wyrowski (eds), Akademie Verlag, Berlin, Germany (1997), 165–187.

[44] A. Vasara, M. R. Taghizadeh, J. Turunen, J. Westerholm, E. Noponen, H. Ichikawa, J. M. Miller, T. Jaakkola and S. Kuisma, "Binary surface-relief gratings for array illumination in digital optics", *Appl. Opt.* **31** (1992), 3320–3336.

[45] H. Machida, J. Nitta, A. Seko and H. Kobayashi, "High efficiency fiber grating for producing multiple beams of uniform intensity", *Appl. Opt.* **23** (1984), 330–332.

[46] N. Streibl, U. Nölscher, J. Jahns and S. J. Walker, "Array generation with lenslet arrays", *Appl. Opt.* **30** (1991), 2739–2742.

[47] S. J. Walker and J. Jahns, "Optical Clock Distribution Using Integrated Free-Space Optics", *Opt. Comm.* **90** (1992), 359–371.

[48] F. B. McCormick, "Generation of large spot arrays from a single laser beam by multiple imaging with binary phase gratings", *Opt. Eng.* **28** (1989), 299–304.

[49] A. W. Lohmann, "Array illuminators and complexity theory", *Opt. Comm.* **89** (1992), 167–172.

10 Microoptical components for beam shaping

In this chapter, we shall address the use of microoptics for generalized beam shaping. Multiple beam splitting, as discussed in the previous chapter, can be considered as a special case of beam shaping. However, beam shaping comprises many other cases so that it is appropriate to discuss it in a separate chapter [1]. Table 10.1 is an attempt to classify the different techniques. Before we start to discuss the contents of the table, it should be noted that it is not necessarily complete in the sense that it lists all possibilities one might be able to think of. We shall return to this point later.

We assume that, in general, a laser or laser array is used as the source of radiation. Thus, it is appropriate to distinguish between beam shaping of a single beam and multiple beam shaping. A single beam might be modified in its spatial and temporal shape. Furthermore, spatial beam shaping can be subdivided further into lateral and longitudinal (or axial) beam shaping. This distinction obviously relates to whether the profile is modified in the lateral direction(s) or along the optic axis. The conversion of a beam with a Gaussian amplitude profile to a flat-top profile would be an example for lateral beam shaping. The generation of "non-diffracting" beams using elements such as axicons is an example for axial beam shaping. We include temporal beam shaping in our scheme, since it is accomplished in space after time-to-space conversion. Similar to other beam-shaping techniques, diffractive elements can be used as elements for temporal beam shaping. Multiple beam shaping includes the situation where a single input beam is split into an array of output beams ("multiple beam splitting") and the "inverse" case, where beams from several sources are coupled into a single output beam ("beam combination"). Hence, all the techniques discussed in Chapter 9 for multiple beam splitting are also of interest for multiple beam shaping. So far, techniques known for multiple beam shaping only work on the lateral beam profile. On the other hand, it would be conceivable to think of an application where beams from multiple sources are combined and also modified in their axial or temporal behaviour.

Furthermore, there are reports on a number of techniques where, for example, diffraction gratings are used inside the laser cavity to modify the beam characteristics. Therefore, we distinguish between techniques, where the beam-shaping element is located outside the laser cavity ("extra-cavity beam shaping") or inside ("intra-cavity beam shaping"). Intra-cavity beam shaping techniques have already been used for a long time for the design of laser cavities with high efficiency or increased stability. Only recently with the introduction of diffractive optics in this field has the complex optimization of the lateral mode profile become feasible.

<table>
<tr><th colspan="2">extra-cavity</th></tr>
<tr><th>single beam</th><th>multiple beam</th></tr>
<tr><td>lateral
e.g., Gauss to flat-top

axial
e.g., non-diffracting beams
3D wavefront shaping

temporal
e.g., pulse compression,
time lens</td><td>multiple beam splitting
e.g., array illumination

beam combination
e.g., aperture filling,
superposition</td></tr>
<tr><th colspan="2">intra-cavity</th></tr>
<tr><th>single beam:</th><th>multiple beam:</th></tr>
<tr><td>mode shaping
e.g., mode selection

resonator optimization
e.g., stability,
efficiency</td><td>mode locking
e.g. coherent LD arrays

beam combination
e.g., aperture filling,
superposition</td></tr>
</table>

Tab. 10.1: Summary of approaches and applications for beam shaping.

Before we address specific examples we shall discuss "beam shaping from a general perspective". The main part of the section on "lateral beam shaping" will focus on the "collimation of astigmatic beams", which is an important task due to the emission characteristics of edge-emitting laser diodes. In the subsections on "beam shaping for laser machining" and "beam shaping for LIDAR" we address two further applications which require radiation with a specific lateral beam profile. Subsections on "axial beam shaping" and "temporal beam shaping" conclude the section on single source extra-cavity beam shaping. The most important techniques for "multiple beam shaping" are "beam combination" and multiple beam splitting, which has already been discussed in the previous chapter. Many of the techniques discussed for extra-cavity beam shaping can also be applied to "intra-cavity beam shaping", providing elegant solutions to the beam-shaping problem. This final section of the chapter is further subdivided into subsections on "intra-cavity beam shaping of individual laser beams" and "intra-cavity beam shaping of arrays of laser beams".

10.1 Beam shaping from a general perspective

In general terms, the purpose of a beam-shaping component is to transform an incident wavefront $U_i(x, -0)$ into a desired complex wavefront $U_2(x, z_0)$ at a distance z_0 from the component plane (Fig. 10.1). The straightforward approach towards designing such a beam-shaping component is to calculate the inverse propagation from the output plane to the component at $z = 0$. This is possible, for example, by calculating the inverse Fresnel transform of $U_2(x, z_0)$ for a near-field relationship or the inverse Fourier transform for far-field beam shaping. The complex amplitude $U_C(x)$ of the component is then calculated simply as the difference in amplitude and phase between the incident wavefront $U_i(x, -0)$ and the propagated output wavefront $U_1(x, +0)$. Two problems become obvious from these basic design considerations. Firstly, in general, the beam-shaping component needs to influence both amplitude and phase of the wavefront. This is not possible with the conventional multilevel phase components often discussed as microoptical elements. However, thick holographic optical elements (HOEs), detour-phase computer-generated holograms or specifically blazed holograms are capable of implementing arbitrary complex wavefronts [2]. The second problem affects the efficiency of the beam-shaping component. In principle a uniform beam can be deflected by a thick HOE or a blazed DOE with an efficiency of 100%. However, the efficiency generally is significantly lower for HOEs applied to generalized beam shaping. Although the HOE is a phase-only element, it provides amplitude weighting by variable diffraction of the light into higher diffraction orders. These orders do not contribute to the signal. This is the reason why, for the generation of arbitrary wavefronts, the diffraction efficiency η_{BS} is significantly lower that 100%. According to Horner [3], η_{BS} can be calculated as:

$$\eta_{BS} = \eta_S \cdot \eta_C = \eta_S \cdot \frac{\int |A(x) \otimes A_1(x)|^2 dx}{\int |A(x)|^2 dx} \tag{10.1.1}$$

Here, η_S is the maximum efficiency which can be achieved with the technological implementation of the HOE. $A(x)$ and $A_1(x)$ denote the amplitudes of the incident wavefront $U_i(x, -0)$ and the transformed wavefront $U_1(x, +0)$. $\otimes$ denotes the mathematical correlation between the amplitude distributions:

$$A(x) \otimes A_1(x) = \int A(x')A'(x' - x)dx' \tag{10.1.2}$$

Eq. 10.1.1 states that the efficiency of a beam-shaping element can only be high if the amplitude distribution of the incident and the transformed wavefront are highly correlated. However, this is generally not the case for arbitrary beam-shaping applications.

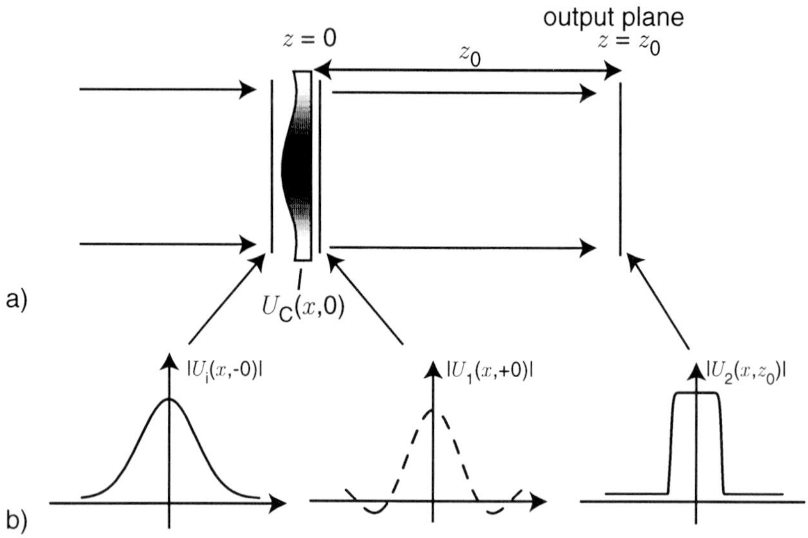

Fig. 10.1: The general situation for arbitrary beam shaping: a) geometry; b) example of the amplitude of the incident wavefront $U_i(x,-0)$ just in front of the beam-shaping element, the amplitude $U_1(x,+0)$ just behind the element, and the desired amplitude $U_2(x,z_0)$ in the observation plane. The beam-shaping element is described by the complex transmission function $U_C(x,0)$.

In order to escape from this dead end, the use of tandem components has been suggested for arbitrary beam shaping [4]. The first component is used to transform the incident complex wavefront into a complex amplitude distribution with arbitrary phase profile but with the desired amplitude distribution. This is possible using a single phase-only element. If blazed or holographic elements are used, the efficiency can be close to $\eta_1 \approx 100\%$ for this element. The phase profile of the desired complex wavefront can be generated with the second phase-only component. Again, efficiencies close to $\eta_2 \approx 100\%$ are possible. Thus, by splitting the beam-shaping task between two phase-only filters, no amplitude weighting is necessary. The two phase filters provide the necessary design freedom to shape both the amplitude as well as the phase of the wavefront. The efficiency of the overall beam-shaping system can be very high and is calculated by means of:

$$\eta_{BS} = \eta_1 \cdot \eta_2 \qquad (10.1.3)$$

The second phase filter can be positioned either in the near field or in the far field behind the first element. The freedom in the system geometry is only restricted by the fact that effective redistribution of the light amplitude has to be possible from the first to the second component. The larger the similarity between the incident and the desired amplitude distribution, the smaller is the distance necessary between the two filter planes (Fig. 10.2).

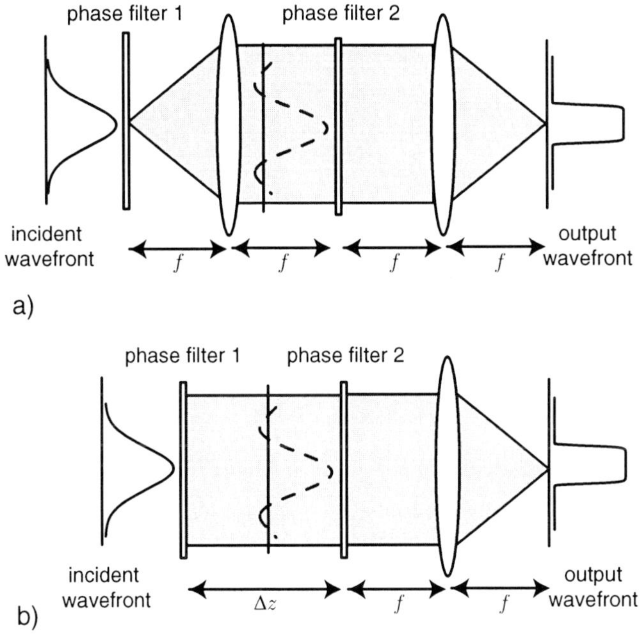

Fig. 10.2: Two possible configurations for beam shaping with two phase filters; a) filters located in Fourier reciprocal planes (far field); b) filters located in Fresnel planes (near field).

For the design of the element which generates the desired amplitude distributions any of the design approaches introduced in Chapter 5 can be applied. Iterative Fourier transform approaches have been suggested. Alternative design approaches are based on space-variant imaging configurations, generating so-called geometrical transforms of the incident beam [5, 6]. The design approach commonly applied here is closely related to the ray-tracing approach (Chapter 5). In order to describe the redistribution of the light amplitude by the geometrical transform, the redistribution of the rays is calculated. Each ray is supposed to carry the same amount of energy. The intensity distribution of both the incident optical field as well as the transformed field, are subdivided into equal energy intervals A_i, each of which is the origin or end of an optical ray (Fig. 10.3) [7, 8].

$$\int_{A_i} |U(x)|^2 dx = \int_{A_i'} |U'(x')|^2 dx' = \frac{1}{I}; \quad i = 1, 2, \ldots, I \tag{10.1.4}$$

where $U(x)$ and $U'(x)$ are the incident and the transformed amplitude distribution with normalized intensity ($|U|^2 = 1$). From geometrical considerations we find the correlation between the ray directions and the optimized phase function $\phi(x)$ (Eq. 2.1.7):

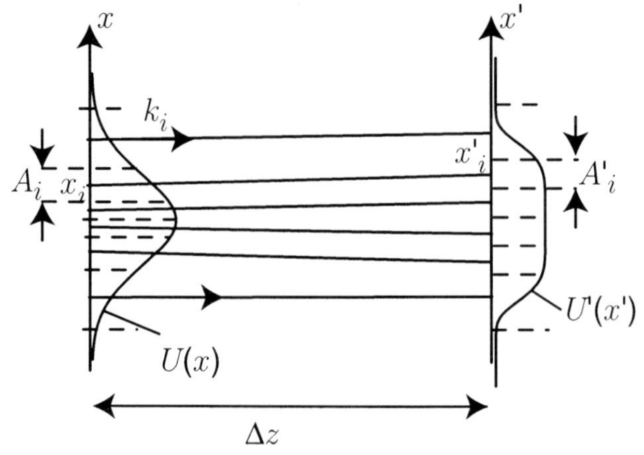

Fig. 10.3: Geometrical transform described by a change in density of optical rays.

$$x_i' = x_i + \frac{\Delta z \lambda}{2\pi} k_x = x_i + \frac{\Delta z \lambda}{2\pi} \frac{d\phi(x)}{dx} \tag{10.1.5}$$

Thus, the k-vectors of the individual rays can be calculated. By partial integration of the k-vectors we achieve the phase distribution of the transforming element (Chapter 2.). For some specific situations this design approach yields analytic solutions for the phase profile of the beam-shaping element [5, 9, 10]. A variety of numerical approaches for a more general solution of the problem have been suggested [11, 12].

Having in mind the trade-offs described above, we now wish to discuss a variety of approaches for beam shaping. We will find that for some specific situations it is possible to achieve good quality beam shaping even with a single element. This is the case, for example, if the shape of the amplitude profile of the incident wavefront is sufficiently close to the desired amplitude distribution.

10.2 Lateral laser beam shaping

10.2.1 Collimation of astigmatic beams

One of the basic and most widespread beam-shaping tasks is to transform a non-symmetrical astigmatic laser beam into a uniform Gaussian beam. This is typically necessary for the collimation of edge-emitting laser diodes. Due to their compact and cheap manufacturability, edge-emitting laser diodes are important optical sources for a large variety of applications. For many applications, however, the astigmatic characteristics of the emitted radiation causes some problems. Significant effort is, therefore, devoted to the transformation of the astigmatic and non-symmetrical beam into a symmetrical, collimated beam. The origin of the beam characteristics can be found in the structure of the edge-emitting laser diodes. They consist of a layered stack of semiconductor materials with different doping (p-n junctions). The resonator structure is oriented parallel to the layers and generally has a highly asymmetrical shape. The layer thickness representing the height of the resonator is of the order of fractions of one micron, O(μm), while the width is in the order of several microns, O(10 μm). This asymmetry of the laser resonator is the reason for an asymmetrical beam shape which upon propagation generates astigmatism due to different divergence angles caused by diffraction (Fig. 10.4). Depending on the type of laser diode, an additional source of astigmatism stems from different guiding mechanisms in the waveguide structure (e.g., in gain-guided laser diodes). In the most general case, the shaping of such a beam into a symmetrical, collimated beam requires both amplitude and phase shaping. In order to generate a symmetrical amplitude distribution often anamorphic imaging systems are used which perform the necessary geometrical transformation. Additional cylindrical lenses are necessary to compensate for the different divergence angles in x and y [13].

Large angles of diffraction occur in the direction of the small output window (typical values: $30° < \alpha_\perp < 60°$). The divergence in the perpendicular direction is much smaller due to the larger output window (typical values: $\alpha_\| \approx 10°$). Consequently, at a specific distance z_0 the beam has a symmetrical intensity distribution. According to the considerations of the previous section, efficient shaping of the laser beam is possible with a single microoptical element aligned at z_0 which compensates for the different phase distributions. In the following, we briefly outline the design process for such a component [14].

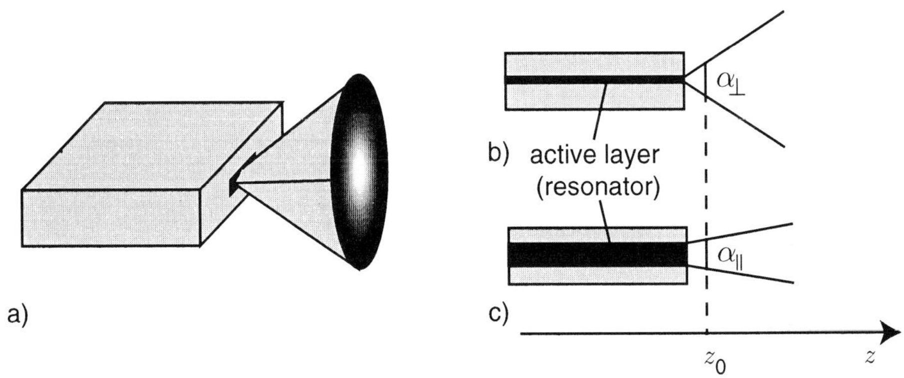

Fig. 10.4: Beam asymmetry in the radiation of edge-emitting laser diodes: a) pseudo-3D illustration of the beam emission; b) side view of the diode; c) top view of the diode.

For the design it is sufficient to assume a constant phase distribution at the output window of the diode. The beam profile in each dimension can be approximated by a Gaussian function with widths $\sigma_{x,0}$ and $\sigma_{y,0}$, respectively. Since we assume constant phase over the output window, $\sigma_{x,0}$ and $\sigma_{y,0}$ are real quantities:

$$u(x,y) = A_0 e^{-\left(\frac{x^2}{2\sigma_{x,0}^2}+\frac{y^2}{2\sigma_{y,0}^2}\right)} \tag{10.2.1}$$

The propagation of the beam (wavelength λ) along the z-axis can be calculated by a Fresnel transform. At any plane z behind the laser diode we obtain a beam which still has Gaussian shape. However, since the phase distribution is no longer constant over the beam diameter the Gaussian widths are now complex quantities. For the real part of the beam width $\sigma_x(z)$ at a distance z from the diode we find (see Exercises):

$$\mathrm{Re}\left[\frac{1}{\sigma_x^2(z)}\right] = \frac{4\pi^2\sigma_{x,0}^2}{4\pi^2\sigma_{x,0}^4+\lambda^2 z^2}; \qquad \mathrm{Re}\left[\frac{1}{\sigma_y^2(z)}\right] = \frac{4\pi^2\sigma_{y,0}^2}{4\pi^2\sigma_{y,0}^4+\lambda^2 z^2} \tag{10.2.2}$$

The plane z_0 where we wish to position the microoptical element is chosen in such a way that the real parts of the Gaussian widths are equal $\sigma_x(z_0) = \sigma_y(z_0)$. We obtain for z_0:

$$z_0 = \frac{2\pi}{\lambda}\sigma_{x,0}\sigma_{y,0} \tag{10.2.3}$$

The phase profile of the beam is represented by the imaginary part of the Gaussian widths which result from the Fresnel transform (see Exercises):

$$\phi_{\mathrm{beam}}(x,z) = \frac{8\pi^3\sigma_{x,0}^4}{4\pi^2\sigma_{x,0}^4\lambda^4 z^4} + \frac{\pi x^2}{\lambda z} = \pi x^2 \frac{\lambda z}{4\pi^2\sigma_{x,0}^4+\lambda^2 z^2} \tag{10.2.4}$$

$\phi_{\mathrm{beam}}(y,z)$ is found analogously. The beam-shaping component at z_0 is designed to compensate for the phase profile of the beam at this location ($\phi_{\mathrm{beam}}(x,z_0)$). Collimation of the astigmatic beam is possible with an astigmatic microlens. Such a lens is described by two different focal lengths f_x, f_y in the two axial directions. The phase terms introduced by the lens are written as:

$$\phi_{\mathrm{lens}}(x) = -\frac{\pi x^2}{\lambda f_x}; \qquad \phi_{\mathrm{lens}}(y) = -\frac{\pi y^2}{\lambda f_y} \tag{10.2.5}$$

The two jobs of the astigmatic microlens are to compensate for the difference in the beam divergence as well as to collimate the beam. It can be easily verified that a constant phase results for an astigmatic lens with the focal lengths:

$$f_x = z_0 + \frac{4\pi^2\sigma_{x,0}^4}{\lambda^2 z_0}; \qquad f_y = z_0 + \frac{4\pi^2\sigma_{y,0}^4}{\lambda^2 z_0} \tag{10.2.6}$$

The microlens described by Eq. 10.2.6 generates a flat phase profile over the beam diameter in the plane z_0. From Gaussian beam relaying we know that in terms of collimation this is not the ideal situation (Chapter 7). It might be advantageous with respect to the longest transmission distance to relay the beam, i.e., to generate a slightly converging beam at z_0. Eq. 10.2.6 can be easily adjusted to this situation.

For a numerical example, we consider a laser diode with an output window of $0.5\ \mu m \times 5\ \mu m$. In this case we find a symmetrical intensity profile of the beam at $z_0 = 19.8\ \mu m$. At this point the beam has a diameter of $\sigma_{x,z_0} = \sigma_{y,z_0} \approx 12.6\ \mu m$. For the astigmatic lens Eq. 10.2.6 yields focal lengths of $f_x = 19.8\ \mu m$ and $f_y = 1986.6\ \mu m$.

For the fabrication of such an astigmatic microlens several of the different technological approaches discussed in Chapters 3–5 can be used. From the point of view of design freedom, a diffractive realisation is the most desirable [15]. However, it is limited in efficiency as well as in numerical aperture. Promising alternatives are fabrication techniques for refractive elements such as reflow and mass transport after binary preshaping of the elements [16, 17]. The fabrication of astigmatic GRIN elements has also been suggested and demonstrated for this application [14, 18, 19]. With respect to the integration of the beam-shaping system with the laser diode the possibility of fabricating the microoptical element within the laser diode substrate is very interesting. This is possible in GaAs using etching techniques (see Chapters 4 and 5). Modular approaches which mechanically integrate microoptical components have been demonstrated using cylindrical [20] as well as GRIN rod lenses [21, 22].

10.2.2 Beam shaping for laser machining

Laser machining is an important method with applications in many industrial areas. For the sake of efficiency and precision, precise control of the energy deposition in the processed material is important. To this end, laser beams with sharply defined intensity profiles are necessary. The phase distribution over the beam diameter, on the other hand, is rather unimportant in this case. Thus, single optical elements performing geometrical transformations are sufficient to perform the beam shaping. With an incident Gaussian beam, a good approximation of a flat intensity profile can be generated by aperture splitting and overlaying the different parts in the near field with a relative lateral shift [23]. This can be performed, e.g., with a number of properly aligned prisms. Beam-shaping systems working according to this principle have been demonstrated as discrete mechanically aligned versions as well as semi-integrated versions, where all necessary prism components were integrated into one element. Fig. 10.5 shows the basic configuration as well as a simulated intensity distribution of the beam behind the component. With a radially symmetrical component consisting of four prisms which perform the shifted overlay it is possible to achieve a uniformity of 3% in the flat-top region and an efficiency of 94%. Here, an incoherent overlay has been assumed. This is valid only for laser radiation with short coherence length, e.g., from pulsed excimer lasers. For radiation with longer coherence length high frequency interference fringes can be observed in the beam profile. This causes problems for micromachining with high lateral resolution. For such high resolution applications it is necessary to prevent the interference fringes. Diffusers or polarisers are used to destroy the coherence of the beam. For many other applications in industry, however, interference fringes do not harm the performance of the laser machining system. Aperture splitting can also be performed by kinoform elements. In combination with microlens arrays, good beam-shaping functionality is achieved [24].

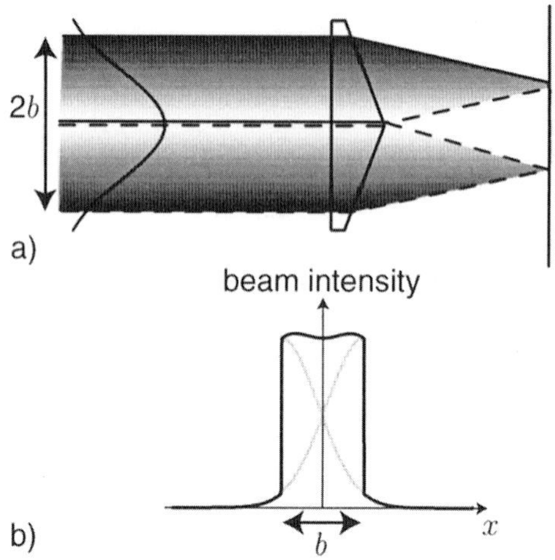

Fig. 10.5: a) Laser beam shaping using beam splitting through multiple prisms and subsequent overlaying of the shifted beams. b) Simulation of the beam shape after incoherent overlaying of the two shifted parts of the Gaussian beam.

It is possible to perform more complex beam shaping by using several microprisms . For example, a refractive optical element which generates a cross-like near field intensity profile has been demonstrated [25]. For better performace, it is also possible to combine beam-shaping microoptics with beam-splitting components, e.g., of the Fresnel- or Fourier-type. Enhanced uniformity in the spot array generated by a Fresnel-type array illuminator can be achieved by increasing the uniformity of the illumination with a beam-shaping element. As we have seen in the previous chapter, one property of Fourier array generators is that the shape of the illuminating beam appears in the shape of the individual spots. Thus, with flat-top illumination, it is possible to generate an array of beamlets with flat-top intensity profiles.

For specific material processing applications using laser radiation, e.g., material hardening, specific nonuniform intensity profiles are required. For application in the far field, the phase functions necessary to generate those profiles can be designed, e.g., with the iterative Fourier transform algorithms. Haupt et al. [26] designed such a hologram for the generation of a nonuniform rectangular intensity profile in the far field. Due to the specific configuration of their optical system for material processing, the DOE was optimized for oblique illumination. In order to be able to use high power laser radiation of up to 2 $\frac{\mathrm{kW}}{\mathrm{cm}^2}$, the DOE was fabricated as an efficient eight phase level reflection hologram in silicon. It is possible to access the DOE for cooling purposes. Fins or microchannels for air or liquid cooling are fabricated on the back of the silicon substrate by anisotropic etching.

10.2.3 Beam shaping for LIDAR

For some applications, such as laser radar, it is desirable to determine the beam shape in the far field. In this case, we again have to determine whether only the intensity distribution is of significance or if the phase distribution is important as well. The former case can be solved using a phase filter which is optimized with basically the same design procedure as described previously for Fourier array generators. In an iterative Fourier transform algorithm (Chapter 5) we use the phase freedom in the spatial frequency domain to optimize a pure phase filter with the desired diffraction amplitudes [27].

For laser radar (LIDAR), however, laser beams are necessary which have a flat-top intensity profile not only in one particular plane, but remain quasi-collimated over a reasonable distance. To this end we need to generate a flat phase simultaneously with the flat-top amplitude distribution. As discussed above, this can be achieved with a second phase-only component. However, a simple trick can be exploited to reduce the number of degrees of freedom necessary for such a beam-splitting component. In the area of the main lobe, the amplitude profile of a Gaussian beam is very similar to that of a sinc-function (Fig. 10.6). We know from Fourier optics that the sinc-profile generates a rectangular distribution in the far field. According to Veldkamp [28] the similarity between the Gaussian and the sinc-function is good enough to generate a nearly flat-top profile. For this purpose only the phase of the Gaussian beam needs to be adjusted. This is possible with high efficiency using a single binary phase grating. Thus, efficient single element beam shapers approximating 1D sinc-shape or 2D Bessel-shape complex amplitude distributions can be fabricated. In the far field, 1D rectangular and 2D circular flat-top profiles in amplitude and phase result. The efficiency of such a beam-shaping component can be further improved using an interlaced phase grating to adjust the Gaussian amplitude even better to the sinc-shape. Effectively, the interlaced grating weights the local amplitude values by a local variation of the diffraction efficiency. With this approach, efficiencies as high as 74% could be achieved [29].

The problem of generating a 2D rectangular flat-top profile, however, is more complex. Due to the radial symmetry, the incident 2D Gaussian amplitude distribution does not approximate the necessary $\mathrm{sinc}(x) \cdot \mathrm{sinc}(y)$ distribution. The

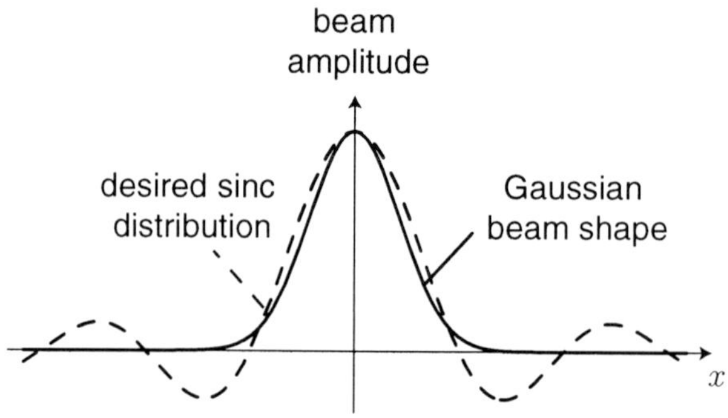

Fig. 10.6: Illustration of the similarity between the Gaussian beam shape and the sinc-distribution, desired for flat-top generation in the far field.

$\text{sinc}(x) \cdot \text{sinc}(y)$ distribution has the light concentrated much more on the x- and y-axes. Thus, the approach for adjusting amplitude and phase in a single beam-shaping element, which worked for the 1D case, leads to very low efficiencies. Configurations using two phase elements have been shown to work for this purpose. The first phase element implements a geometrical transformation which converts the incident Gaussian beam into a 2D sinc distribution. The purpose of the second element is to adjust the phase in a similar way. Eismann et al. [30] demonstrated such a two element beam shaper using holographic optical elements.

10.3 Axial beam shaping

In the previous section we described the shaping of the lateral profile of the complex amplitude of an optical beam. For many applications it is, however, also desirable to influence the axial light distribution of a wavefield. Specifically the generation of beams with extended focal length is interesting in numerous applications such as optical sensing, data storage, laser beam cutting, welding and micromachining. A variety of names has been coined for this type of optical beams, such as non-diffracting beams, Bessel beams or light bullets. Bessel beams can be generated using so-called axicons or lithographically fabricated generalized axicons [31, 32].

The field distribution $E(\mathbf{r}, t)$ of the fundamental non-diffracting Bessel beam is described by a superposition of plane waves [33, 34]:

$$E(\mathbf{r}, t) = e^{i(\beta z - \omega t)} \int_0^{2\pi} \frac{1}{2\pi} e^{i\alpha(x\cos(\Phi) + y\sin(\Phi))} d\Phi = e^{i(\beta z - \omega t)} J_0(\alpha\varrho) \qquad (10.3.1)$$

where $\varrho = \sqrt{x^2 + y^2}$ and J_0 denotes the zeroth-order Bessel function. β and α

denote the components of the wavevector $\mathbf{k}$ in the z-direction as well as in the x-y plane, respectively:

$$\begin{aligned}\beta &= k_z \\ \alpha &= \sqrt{k_x^2 + k_y^2}\end{aligned} \tag{10.3.2}$$

$E(\mathbf{r}, t)$ is an exact solution of the wave equation with $\beta^2 + \alpha^2 = k^2$. As long as β is real-valued, the intensity profile of the wavefield described by Eq. 10.3.1 remains constant along the optic axis. Fig. 10.7 shows a density plot as well as the profile of the intensity distribution in the x-y plane of the Bessel beam. The interesting feature is that the (non-diffracting) Bessel beam exhibits a sharp central lobe which remains localized during beam propagation:

$$I(x, y, z = 0) = \frac{1}{2}|E(x, y, z = 0, t)|^2 = I(x, y, z \geq 0) \tag{10.3.3}$$

In contrast to what we are used to, e.g., with regard to Gaussian beams, Bessel beams do not exhibit any lateral spread due to diffraction. The effective width of the central lobe is determined by α and is smallest for $\alpha = \frac{\omega}{c} = \frac{2\pi}{\lambda}$. In this case the width w of the central lobe is approximately $w = \frac{3\lambda}{4}$. For the ideal Bessel beam, being the superposition of plane waves with infinite extension, the energy content is infinite. Each of the rings of the pattern in Fig. 10.7 contains the same amount of energy. Ideally, the number of rings is infinite. Such ideal Bessel beams cannot be realized in practice. However, it is possible to realize approximately non-diffracting beams with extended focal length.

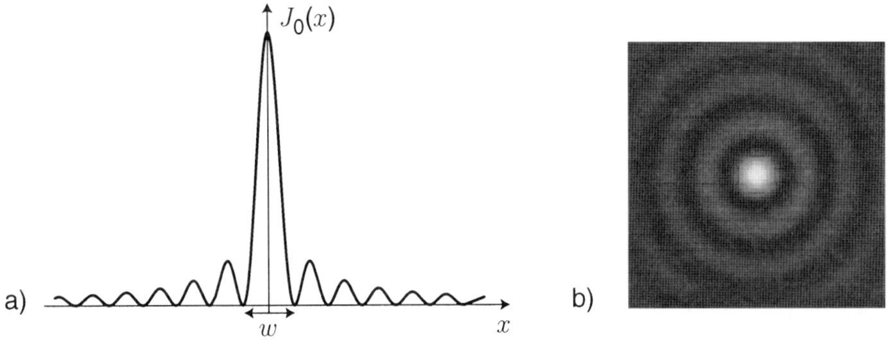

Fig. 10.7: a) Lateral intensity distribution of a Bessel beam; b) density plot of the intensity of the Bessel beam in the x-y plane.

In a simple optical setup to generate a Bessel beam with extended focal length an annular ring aperture is illuminated by a plane wave. Diffraction at the ring-shaped slit generates beams which interfere on the optic axis to form the Bessel beam (Fig. 10.8a). In order to avoid absorption losses at the aperture, it is possible to use so-called axicons which are sometimes also called conical lenses. Refractive axicons are biprisms with circular symmetry which split and deflect the incident beam (Fig. 10.8b). The Bessel beam is generated in the region of superposition of the individual beams. For infinitely large axicons and beam diameters, ideal Bessel beams with infinite propagation length would be generated. Due to the limited width of the incident beam as well as the axicon, however, the propagation distance $z_{\max}$ is limited. From geometrical consideration we can calculate $z_{\max} = \frac{D}{2\tan(\theta)}$ and for the propagation angle we find:

$$\sin(\theta) = \frac{\sqrt{k_x^2 + k_y^2}}{k^2} = \frac{\alpha\lambda}{2\pi} \tag{10.3.4}$$

Thus, $z_{\max}$ is given by:

$$z_{\max} = \frac{D}{2}\sqrt{\frac{1}{\sin^2(\theta)} - 1} = \frac{D}{2}\sqrt{\left(\frac{2\pi}{\alpha\lambda}\right)^2 - 1} \tag{10.3.5}$$

For the case of refractive axicons or diffractive optical elements, as illustrated in Fig. 10.8b and c, the region where the Bessel beam is formed ranges from $[-z_{\max}, +z_{\max}]$. Beyond $z_{\max}$ the axial beam intensity decreases rapidly, proportional to $\frac{1}{\alpha\varrho}$. From Eq. 10.3.1 we find the condition for calculating the effective width $w = 2\varrho$ of the central lobe of the Bessel beam:

$$\varrho\alpha = \frac{w}{2}\alpha \approx 2.4 \quad \longrightarrow \quad w \approx 0.766\frac{\lambda}{\sin(\theta)} \tag{10.3.6}$$

w is proportional to the numerical aperture of the lens used to generate the Bessel beam (Fig. 10.8a), a result which is expected intuitively.

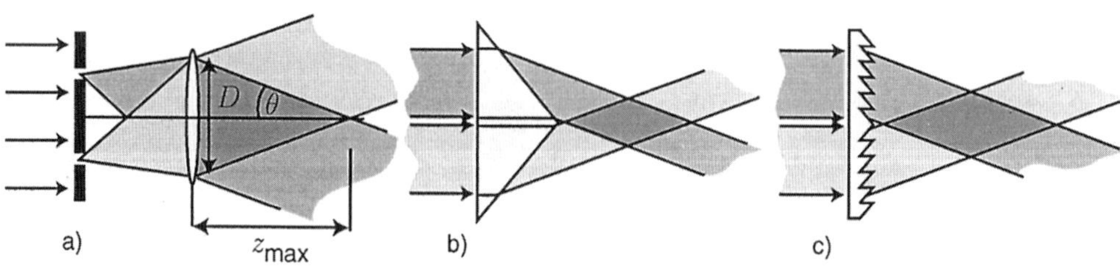

Fig. 10.8: Optical setups for generating Bessel beams: a) annular ring aperture followed by a lens; b) classical refractive axicon; c) blazed diffractive axicon.

Lithographically fabricated microoptical elements have recently been investigated for the generation of non-diffracting beams. The design freedom offers several possibilities for improved axial beam shaping. A variety of diffractive optical elements have been suggested for the generation of higher order Bessel beams [35-38].

Although in the propagation region of $2z_{\mathrm{max}}$ the shape of the lateral intensity distribution does not change, the axial intensity varies significantly for non-ideal Bessel beams. This is due to diffraction at the edges of the finite plane waves which interfere to form the Bessel beam. Apodization filters [39] which gradually decrease the transmission towards the edges of the optical elements can be used to avoid such fluctuations of the axial intensity [40, 41]. Alternatively, modified phase filters with circular symmetry can be fabricated. These phase filters provide enhanced uniformity in the axial intensity distribution [42, 43]. More general shaping of the 3D distribution of a wavefield is possible with more complex design techniques [44, 45, 46]. For example, iterative techniques have been suggested for designing the wavefield in two or more axial locations behind a diffractive optical element [27, 47].

10.4 Temporal beam shaping

The dispersive properties of a diffraction grating open up possibilities for another type of beam shaping, namely time domain pulse shaping [48, 49]. This is possible since the wavelength spectrum $S(\lambda)$ of an optical pulse is determined from the Fourier transform of its temporal shape $s(t)$:

$$S(\lambda) = \int s(t) e^{2\pi i \nu_t t} dt \tag{10.4.1}$$

Optical spectrometers are used to convert the wavelength spectrum $S(\lambda)$ into a spatial distribution $S(x)$. For example, the dispersion of prisms or diffraction gratings is exploited for this purpose. Thus, spectrometers can be used to display the Fourier transform of the temporal shape of an optical pulse in the form of a spatial light distribution. Consequently, with the techniques described for lateral shaping of the spatial distribution of optical wavefronts, it is also possible to shape the temporal shape of an optical pulse. The duality between space and time has been recognised a long time ago and was applied very early on for the compression of optical pulses using electro-optically generated chirped phase modulators [50]. From the point of view of signal theory, the performance of a diffraction grating can be compared to a finite impulse response filter.

Fig. 10.9 shows an optical setup for time-domain pulse shaping. The main parts of the system are two conventional high resolution optical spectrometers for spatial-temporal conversion. The wavelength spectrum $S(\lambda)$ of an incident optical wave is displayed as a spatial light distribution in the Fourier plane of the diffraction grating. The spatial filter $F(x)$ can thus be used to influence the temporal shape of the pulse in the output. We imagine an ideal continuous monochromatic plane wave incident onto the spectrometer. In the filter plane we find a single spatial

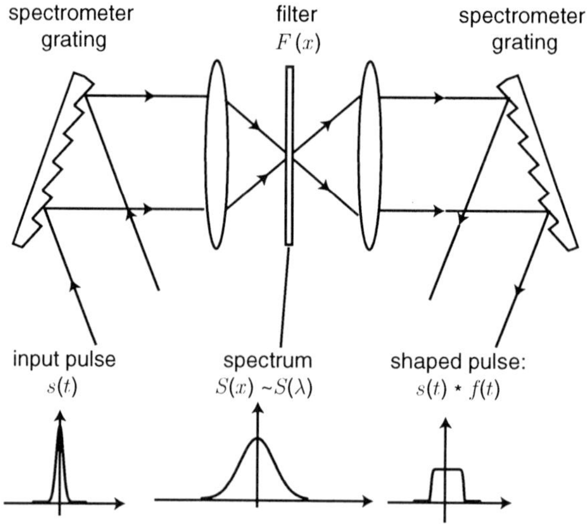

Fig. 10.9: The optical setup for time domain pulse shaping using spatial filtering.

$\delta(x)$-pulse which is located at the position corresponding to the wavelength of the incident wavefront. The $\delta(x)$-shape corresponds to the Fourier transform of the continuous, i.e., temporally uniform pulse. However, if we now change to an ultrashort optical pulse (pulse length $\approx 10^{-12}$–10^{-15} s) in the input (approximately described by $\delta(t)$), in the output of the spectrometer we obtain a continuous spectrum distributed laterally. The width of the spatial distribution of the spectrum in the filter plane is determined by the spectral resolution of the spectrometer setup, which is determined by :

$$\frac{\Delta x}{\Delta \lambda} = \frac{f}{p \cdot \cos(\theta)} \tag{10.4.2}$$

where p is the grating period and θ the diffraction angle of the grating. For example, we calculate the spectral width of a 100 fs optical pulse at $\lambda_0 = 850$ nm to be:

$$\Delta\lambda = \frac{\Delta\nu}{c}\lambda_0^2 = 24 \text{ nm} \tag{10.4.3}$$

According to Eq. 10.4.2 a spectrometer with a period $p = 0.8$ μm and lenses with focal length $f = 100$ mm distributes this wavelength range spatially over a distance of $\Delta x = 3$ mm. The number N of grating periods illuminated by the incident beam determines the spectral resolution of the setup and the space-bandwidth product available for the spatial filter [51].

Based on the concepts introduced for lateral beam shaping, spectrometer systems for space-time conversion have been applied to the generation of a large variety of temporal pulses. Correlators [52], holographical systems [53] as well as lenses working in the temporal domain to overcome temporal dispersion have been suggested and demonstrated successfully [54, 55]. Microoptical elements are important components in such systems. High frequency spectrometer gratings are equally important as efficient spatial filters for the adjustment of the spectral distribution of the pulses. An integrated microoptical spectrometer system for temporal beam shaping has been suggested recently [56].

10.5 Multiple aperture beam shaping

Laser beam combination

For laser machining applications high power laser beams are needed. Single lasers are limited in the maximum power output by physical constraints. An interesting task for passive microoptical components is the combination of an array of mutually coherent laser beams. Accordingly it is possible to use the same kind of systems as those previously discussed for beam splitting. Fourier beam-splitter gratings, phase-contrast setups, Talbot phase gratings and microlens arrays have been demonstrated for high efficiency beam combination [57]. For efficient beam combination the optical setups described in the previous chapter are simply used with light propagating in the opposite direction. Analogous to the Fourier-type beam splitter, for beam combination a set of multiple beams is incident on the optimized phase grating and converted into a single diffraction order (Fig. 10.10a). A necessary condition is mutual coherence between all the incident beams. Leger et al. showed that the efficiency of the beam combination in this approach depends on the phase relation between the beams [58]. For binary phase gratings, theoretical efficiencies of 80% and higher have been demonstrated. Just as in the beam-splitting configuration the use of multilevel or continuous phase gratings yields even higher coupling efficiencies up to theoretically 97.3% for a 10 element laser array [59].

An alternative to the combination of laser beams, e.g., through Fourier phase gratings, is so-called aperture filling. This is the inverse process to aperture division frequently applied to beam splitting. To this end, e.g., a lenslet array or a Talbot phase grating is placed behind the laser array in such a way that the individual beams are densely packed to form a single wide beam. In this case the beam superposition takes place in the Fourier plane (i.e., in the far field). Methods similar to the phase contrast beam splitter have also been suggested for aperture filling [60, 61]. The efficiency of beam combination through aperture filling depends on the fill factor of the incident array of beamlets as well as on the efficiency of the phase grating. As we discussed in the context of beam splitting devices, e.g., for the phase contrast method tight constraints on the incident beam array have to be fulfilled in order to achieve maximum efficiency (i.e, fill factor) in the resulting superposed beam. For a fill factor of 25% in the incoming array a transfer of 92%

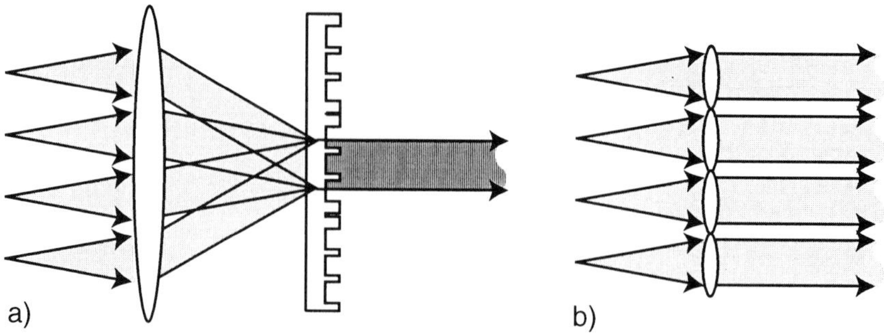

Fig. 10.10: Examples for beam combination: a) by superposition; b) by aperture filling

of the total energy into the center lobe of the output beam has been obtained [60].

As mentioned previously, for efficient beam combination of laser diode arrays mutual coherence (i.e., a fixed phase relation) between the incident beams is necessary. In this case we can get constructive interference in the main mode of the combined beam. The maximum achievable intensity in the combined beam is given by:

$$I_{\text{total}} = N \cdot I_0 \qquad (10.5.1)$$

where N is the number of laser diode sources and I_0 is the intensity of the individual sources. For incoherent beams no increase in beam intensity can be achieved through superposition [57].

10.6 Intra-cavity beam shaping

The systems discussed in the previous sections used optical components located outside the active laser cavity for the shaping of a coherent laser beam or beam array. Alternatively it is possible to integrate similar optical systems inside the laser cavity [1]. The specific beam shape is then generated by a discrimination of the modes which are amplified in the cavity. Intra-cavity modal discrimination can be achieved for single laser beams as well as for laser diode arrays (Figs 10.11 and 10.13).

10.6.1 Intra-cavity beam shaping of individual laser beams

Mirror shaping is one of the well established techniques for influencing the stability and mode selectivity of conventional laser resonators. Generally, for low gain laser materials it is advantageous to use spherical mirrors which form a so-called stable

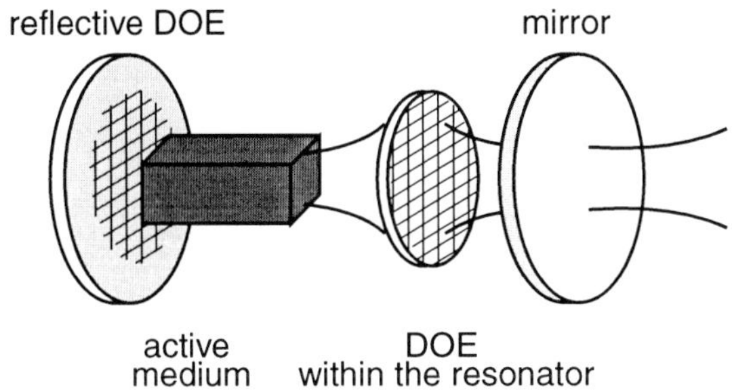

Fig. 10.11: Intra-cavity beam shaping using microoptically structured mirrors and phase plates.

resonator [62]. The mirrors are arranged such that they perform a relaying of the Gaussian beams onto the surface of the opposite mirror. The supported eigenmodes of such a resonator are Hermite-Gaussian modes. With sufficiently large mirror sizes the diffraction losses in this cavity can be minimized.

Enhanced mode discrimination and beam shaping can be achieved with more complex, stable or unstable resonators [63]. In this case, however, diffraction losses have to be accepted. These types of resonators are only useful for high gain laser materials. In order to achieve a desired shape of the supported modes in such resonators, specifically optimized mirror structures have been suggested. Variable reflectivity mirrors have been shown to reduce diffraction losses at the sharp edges of conventional mirrors [64, 65]. Losses due to the locally reduced reflectivities can be avoided if phase elements are used for the mode discrimination. This leads to aspherical mirror structures with complex shape [66, 67, 68]. Such mirrors are also called "graded-phase mirrors" or "mode-selective mirrors". Due to the difficult fabrication process for mirrors which introduce optimized continuous phase profiles, diffractive optics has been suggested for this application [69]. The available design freedom of diffractive optics in combination with lithographic fabrication techniques allows one to enhance the characteristics of the laser cavities significantly. Fig. 10.12 shows the beam profile resulting from intra-cavity beam shaping designed to generate a super-Gaussian beam.

The design procedure starts by assuming the desired complex amplitude distribution at the output mirror just inside the cavity. By non-paraxial beam propagation it is possible to calculate the amplitude distribution at the second mirror. In order to support the desired beam profile as an eigenmode of the resonator we require the second mirror to phase conjugate the beam. In this case the initial amplitude distribution is reconstructed at the output mirror. In order to perform the complex conjugation, phase-only filters are necessary. Reflective diffractive optical elements fabricated lithographically at the surfaces of conventional mirrors provide an efficient solution to this problem. Since the mirrors are shaped through

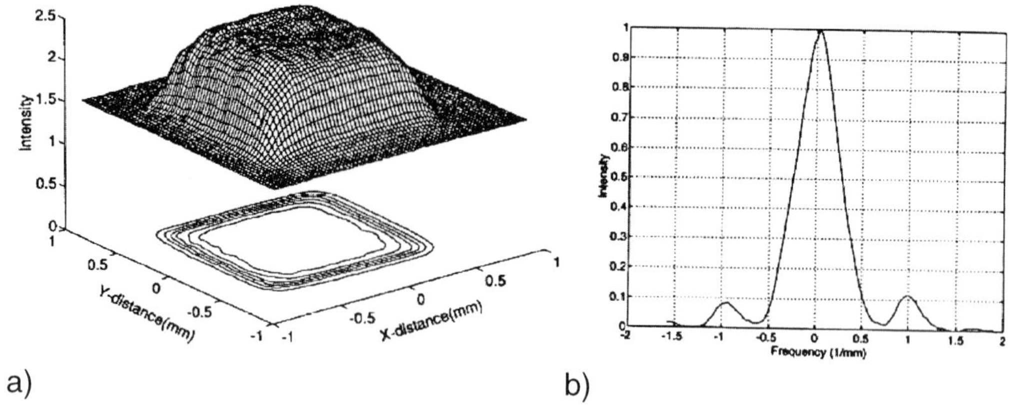

Fig. 10.12: Intra-cavity beam shaping with diffractive optics: a) near field beam profile resulting from a laser cavity with "mode-selective mirrors". The profile corresponds to a 14th-order super-Gaussian beam; b) focus distribution of this beam in the far field (Fourier plane). (pictures courtesy of Prof. J. Leger, University of Minnesota.)

the DOE such that a specific mode profile is selected by the resonator the term diffractive mode-selective mirror (DMSM) has been coined for reflective DOEs for this application [69].

The basic form of a beam-shaping resonator contains a conventional flat mirror in combination with a reflective DOE. With this configuration it is possible to design the laser cavity to support an arbitrary real valued mode. In the same way as for extra cavity beam shaping we need a second phase element if the complex amplitude of the supported fundamental mode is to be shaped. High efficiencies can be achieved with multilevel phase elements with sufficiently large diameters. However, the losses of higher order modes are also rather small which reduces the mode selectivity. In order to achieve single mode operation, an aperture stop can be introduced in the cavity. Such an absorbing aperture increases the loss for higher order modes. Alternatively, additional phase gratings have been demonstrated to increase the modal discrimination [70]. Sinusoidal phase, square phase and pseudo-random phase plates have been investigated for this purpose. For a sinusoidal phase plate it is possible to find an optimum frequency, phase and modulation depth. For square wave and random phase plates the requirements on the minimum feature size can easily become rather stringent if large mode discrimination is desired [69].

10.6.2 Intra-cavity beam shaping of arrays of laser beams

Intra-cavity mode-shaping techniques can be applied for mode coupling of laser diode arrays (Fig. 10.13). In order to gain access into the resonator, the original output windows of the array are anti-reflection coated in order conveniently to add microoptical components off-chip. An off-chip mirror can be shaped to select the desired mode. A mathematical description of this situation is possible using coupled mode theory (Chapter 5). The complex amplitude of the laser array is described by a linear superposition of the individual laser modes. In coupled mode theory the amplitude coefficients of the lasers are related in an eigenvalue equation to a coupling matrix which contains the coupling coefficients between the modes.

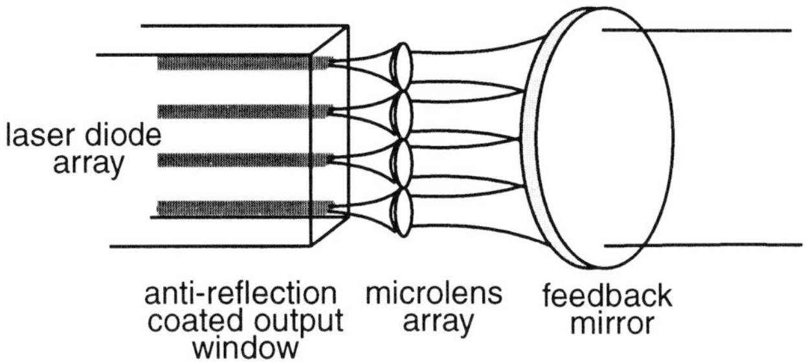

Fig. 10.13: Intra-cavity beam shaping of laser arrays using microoptical beam splitter configurations.

A suitable positioning of the external mirror allows one to provide mode locking of the individual lasers. This is achieved, e.g., in so-called Talbot cavities where the mirror is placed at a distance of $\frac{z_T}{2}$ from the laser diode array. The periodic laser array is self-imaged and the light efficiently couples back into the waveguiding laser medium. Since the Talbot effect only works with coherent illumination, this Talbot cavity does not support mutually incoherent laser modes. It is an efficient technique for mode locking of laser diode arrays. It has been shown that the modal behaviour of the resulting beam depends on the cavity length, the fill factor and the size of the array [71]. Theoretically, it is found that the maximum modal discrimination in combination with minimum fundamental mode loss occurs if the cavity length is one quarter of the Talbot length. Fig. 10.14 shows a photograph of such a Talbot cavity designed for combining a 1D array of laser diodes. This configuration leads to a single spatial supermode with the phases of the individual lasers alternating between 0 and π. Additional losses to the modes occur due to the finite size of the laser array and disappear for sufficiently large arrays. A trade-off exists between the modal discrimination, which can be increased by a reduction of the laser fill factor and the fundamental mode loss due to edge effects [72]. Additionally, the propagation angles have to be limited in order not to violate the paraxial condition.

This limits the apertures of the individual lasers [1]. An active phase control using phase contrast imaging and active liquid crystal phase modulators inside the laser cavity can help further to improve the spatial mode selection [73].

Fig. 10.14: Talbot cavity for coherent combination of a 1D laser array. The cavity consist of a cylindrical lens element with a thickness of 1/4th of the Talbot length. Diffractive optics for beam combination is contained on the flat side of the lens element. (photograph courtesy of Prof. J. Leger, University of Minnesota.)

10.7 Glossary

- **aperture filling**
  technique for combining an array of optical beams by densely packing the beams to fill the aperture of an optical system. The individual beams are arranged to propagate in parallel (i.e., in the same direction) but do not occupy the same location. Aperture filling is a technique for laser beam combination which is Fourier reciprocal to the beam superposition technique.

- **beam superposition**
  combination of an array of laser beams by directing them to the same location, where they overlap and are redirected, e.g., by a diffraction grating to form a single beam.

- **LIDAR**
  laser radar: optical technique which uses the radar principle. A focused high power laser beam is used for scanning the target volume, e.g., a section of the atmosphere. The backscattered light is detected and evaluated using specific filters and detection systems.

- **mode selective resonator**
  laser resonator which supports modes with specific beam profiles. This can be achieved with specifically shaped reflective ("mode selective mirrors (MSMs)") or transmissive diffractive optical elements inside the cavity.

- **super-Gaussian beam**
  laser beam with an amplitude distribution:

  $$a(x,y) = A_0 e^{-\left(\frac{x^2}{\sigma_x^2}\right)^n} e^{-\left(\frac{y^2}{\sigma_y^2}\right)^n}$$

  For $n \gg 1$ this distribution approaches a rectangular shape.

- **Talbot cavity**
  laser resonator which uses Fresnel plane phase filters to perform intra-cavity beam shaping; specifically a cavity with Talbot phase gratings similar to Talbot array illuminators which are used for mode locking of laser arrays.

10.8 List of new symbols

α:	components of the k-vector of a Bessel beam in a plane perpendicular to the propagation direction
β:	propagation constant of a Bessel beam
$F(\lambda)$:	spatial filter for temporal pulse shaping
J_0:	zeroth order Bessel function
σ_x, σ_y:	the width of the Gaussian amplitude distributions; generally these widths (waists) are measured as the 1/e diameters of the beam
$S(\lambda)$:	wavelength spectrum of a temporal pulse
$s(t)$:	shape of a temporal pulse
ϱ:	radial coordinate in the x-y plane
w:	width of the center lobe of a Bessel beam

10.9 Exercises

1. **Propagation of Gaussian beams; design of an astigmatic element for laser diode collimation**
   Eq. 10.2.1 describes a Gaussian beam where w is real-valued, i.e., with a constant phase over the beam diameter. The beam widths, however, are different in the two lateral directions. Calculate the (complex-valued) beam widths after a propagation distance z. Derive the distance z_0 where the diameters in both lateral directions are equal (Eq. 10.2.3).

2. **Bessel beams**
   What is the diameter of a Bessel beam generated according to Fig. 10.8 for a lens with an f-number $f/\# = 30$? How long is the propagation distance z_{max} of the Bessel beam for the diameter of an incident plane wave $D = 1$ cm? (Assume paraxial approximation.)

3. **Super-Gaussian beam shape**
   Illustrate the shape of super-Gaussian beams (see Glossary) for different parameters n and compare it to a rectangular beam profile. How efficiently is the beam confined to the area with the width of 2σ if $n = 5$ ($\lambda = 633$ nm)?

References

[1] J. R. Leger, "Laser beam shaping", in Microoptics: elements, systems, and applications, H.-P. Herzig (ed.), Taylor & Francis, London, UK (1997), 223–257.

[2] Y. Danziger, E. Hasman, A. Friesem and A. W. Lohmann, "Multilevel diffractive elements for generalized wavefront shaping", *Opt. Eng.* **35** (1996), 2556–2565.

[3] J. L. Horner, "Light utilization in optical correlators", *Appl. Opt.* **21** (1982), 4511–4514.

[4] H. O. Bartelt, "Computer-generated holographic component with optimum light efficiency", *Appl. Opt.* **23** (1984), 1499–1502.

[5] O. Bryngdahl, "Geometrical transformations in optics", *JOSA* **64** (1974), 1092–1099.

[6] F. S. Roux, "Branch-point diffractive optics", *JOSA A* **11** (1994), 2236.

[7] J. Turunen and F. Wyrowski, "Introduction to diffractive optics", in Diffractive optics for industrial and commercial applications, J. Turunen and F. Wyrowski (eds), Akademie Verlag, Berlin (1997), 1–50.

[8] Y. Arieli, N. Eisenberg, A. Lewis and I. Glaser, "Geometrical-transformation approach to optical two-dimensional beam shaping", *Appl. Opt.* **36** (1097), 9129–9131.

[9] J. Cederquist and A. M. Tai, "Computer-generated holograms for geometric transformations", *Appl. Opt.* **23** (1984), 3099–3104.

[10] N. C. Roberts, "Multilevel computer-generated holograms with separable phase functions for beam shaping", *Appl. Opt.* **31** (1992), 3198–3200.

[11] T. Dresel, M. Beyerlein and J. Schwider, "Design of computer-generated beam-shaping holograms by iterative finite-element mesh adaptation", *Appl. Opt.* **35** (1996), 6865.

[12] H. Aagedahl, F. Wyrowski, M. Schmid, "Paraxial beam shaping and splitting", in Diffractive optics for industrial and commercial applications, J. Turunen, F. Wyrowski (eds), Akademie Verlag, Berlin, Germany (1997), 165–188.

[13] T. Tschudi and L. Wang, "Diffractive optics for semiconductor lasers", in Diffractive optics for industrial and commercial applications, J. Turunen and F. Wyrowski (eds), Akademie Verlag, Berlin, Germany (1997), 217–241.

[14] S. Sinzinger, K.-H. Brenner, J. Moisel, T. Spick and M. Testorf, "Astigmatic gradient-index elements for laser diode collimation and beam shaping", *Appl. Opt.* **34** (1995), 6626–6632.

[15] J. R. Leger, M. L. Scott and W. B. Veldkamp, "Coherent addition of AlGaAs lasers using microlenses and diffractive coupling", *Appl. Phys. Lett.* **52** (1988), 1771–1773.

[16] Z. L. Liau, J. N. Walpole, D. E. Mull,C. L. Dennis and L. J. Missaggia, "Accurate fabrication of anamorphic microlenses and efficient collimation of tapered unstable-resonator diode lasers", *Appl. Phys. Lett.* **64** (1994), 3368–3370.

[17] J. S. Swenson, R. A. Fields and M. H. Abraham, "Enhanced mass-transport smoothing of $f/0.7$ GaP microlenses by use of sealed ampoules", *Appl. Phy. Lett.* **66** (1995), 1304–1306.

[18] Y. Asahara, S. Omi, H. Sakai, S. Nakayama and Y. Yoneda, "Gradient refractive index type anamorphic planar microlens and method of producing the same", US patent 4805997 (30 May 1986), see also *Appl. Opt.* **25** (1989), 2625.

[19] D. D'Amato and R. Centamore, "Two applications for microlens arrays: detector fill factor improvement and laser diode collimation", *Proc. SPIE* **1544**, Miniature and micro-optics: fabrication and systems applications, San Diego, CA, USA (1991), 166–177.

[20] J. J. Snyder, P. Reichert and T. M. Baer, "Fast diffraction limited cylindrical microlenses", *Appl. Opt.* **30** (1991), 2743–2747.

[21] B. Messerschmidt, T. Poßner and R. Goering, "Colorless gradient-index cylindrical lenses with high numerical apertures produced by silver-ion exchange", *Appl. Opt.* **34** (1995), 7825–7830.

[22] T. Poßner, B. Hoefer, G. Leibeling, B. Messerschmidt and G. Tittelbach, "Fabrication of optical sub-systems for hybrid integration in microsystems", *VDE-Verlag* , MICRO SYSTEM Technologies, Potsdam (1996), 579–584.

[23] Y. Kawamura, Y. Itagaki, K. Toyoda and S. Namba, "A simple optical device for generating square flat-top intensity irradiation from Gaussian laser beams", *Opt. Comm.* **48** (1983), 44–46.

[24] A.-K. Holmer and S. Hård, "Laser-machining experiment with an excimer laser and a kinoform", *Appl. Opt.* **34** (1995), 7718–7723.

[25] A. Bewsher and W. Boland, "Design of a single-element laser-beam uniform cross projector", *Appl. Opt.* **33** (1994), 7367–7370.

[26] C. Haupt, M. Pahlke, R. Krupka and H. J. Tiziani, "Computer-generated microcooled reflection holograms in silicon for material processing with a CO_2 laser", *Appl. Opt.* **36** (1997), 4411–4418.

[27] V. Soifer, V. Kotlyar and L. Doskolovich, "Iterative methods for diffractive optical elements computation", Taylor & Francis, London (1997).

[28] W. B. Veldkamp, "Laser beam profile shaping with binary diffraction gratings", *Opt. Com.* **38** (1981), 381–386.

[29] W. B. Veldkamp, "Laser beam profile shaping with interlaced binary diffraction gratings", *Appl. Opt.* **21** (1982), 3209–3212.

[30] M. T. Eisman, A. M. Tai and J. N. Cederquist, "Iterative design of a holographic beamformer", *Appl. Opt.* **28** (1989), 2641–2650.

[31] J. H. McLeod, "The axicon: a new type of optical element", *JOSA* **44** (1954), 592–597.

[32] J. H. McLeod, "Axicons and their uses", *JOSA* **50** (1960), 166–169.

[33] J. Durnin, "Exact solution for nondiffracting beams. I. The scalar theory", *JOSA A* **4** (1987), 651–654.

[34] J. Durnin, J. J. Miceli and J. H. Eberly, "Diffraction-free beams", *Phys. Rev. Lett.* **58** (1987), 1499–1501.

[35] I. A. Mikhaltsova, V. I. Navivaiko and I. S. Soldatenkov, "Kinoform axicons", *Optik* **67** (1984), 267–278.

[36] J. Turunen, A. Vasara and A. T. Friberg, "Holographic generation of diffraction-free beams", *Appl. Opt.* **27** (1988), 3959–3962.

[37] A. Vasara, J. Turunen and A. T. Friberg, "Realization of general nondiffracting beams with computer-generated holograms", *JOSA A* **6** (1989), 1748–1754.

[38] A. T. Friberg and S. Y. Popov, "Radiometric description of intensity and coherence in generalized holographic axicon images", *Appl. Opt.* **35** (1996), 3039–3046.

[39] J. Ojeda-Castañeda, R. Ramos and A. Noyola-Isgleas, "High focal depth by apodization and digital restorating", *Appl. Opt.* **27** (1988), 2583–2586.

[40] A. J. Cox and C. Dibble, "Nondiffracting beam from a spatially filtered Fabry-Perot resonator", *JOSA A* **9** (1992), 282–286.

[41] R. M. Herman and T. A. Wiggins, "Apodization of diffractionless beams", *Appl. Opt.* **31** (1992), 5913–5915.

[42] N. Davidson, A. A. Friesem and E. Hasman, "Holographic axilens: high resolution and long focal depth", *Opt. Lett.* **16** (1991), 523–525.

[43] L. Niggl, T. Lanzl and M. Maier, "Properties of Bessel beams generated by periodic gratings of circular symmetry", *JOSA A* **14** (1997), 27–33.

[44] J. Rosen and A. Yariv, "Synthesis of an arbitrary axial field profile by computer holograms", *Opt. Lett.* **19** (1994), 843–845.

[45] R. Piestun, B. Spektor and J. Shamir, "Wave fields in three dimensions: analysis and synthesis", *JOSA A* **13** (1996), 1837–1848.

[46] V. Kettunen and J. Turunen, "Propagation invariant spot arrays", *Opt. Lett.* **23** (1998), 1247–1249.

[47] R. G. Dorsch, A. W. Lohmann and S. Sinzinger, "Fresnel ping-pong algorithm for two plane CGH display", *Appl. Opt.* **33** (1994), 869–875.

[48] C. Froehly, B. Colombeau and M. Vampouille, "Shaping and analysis of picosecond pulses", in Progress in Optics **XX** (1983), E. Wolf (ed.), North-Holland, Amsterdam, 65–153.

[49] A. M. Weiner, J. P. Heritage and E. M. Kirschner, "High resolution femtosecond pulse shaping", *JOSA A* **5** (1988), 1563–1572.

[50] J. A. Giordmaine, M. A. Duguay and J. W. Hansen, "Compression of optical pulses", *IEEE J. Quant El.* **QE 4** (1968), 252–255.

[51] M. C. Nuss and R. L. Morrison, "Time domain images", *Opt. Lett.* **20** (1995), 740–742.

[52] M. C. Nuss, M. Li, T. H. Chiu and A. M. Weiner, "Time-to-space mapping of femtosecond pulses", *Opt. Lett* **19** (1994), 664–666.

[53] A. M. Weiner, D. E. Leaird, D. H. Reitze and E. G. Paek, "Femtosecond spectral holography", *IEEE J. Quant Electr.* **28** (1992), 2251–2261.

[54] B. H. Kolner and M. Nazarathy, "Temporal imaging with a time lens", *Opt. Lett.* **14** (1989), 630–632.

[55] A. W. Lohmann and D. Mendlovic, "Temporal filtering with time lenses", *Appl. Opt.* **31** (1992), 6212–6219.

[56] M. Testorf and U. Österberg, "Planar-integrated systems for pulse shaping", *OSA Techn. Digest* , **10**, Diffractive Optics and Microoptics, Kona, Hawaii (1998), 62–64.

[57] J. R. Leger, "External methods of phase locking and coherent beam addition of diode lasers", in Surface emitting semiconductor lasers and arrays, G. A. Evans and J. M. Hammer (eds), Academic Press, Boston, Mass., USA (1993), 379–433.

[58] J. R. Leger, G. J. Swanson and W. B. Veldkamp, "Coherent laser addition using binary phase gratings", *Appl. Opt* **26** (1987), 4391–4399.

[59] P. Ehbets, H.-P. Herzig, R. Dändliker, P. Regnault and I. Kjelberg, "Beam shaping of high power laser diode arrays by continuous surface-relief elements", *J. Mod. Opt.* **40** (1993), 637–645.

[60] G. J. Swanson, J. R. Leger and M. Holz, "Aperture filling of phase-locked laser arrays", *Opt. Lett.* **12** (1987), 245–247.

[61] L. Liu, "Talbot and Lau effect on incident beams of arbitrary wavefront, and their use", *Appl. Opt.* **28** (1989), 4668–4678.

[62] A. E. Siegman, "Lasers", University Science Books, Mill Valley, California (1986).

[63] A. N. Chester, "Mode selectivity and mirror misalignment effects in unstable laser resonators", *Appl. Opt.* **11** (1972), 2584–2590.

[64] H. Zucker, "Optical resonators with variable reflectivity mirrors", *Bell Sys. Techn. J.* **49** (1970), 2349–2376.

[65] A. Yariv and P. Yeh, "Confinement and stability in optical resonators employing mirrors with Gaussian reflectivity tapers", *Opt. Comm.* **13** (1975), 370–374.

[66] P. A. Bélanger and C. Paré, "Optical resonators using graded-phase mirrors", *Opt. Lett.* **16** (1991), 1057.

[67] P. A. Bélanger, R. L. Lachance and C. Paré, "Super-Gaussian output from a CO_2 laser by using a graded-phase mirror resonator", *Opt. Lett.* **17** (1992), 739–741.

[68] V. Kermene, A. Saviot, M. Vampouille, B. Colombeau and C. Froehly, "Flattening of the spatial laser beam profile with low loss and minimal beam divergence", *Opt. Lett.* **17** (1992), 859–861.

[69] J. R. Leger, D. Chen and G. Mowry, "Design and performance of diffractive optics for custom laser resonators", *Appl. Opt.* **34** (1995), 2498–2509.

[70] J. R. Leger, D. Chen and K. Dai, "High modal discrimination in a Nd:YAG laser resonator with internal phase grating", *Opt. Lett.* **19** (1994), 1976–1978.

[71] J. R. Leger, G. Mowry and D. Chen, "Modal analysis of a Talbot cavity", *Appl. Phys. Lett.* **64** (1994), 2937–2939.

[72] E. Ho, F. Koyama and K. Iga, "Effective reflectivity from self-imaging in a Talbot cavity and its effect on the threshold of a finite 2D surface emitting laser array", *Appl. Opt.* **29** (1990), 5080–5085.

[73] J. C. Ehlert, W. Cassarly, S. H. Chahmakijan, J. M. Finlan, K. M. Flood, R. G. Waarts, D. Nam and D. F. Welch, "Automated phase sensing and control of an external Talbot cavity laser with phase contrast imaging", *Appl. Opt.* **33** (1994), 5550–5556.

11 Microoptics for optical information technology

The handling of information involves three aspects: processing, communication and storage. All three are present in any complex information system such as a computer or a communications network. Optical information technology covers all three aspects named above. The most prolific example is fiber optics which has fully replaced electric communication for long-haul transmission and partly also for local area networks. For short and very short distances, as they occur in computer and switching systems, optical interconnections are currently being investigated. While for board-level interconnections waveguide solutions appear to be more suited, for chip-to-chip communication the parallelism of free-space optics is of interest for alleviating the existing communication problems in electronic computers. This approach requires a suitable microoptical hardware that fits into the highly integrated world of electronics. The challenge of packaging optical systems is also of importance when optics is used for data storage. Optical pick-ups for CD-ROMs are examples for the trend from miniaturized, discrete optics to integrated microoptics. Finally, optical information processing involves areas such as the optical processing of broadband radar signals and real-time pattern recognition based on optical correlation. For the latter application, computer-generated phase-only filters are being used as another example of microoptical elements.

The purpose of this chapter is not to give an overview of optical information technology, since this is not within the scope of this book. The interested reader might refer to other literature [1–7]. Here, we rather wish to consider the use of microoptics in processing, interconnections and optical storage.

11.1 Optical information processing

11.1.1 Analog information processing

Optical systems for analog information processing are based on spatial filtering techniques. Consequently, the basic optical setup of an optical processor is the so-called $4f$-setup shown in Fig. 11.1. In general, such a system can be used for correlating an input complex amplitude distribution $u(x', y')$ with a filter function $\tilde{F}(\nu_x,\ \nu_y)$ located in the Fourier plane. The functionality and the quality of the optical processing system depend on the potential for implementing the filter functions. Many tasks which are useful for image processing can be performed with low complexity filters. Important examples are dark-field filters for edge recognition, inverse filtering for image restauration, apodisation and the phase-contrast technique discussed in Chapter 9 in the context of array illumination. Matched complex filters implementing amplitude as well as phase values are interesting for optical pattern recognition or feature extraction through correlation. For this purpose holographic techniques (e.g., Fourier holography and computer-generated detour-phase holography) have been developed (see Chapter 5). The concept of pattern recognition can be understood if we consider the case where the filter function $\tilde{F}(\nu_x,\ \nu_y)$ is identical to the complex conjugate of the Fourier transform of the incident object wavefront. Such a filter is called a "matched filter" :

$$\tilde{F}(\nu_x, \nu_y) = \tilde{U}^*(\nu_x, \nu_y) \tag{11.1.1}$$

In this case, in the correlation plane we obtain the autocorrelation AC which exhibits an intense "correlation peak" at the origin:

$$\mathrm{AC} = \iint u(x', y')u^*(x' - x, y' - y)\, dx'dy' \tag{11.1.2}$$

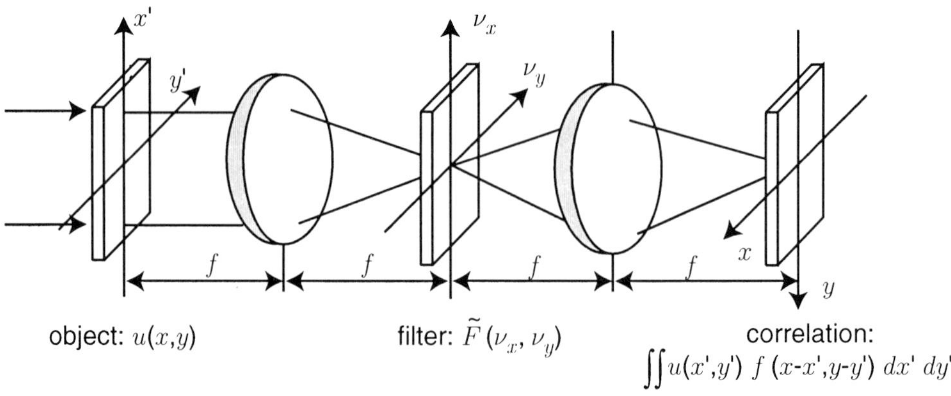

Fig. 11.1: $4f$-setup for optical correlation.

Real-time image processing (e.g., of synthetic aperture radar images), associative memories and neural networks are the most important application areas for analog optical processors [8–11]. The use of microoptical techniques in these fields is mainly focused on the implementation of complex filter functions. The design freedom of lithographically fabricated optical elements, e.g., allows one to implement complex spatial filters for both spatially and rotationally invariant correlation systems. In this case the basic design considerations are analogous to the design of beam-shaping optical elements described in Chapter 10. In fact, beam shaping and the design of optical correlation filters can be considered as inverse problems. The goal for the design of beam-shaping elements is the generation (i.e., the synthesis) of a specific complex optical wavefront. In optical correlators, on the other hand, spatial filters are used for analysing an optical wavefront with respect to specific features. For a discussion of the design methods for correlation filters, the reader can thus be referred to the previous chapter.

Increased stability and a reduction of alignment problems for optical correlator setups can be achieved with the joint transform correlator [12]. In this configuration, object and filter are located in the same physical plane. This can be considered as the first step towards the integration of the optical system. Integration of optical correlators for image-processing applications, however, is facing the challenging need for large space-bandwidth products. From our discussion of the scaling behaviour of lenses (Chapter 2), we know that microlenses cannot support such large space-bandwidth products. Here, the hybrid techniques discussed in Chapter 7 again might prove useful. Integrated optical correlator systems for low space-bandwidth applications have been demonstrated recently [13, 14, 15].

11.1.2 Digital optical information processing

Optical correlators represent special purpose processors which are capable of performing specific operations very efficiently. Analog optical computing, however, is usually restricted to linear operations such as a Fourier transformation and correlation. General purpose computing requires the realization of nonlinear operations. The use of highly parallel free-space optical interconnections and optoelectronic switching devices has been investigated intensely during the past years [4]. The fine-grain (often gate-level) architectures, however, did not turn out to be promising. Nonetheless, out of these efforts the concept of "smart pixel devices" (Chapter 7) has emerged. In general, the smart pixel concept involves some amount of information processing. Often, however, a smart pixel may simply be an optical input/output device on an electronic chip consisting of an emitter/detector pair. As such, smart pixel arrays and three-dimensional free-space optical interconnections are of growing interest for VLSI.

11.2 Optical interconnects

The use of optics as an interconnection technology for electronic systems is motivated by the observation of a variety of advantages over electronic interconnections. Above all the 3D nature of free-space optics has been listed frequently as the major advantage, leading to the potential for interconnecting planar arrays of logic components in parallel by means of the third dimension. This is in contrast to the planar nature of electronic interconnections, which results in a limited number of interconnects (pins) with the origins at the chip boundaries (Fig. 11.2). In particular, the unfavourable scaling behaviour is a major problem in this case. With rapidly proceeding integration technology, the number of electronic logic circuits is constantly increasing proportional to the chip area A, while the number of electronic interconnects only scales as $\sqrt{A}$. The performance of electronics today is limited by the interconnect technology rather than by the switching speed of electronic devices. The topological advantage is complemented by physical as well as architectural reasons in favour of optical interconnections in VLSI systems [16–22].

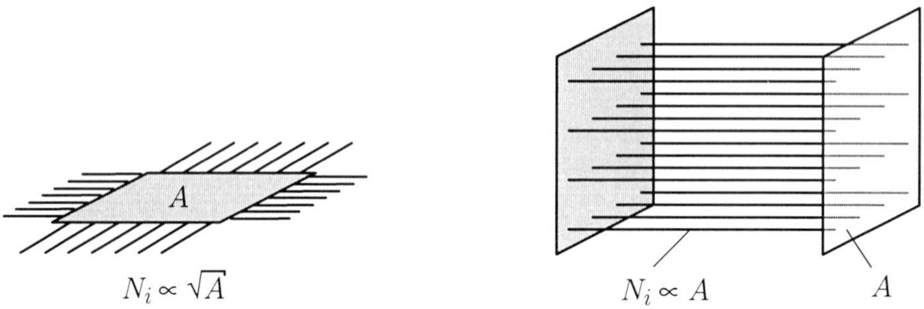

Fig. 11.2: Interconnect topologies and scaling of the number of interconnects: a) planar interconnections; b) 3D interconnections between 2D data planes.

Optical interconnections behave very much like perfectly matched transmission lines. The propagation speed in a transmission line is the same as in an optical fiber, for example. However, for electrical interconnections to achieve transmission line speeds, they must be driven by low impedance drivers and be terminated at the end of the wires in order to avoid reflections. Optical drivers always behave as if they were driven by a low impedance source and optical reflections at detectors are usually not a problem [23]. Optical interconnections also offer the advantage of a bandwidth-independent low absorption and they can support larger fanouts [24]. Another major advantage is that they lack mutual coupling effects, which is why optical signals can cross through each other without the generation of noise or loss of information.

11.2.1 Terminology

Space-invariant and space-variant interconnects

Interconnections for discrete input and output arrays can be represented as so-called bipartite graphs (Fig. 11.3). Both the input and output are represented by a regular array of positions that are denoted by characters (e.g., a, b, c, ...) or numbers (0, 1, 2, 3, ...). Very often the number of positions is a power of 2.

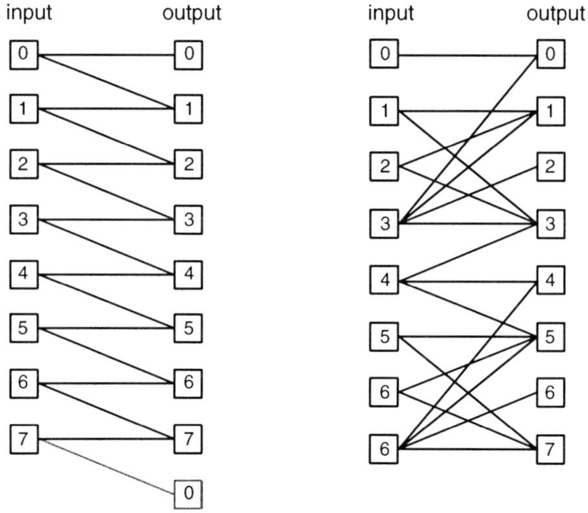

Fig. 11.3: Bipartite graph representing interconnection networks. a) space-invariant interconnects; b) space-variant interconnects.

Linear optical systems can be characterized as either space-invariant or space-variant. A system is called space-invariant if every input position generates the same output pattern (Fig. 11.3a). In a bipartite graph this means that the number and direction of the lines emerging from all input positions are the same. For arrays of finite size, space-invariance is not strictly possible since some lines would not connect to a position in the output array. Occasionally, the lines are wrapped around in a cyclical fashion. A system is called space-variant if the array of lines emerging from one input position to the next varies. This case is represented by Fig. 11.3b. Even though the space-variant interconnect scheme varies from one input to the next this does not necessarily mean that the interconnections are irregular. Therefore we also distinguish between regular and irregular interconnects.

Regular and irregular networks

These terms are not precisely defined since they are used according to the general understanding of what regular and irregular mean. As pointed out above, space-variance does not always mean irregularity. Certain space-variant interconnections exhibit a high degree of regularity. On the other hand, space-invariant interconnection networks are always regular. Actually, the interconnection patterns for space-variant multistage networks can be explicitly expressed in terms of a mathematical mapping of input to output positions. As we will see later on in this chapter, free-space optics can be used for implementing space-invariant regular interconnects as well as space-variant irregular networks.

Fixed and dynamic interconnections

In electronics, interconnections are usually fixed (with a couple of exceptions, such as switch-boards). For metallic wires, a change of a connection requires a mechanical displacement. As we have seen earlier, (micro)mechanical switches are also used in optics. However, since they are slow (milli- to microseconds) their use is limited. Optical interconnections can also be switched by changing the refractive index of a material. Examples are the directional waveguide coupler and dynamic holograms realized in photorefractive materials. Some of these effects can be very fast (nanoseconds or less) and they offer the possibility for dynamic routing of light signals. The usefulness of dynamic interconnections, however, is not quite clear. In computing, there appear to be only a few interesting applications. However, this could also be because current electronic computers are based on fixed interconnections.

11.2.2 Interconnect hierarchy

The data communication in a computer is subdivided into different levels of interconnections. For each of these levels a suitable technology has to meet specific requirements with respect to interconnection length and density. Generally, the longer the interconnection length, the smaller the number of interconnects. The different interconnect levels are generally called rack-to-rack, board-to-board, chip-to-chip and on-chip or gate-to-gate interconnects (Fig. 11.4). Optics is being considered for each of these levels of interconnections. Between racks, fiber-optical links have already been used for several years, for example, as the backplanes of switching systems. For board-to-board communications, polymer waveguides, fiber bundles and free-space optics can be used. For chip-to-chip interconnections, integrated packaging techniques using either waveguide or free-space optics are being investigated. Highly parallel low level optical interconnections, e.g., between smart pixels, which represent logic gates of variable complexity, have been tested successfully for optical switching networks.

Board-to-board and chip-to-chip interconnects

Due to the low number of interconnections and the long interconnect distances, fiber optics is a suitable technology at the rack-to-rack level. The discussion as to which technology is most appropriate for optical board-to-board interconnections, however, is more controversial [25]. Here, free-space optics as well as fiber and waveguide optical systems are being discussed. Since all of the different approaches are able to provide the necessary performance (e.g., moderate parallelism, interconnection density and distances), the most economical technology will prevail at this level. A fiber optical bus system with 20 parallel channels (bandwidth 500 Mbits/s each) has recently been shown to be feasible [26]. The important aspect in this case is the fact that the estimated manufacturing costs are competitive compared to those of conventional copper-based electronic systems. The applicability

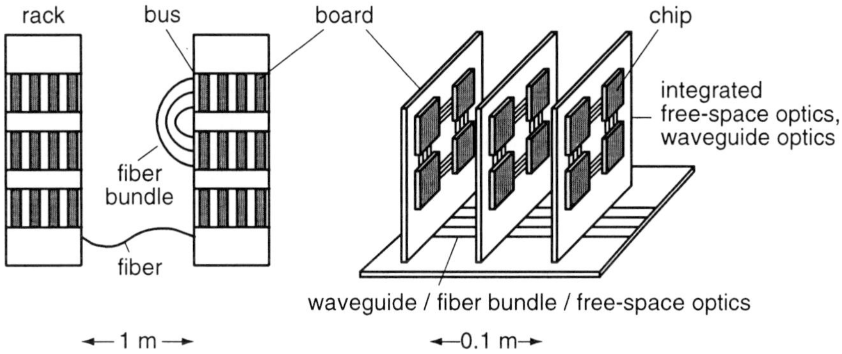

Fig. 11.4: Interconnection hierarchy in a computing system and suitable optics fabrication techniques. Typical transmission distances are on the order of metres for rack-to-rack communication, O(0.1 m) for board-to-board and O(1 cm) for chip-to-chip.

of free-space optics has also been demonstrated for the implementation of optical bus-type interconnections over distances suitable for board-to-board interconnections [27, 28, 29]. The challenge remaining here is the need for cost reduction, e.g., through further microoptical integration of the optical system.

Similarly, for the implementation of chip-to-chip level interconnections, the integration of free-space optical systems is one of the most important tasks [30]. Here, due to the high interconnection densities, waveguide optical approaches are not feasible. In Chapter 7 we have already discussed the trade-offs of the various possible imaging configurations for the implementation of board-to-board or chip-to-chip interconnects. The most compact solution using integrated free-space optics has been introduced in this context [31, 32].

(Micro)optical implementation of interconnection networks

Interconnection networks provide data transmission lines, e.g., between computers, circuit boards, chips or even the individual gates on the chip. They have a variety of applications, for example, in telecommunication and for parallel optical computing. The goal of any interconnection network is to provide an interconnection between each of the input channels and any output channel (point-to-point interconnect) [6, 33]. This goal can be achieved either in one interconnection stage, e.g., in a "Crossbar" interconnect. In this case each input is split and connected to each output channel (Fig. 11.5). In order to be able to control the data flow in this network we require N switches for each input channel, where N is the number of input and output channels. Since cost is proportional to the number of required switches, the Crossbar interconnect is an "expensive network". Additionally, the scaling behaviour $\propto N^2$ is unfavourable. Therefore, the interconnection pattern

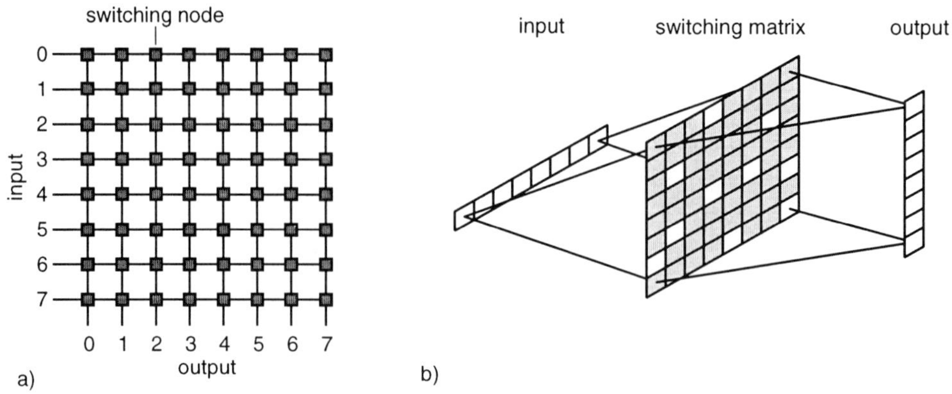

Fig. 11.5: Crossbar interconnect: a) interconnection scheme; b) possible implementation as a vector-matrix multiplier suitable for realisation in free-space optics.

is often implemented in several interconnect stages with less complexity. This results in so-called multi-stage interconnection networks (MINs) [34]. Depending on the interconnect scheme in each stage a variety of different networks may be distinguished, e.g., shuffle, permutation, crossover. The most popular example is the Perfect Shuffle network (Fig. 11.6). Other types are the "Banyan" and the "Crossover" network which are equivalent to the "Perfect Shuffle" interconnect with respect to performance and cost. These interconnection networks are based on 2×2 Crossbar switching nodes and generally consist of $\log_2 N$ interconnect stages to achieve point-to-point routing.

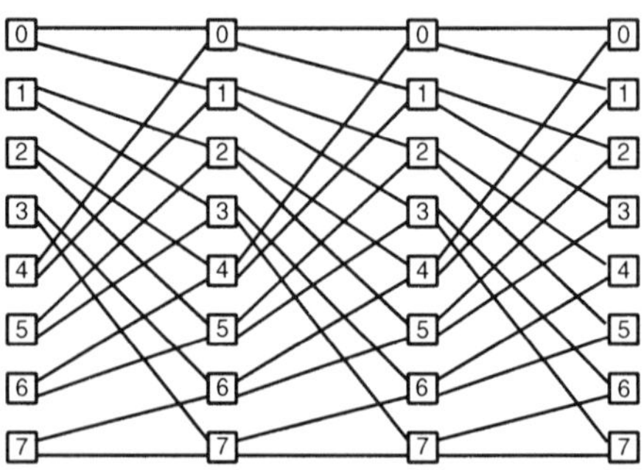

Fig. 11.6: Example of a multistage interconnection network (MIN): Perfect Shuffle interconnect.

In Chapter 6 we discussed the waveguide optical implementation of intercon-

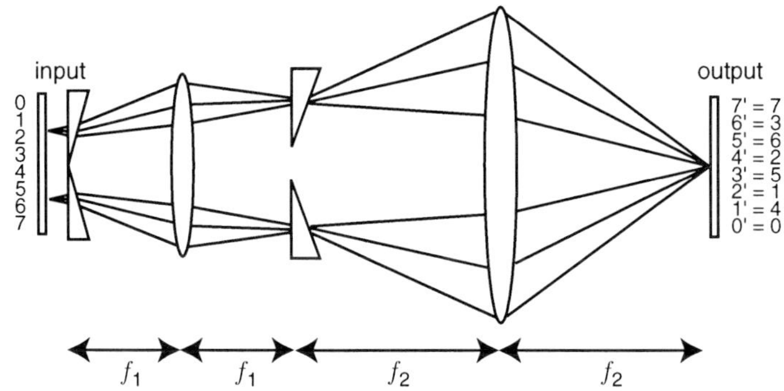

Fig. 11.7: Implementation of a Perfect Shuffle interconnect by aperture splitting using prism arrays.

nection networks. Now we focus on free-space optics. The optical implementation of interconnect networks, as shown in Fig. 11.6, is possible with some basic optical operations. Firstly, the data planes are duplicated in a "copy process". This is followed by a "shifted overlay". A large variety of implementation methods for optical interconnection networks using conventional optical components has been suggested [20]. The beam splitting can be performed by amplitude division (e.g., in interferometric setups using beam splitters or polarising beam splitters)[35, 36, 37] or aperture division, using prisms or lenses which split the aperture (Fig. 11.7) [38, 39, 40]. Alternatively, Mach-Zehnder interferometer setups, tilted mirrors and shifted lenses etc. have been used to produce shuffle interconnects.

Optical Crossbar interconnects can be implemented in waveguide optics using directional couplers or in free-space optics using vector-matrix multiplication as suggested by Goodman et al. (Fig. 11.5b) [41, 42]. Anamorphic imaging systems can be used for expanding the input array over a matrix of switching elements. The beam combination is carried out with a second anamorphic imaging step. Microlens arrays are useful for the implementation of Crossbar interconnects for 2D input arrays [43, 44]. Instead of anamorphic imaging, the broadcast interconnect can also be implemented using beam splitter gratings [45].

The development of lithographically fabricated microoptical components such as diffraction gratings allows one to realize more compact and efficient optical interconnection systems. Diffraction gratings, e.g., have been used as spatial carrier gratings for encoding the different input channels in order to be able to provide space-variant shifts using prisms [46]. Fig. 11.8 shows the use of microprism arrays for the implementation of an optical Crossover network [47]. Efficient diffractive beam splitter gratings offer new design freedom for interconnection networks. For efficiency reasons polarising optics is often used for the necessary beam splitting

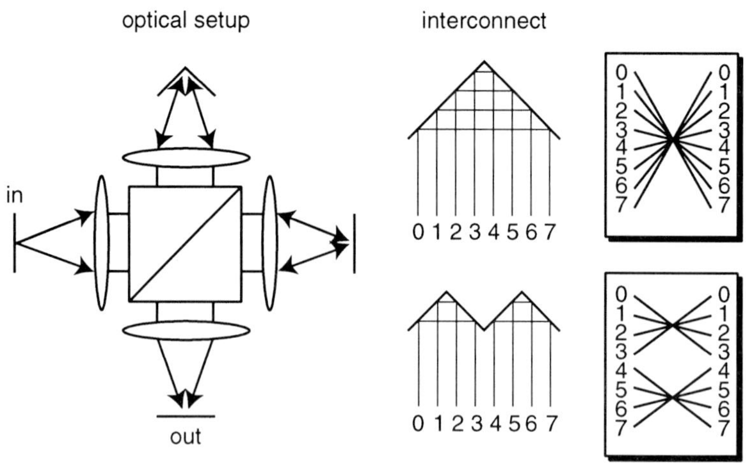

Fig. 11.8: Implementation of a Crossover interconnect using a beam splitter cube and prisms.

and beam combination. Here, polarisation-selective diffractive optical elements [48] or diffractive wave plates consisting of zero-order gratings (Chapter 5) can also be used.

Microlens arrays have been demonstrated for the implementation of a variety of different interconnect schemes [49]. Diffractive or holographic microlenses offer increased design and fabrication flexibility (Fig. 11.9) [50–58]. As discussed in Chapter 7, however, due to diffraction at the apertures (diameter d) of the microlenses or hologram facets, the interconnection distance $l_{\max}$ between two relaying microlenses is limited to $l_{\max} < \frac{d^2}{\lambda}$ [59]. This also leads to a limitation of the size of the input and output arrays if arbitrary interconnects need to be implemented [60] (see Exercises).

Several free-space optical interconnection networks have been designed and demonstrated, which are based on optomechanical alignment and discrete miniature components [61, 62, 63]. Various concepts for fully integrated optical interconnection networks have also been suggested, e.g., for planar optical systems [64, 65]. So far, however, integration of only parts of such networks have been carried out successfully [66, 67]. For example, split and shift modules using diffraction gratings have been integrated into planar optics [68]. Stacked optical systems using microprisms and GRIN microlenses have also been tested [69].

11.2.3 Optical clock distribution

Another important interconnection task in data processing systems is clock distribution [16]. Here, it is necessary to distribute a high frequency central clock

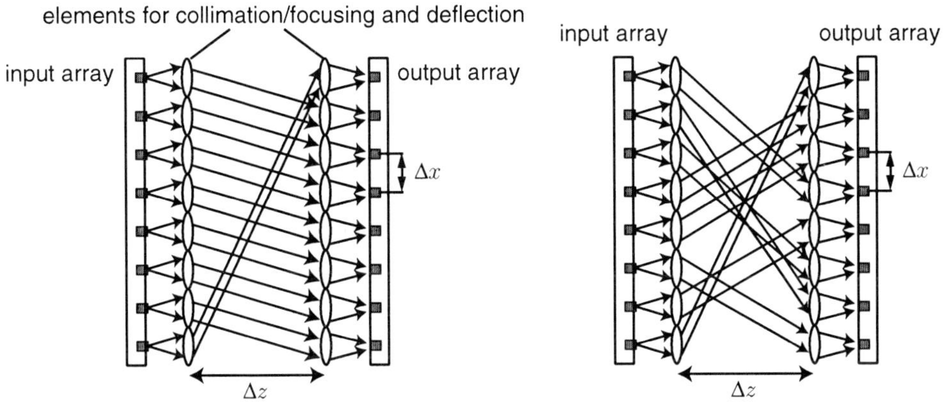

Fig. 11.9: Interconnection network implemented with microlens arrays. a) cyclic shifter; b) arbitrary irregular interconnects.

signal synchronously to a variety of locations on a wafer. With growing wafer dimensions and clock frequencies, this is an increasingly difficult task for electrical wires. Therefore, optical interconnect technology using waveguide or free-space optics has been suggested for this purpose. Efficient beam splitter gratings can be readily implemented in free-space optics for the distribution of the clock signal (see Chapter 8). An integrated optical system which uses a cascade of 1×2 beam splitter and deflection gratings has been demonstrated using planar optics [70].

In summary, the advantages of optical interconnections are not quite clear, as yet. High performance and reliability at low cost are necessary requirements for any kind of technology to be accepted in the competitive world of electronics. This is difficult to achieve, in particular for chip-to-chip interconnections. Here, free-space optics offers the largest potential but also faces a variety of difficulties. Mainly, packaging issues remain to be solved. This is where microoptics can help in future systems. Also, for free-space optical interconnections, the availability and performance of 2D arrays of light sources and detectors is crucial. Both modulator arrays and arrays of vertical cavity surface-emitting lasers (VCSELs) are being considered. For large 2D arrays, addressing mechanisms and thermal issues need to be taken care of. Future computing systems will continue to make use of metallic wires, but the availability of optical interconnections will give system designers alternatives in the engineering process [71–74].

11.3 Microoptics for optical data storage

The increasing processing speed of computers also brings about the need for mass storage systems with growing capacities and data access speed. With the compact disc (CD) system, optically addressed data storage already had its breakthrough about two decades ago [75]. However, since then, significant improvements in optical storage systems have taken place, which are partly due to the development of microoptics. In this section on microoptics for optical data storage, we first discuss some of the "basics of optical data storage", which will help us finding out the aspects where microoptics can help improving the performance of optical storage systems. Specific examples for the use of "microoptics in read/write heads" are discussed in the second subsection. Finally we will focus on the application of microoptics for "volume optical memories".

11.3.1 Basics of optical data storage

The most widespread optical storage system is still analog photography, which is capable of storing several gigabytes of information per picture. Due to the progress in digital computing and data transmission, however, more and more information, even image information, is nowadays available as digital information. This motivated the effort to adjust optical systems to digital data storage. Once again the invention of the laser marks the beginning of the development. The laser is the tool necessary to write digital information into photosensitive material with high contrast and writing density. Just as in the case of laser lithography (Chapter 3), the critical features are tight focusing and high energy density in the focus. Depending on the choice of photosensitive material, significant energy densities may be necessary during the writing process.

In order to calculate the storage density which can be achieved in optical storage systems we approximate the width of a focused laser beam as:

$$\delta x \approx \frac{\lambda}{\mathrm{NA}} \tag{11.3.1}$$

where λ denotes the wavelength and NA is the numerical aperture of the optical system. $\delta A = \delta x^2$ determines the minimum area covered by one optically addressable bit in an optical memory (in the theoretical limit: $\delta A = \lambda^2$). This is independent of the storage technique, i.e., if laser scanning systems are used for information reading/writing or if 2D data planes are written in parallel in holographic memories. For a wavelength of $\lambda \approx 1\ \mu$m and a numerical aperture of NA$= 0.5$ we find a theoretical storage density of $\frac{1}{\delta A} = 2.5 \cdot 10^5 \frac{\mathrm{bit}}{\mathrm{mm}^2}$.

In order to calculate the achievable volume storage density we have to take into account the depth of focus in the data plane:

$$\delta z \approx \frac{\lambda}{\mathrm{NA}^2} \tag{11.3.2}$$

The volume storage density for our example could be as high as $10^8 \frac{\mathrm{bit}}{\mathrm{mm}^3}$. Of course this estimation is based on the assumption that techinques for 3D data storage are available.

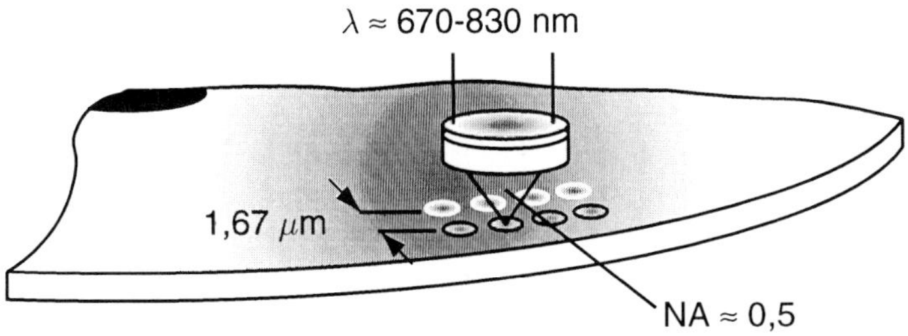

Fig. 11.10: Readout of an optical disc by scanning with a focused laser beam.

However, before we discuss the storage capacity of 3D optical memories any further, we focus on the mechanism of 2D optical disc storage. Such optical storage systems have already been commercially available for many years in the form of the compact disc storage system (CD). More recently, advanced standards with higher storage density have been introduced in the so-called DVD (digital versatile disc). Optical discs for read-only memories (CD ROMs) are mass-fabricated by injection moulding (Chapter 3) of master discs. The masters are fabricated by laser lithography. The bits of information are encoded as tiny pits of variable length and density in the disc surface [7]. The size of the pits is less than 1 μm with a lateral distance of 1.67 μm for standard CDs. Upon readout, the focused laser beam is scanned over the track of pits on the CD and is modulated by the surface profile or the structure of the material (Fig. 11.10).

In order to allow optical recording of the discs by the customer, new photosensitive materials have been exploited. For writeable optical disc systems, complex development processes after illumination have to be avoided. The surface structure or the change of the material properties have to be detectable immediately after illumination. Fig. 11.11 shows a variety of principles which can be used for optical recording [76]. Any of the processes results in changes of the material such that the reflected light of the scanning readout beam is modulated. In the case of the ablative and phase change process, the intensity of the reflected light is modulated. The ablative removal of an absorbing layer of polymer dye leads to a variation of the focus position and thus a modulation of the reflected beam intensity. Changing the recording layer from a crystalline to an amorphous structure results in a phase change and, depending on the layer thickness, in a change of the reflected light intensity. The crystalline/amorphous phase transition can be reversed and leads to rewritable optical discs. The principle of the magneto-optical disc is based upon local heating of the recording material by the laser radiation to temperatures, e.g., above the Curie temperature. In these areas the simultaneously applied magnetic field leads to a change of the magnetic organization. The readout is based on the magneto-optical Kerr effect which changes the polarisation of the reflected light. It is important to note that polarisation-sensitive optics is required for the readout.

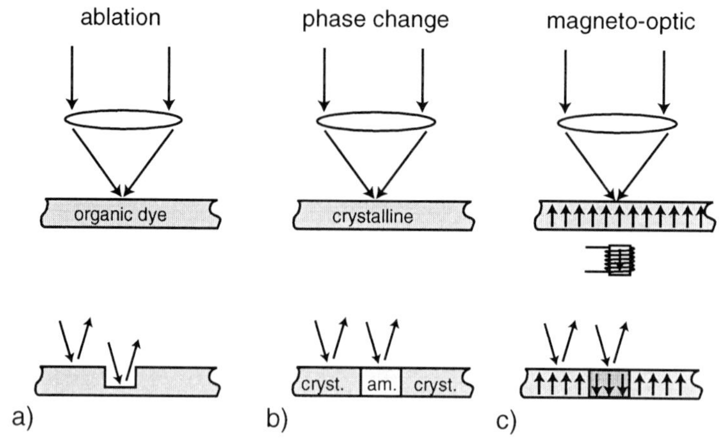

Fig. 11.11: Mechanisms for optical recording on CDs: a) photoablation for write-once-read-many (WORM) memories; b) phase change through crystallization; c) magneto-optical recording.

Rather than going further into the details of the physical effects of recording on optical discs we now wish to switch to the optics involved. We are especially interested in how microoptics can help improving optical disc storage systems. To this end we define the critical parameters which determine the performance of a mass storage system and outline possible improvements by means of enhanced optical systems:

- The **storage capacity** C is the overall amount of data which can be stored in the system. For a general volume data storage system we can calculate C as:

$$C = N_1 N_2 < \frac{A}{\delta A}\frac{t}{\delta z} < \frac{V}{\lambda^3} \tag{11.3.3}$$

  where N_1 is the area storage capacity which is dependent on the minimum area δA required for one bit and the total storage area A. N_2 is the multiplexing factor due to volume storage capacity. Theoretically, N_2 depends on the depth of focus δz and the thickness t of the storing medium. As we will see later on, in real volume storage systems N_2 is limited by the technological approach to volume storage. The considerations lead to a theoretical storing capacity in the total volume V of the medium.

  For increasing the storage capacity per unit area it is possible either to increase the storage density or to use larger areas for data storage. From the considerations above we find that the storage density is determined by the width of focus of the laser. Large NA optical systems lead to smaller focus sizes. However, since the depth of focus is also reduced with increasing NA, the tolerances during the readout are getting very tight. Therefore, currently the numerical apertures of optical read/write heads are limited to NA ≈ 0.5. Alternatively, microoptical components or systems can be helpful to exploit techniques for resolution enhancement, e.g., in the near field or in systems providing optical super-resolution.

The storage capacity can also be increased, e.g., by increasing the data storing area. Although it sounds very straightforward, this approach is not ideal in many cases. Often a large storing area results in long scanning distances of the read/write heads and slow access times. Typical examples are magnetic or optical tape memories. Optical data storage, however, offers the possibility for volume data storage. If the volume is modelled as a densely packed stack of thin storage discs, the increased storage capacity can be understood conceptually as an increased memory area. However, this is only partly true for holographic memories since the 2D information there is stored in the volume rather than in a thin slice of the memory volume. Examples of how microoptics can help implementing optical volume storage systems are discussed later on.

- The **data transfer rate** is defined as the amount of data from the memory which is available to the processor per unit time. The data transfer rate in optical memories is mainly determined by the time it takes to transfer the data to the processor. Here, optical interconnect technology can help to provide parallel high bandwidth access to the memory (see previous section).

- The data **access time** is the time necessary to access one specific piece of information from the mass storage system. Here, the crucial aspect is the localization of the piece of data and the time to access the specific location in the memory. Microoptics can help here to construct compact lightweight optical pick-up heads. Scanning and focusing, thus, becomes easier and faster. The access time is further reduced using optical systems for parallel memory access.

- **cost of the system**: Microoptical systems integration and replication techniques help to reduce the cost for optical pick-up systems significantly.

11.3.2 Microoptics for read/write heads

We shall now focus on some aspects of the application of microoptics for optical disc pick-up systems. We will learn how the design freedom of diffractive optics can significantly help to reduce the complexity of the pick-up head.

Inspite of the rather simple principle of the readout of optical discs illustrated in Fig. 11.10, for real applications the system is rather complex. Fig. 11.12a shows the configuration of a commercial pick-up system assembled from discrete optical

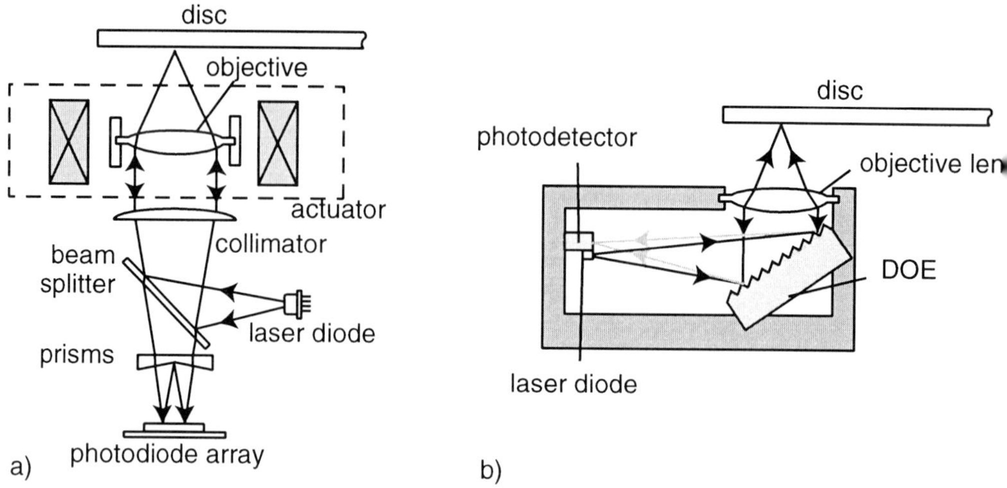

Fig. 11.12: a) Optical pick-up head using discrete optical components for beam focusing and splitting; b) integrated optical disc pick-up using a DOE with complex functionality.

components [77]. The figure illustrates that, due to the necessary beam-splitting elements, the complexity is significantly increased. A beam splitter is necessary for focusing the laser diode on the optics disc and extracting the signal. A second beam-splitting element, realized through aperture-splitting, prisms is used to focus the beam on the different sections of the photodiodes. This is necessary in order to gain information about the focusing and tracking of the beam on the optical disc, which is crucial for high contrast readout. Diffractive optical elements can integrate the functionality of several of the focusing and beam-splitting elements [78]. This helps constructing compact pick-up systems such as the one illustrated in Fig. 11.12b. Here the DOE combines the functionality of the beam splitter as well as the beam-splitting microprisms [79].

DOE optimization for optical disc pick-ups

It is worthwhile to take a closer look at the design of the DOE for this type of optical pick-up head [79]. Fig. 11.13 shows the shape of the DOE as well as a possible arrangement of the detectors for detection of the tracking as well as the focusing error (TE and FE). The DOE is spatially multiplexed to split and deflect the incident wavefront onto the different detector segments. From Fig. 11.12b we find that in the pick-up configuration the propagating wave interacts twice with the DOE, i.e., on the forward path and, after reflection at the disc, on the backward path to the detectors. This double path has to be taken into account for optimization of the DOE. In the forward path, no deflection or splitting of the illuminating beam by the DOE is desired. In other words, we use the non-diffracted zeroth diffraction order of the element. In the backward path, however, we wish to use the first diffraction orders of the different sections of the grating. Thus, we find for the efficiency η_S in the signal order:

$$\eta_S = \eta_0 \eta_1 \tag{11.3.4}$$

where η_0 and η_1 denote the diffraction efficiency of the DOE in the zeroth and first order, respectively.

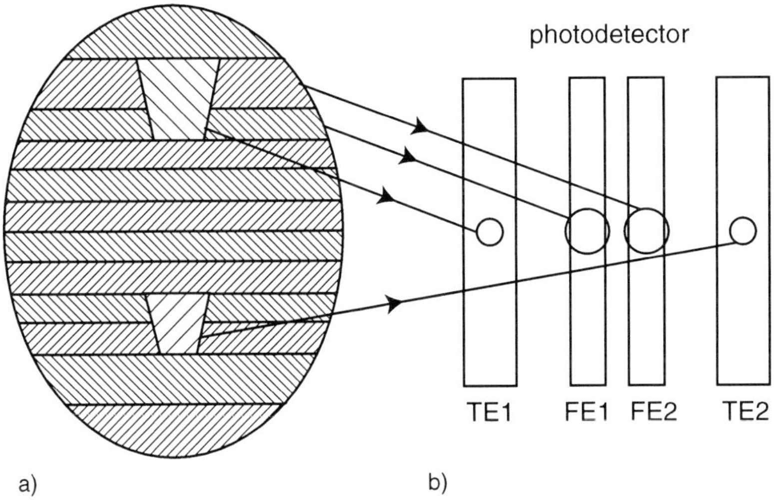

Fig. 11.13: a) Layout of the DOE for an optical disc pick-up performing beam splitting and deflection; b) possible arrangement of the detector segments for detection of the tracking and focusing errors (TE and FE; see text body for detailed explanation).

For best performance of the element we are interested in the optimum signal-to-noise ratio (SNR) which can be obtained. In order to calculate the SNR we identify the most intense noise orders of the DOE which end up at the detector locations. Except for the desired signal (0th order in the forward path and 1st order in the backward path), this is the case for the light which is diffracted to the -1st order during the forward propagation and is not diffracted in the backward path (0th order). A similarly intense component results upon deflection to the +1st order in the forward path and to the +2nd order in the backward path. In this case we took into account the inversion of the propagation direction due to reflection at the disc surface. The most intense noise contribution η_N is thus calculated to be:

$$\eta_N = \eta_{-1}\eta_0 + \eta_{+1}\eta_{+2} \tag{11.3.5}$$

For optimization we assume fabrication with four phase levels. Such a diffractive optical element is fabricated in two etching steps generating phase heights of ϕ_1, ϕ_2 (Fig. 11.14a). We can calculate the diffraction amplitudes a_m of such an element by Fourier-transforming the phase function $\phi(x)$ (Chapter 5):

$$a_m = \frac{1}{p}\int_0^p e^{i\phi(x)} e^{2\pi i m \frac{x}{p}} dx \qquad (11.3.6)$$

For the diffraction amplitudes and efficiencies we obtain:

$$\begin{aligned} a_0 &= \cos\left(\frac{\phi_1}{2}\right)\cdot\cos\left(\frac{\phi_2}{2}\right) \\ a_m &= \frac{4}{\pi m}\cdot\sin\left(\frac{\pi m}{4}\right)\cdot\cos\left(\frac{\phi_1}{2}-\frac{\pi m}{4}\right)\cdot\cos\left(\frac{\phi_2}{2}-\frac{\pi m}{4}\right) \end{aligned} \qquad (11.3.7)$$

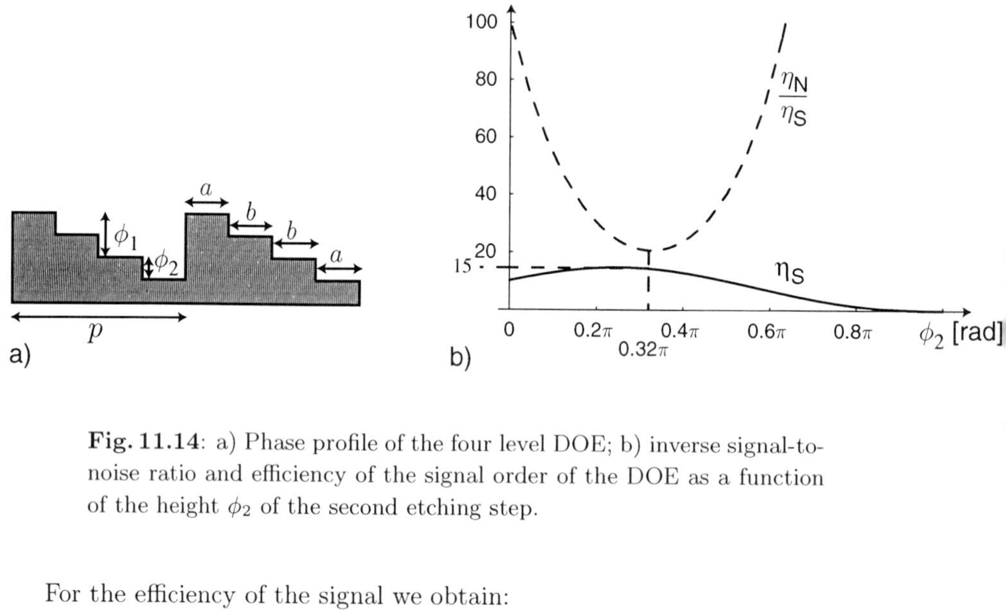

Fig. 11.14: a) Phase profile of the four level DOE; b) inverse signal-to-noise ratio and efficiency of the signal order of the DOE as a function of the height ϕ_2 of the second etching step.

For the efficiency of the signal we obtain:

$$\eta_S = \eta_0\cdot\eta_{+1} = a_0^2\cdot a_1^2 = \frac{2}{\pi^2}\cdot\left[\sin(\phi_1)\cdot\cos\left(\frac{\phi_2}{2}\right)\cdot\cos\left(\frac{\phi_2}{2}-\frac{\pi}{4}\right)\right]^2 \qquad (11.3.8)$$

The inverse signal-to-noise ratio results as:

$$\frac{\eta_N}{\eta_S} = \frac{a_{-1}^2 a_0^2 + a_{+1}^2 a_{+2}^2}{a_0^2 a_{+1}^2} = \left[\tan\left(\frac{\phi_2}{2}-\frac{\pi}{4}\right)\right]^2 + \left[\frac{2}{\pi}\cdot\tan\left(\frac{\phi_2}{2}\right)\right]^2 \qquad (11.3.9)$$

From Eq. 11.3.9 we find that the SNR is independent of ϕ_1. Thus, we choose $\phi_1 = \frac{\pi}{2}$, in order to maximize the light in the signal order according to Eq. 11.3.8. Fig. 11.14b shows the efficiency η_S in the signal order as well as the inverse SNR

$(\frac{\eta_N}{\eta_S})$ as a function of the phase step ϕ_2. We find the best performance, i.e., the best SNR for a phase step height of $\phi_2 = 0.32\pi$, which is close to the maximum of the signal intensity. In this case the element yields an SNR of ≈ 4.8 with an efficiency in the signal of 15%. Similar considerations for a DOE with 8 phase levels show that although the intensity can be increased through the additional etching step, no enhancement of the SNR is possible. For a configuration of the optical system as shown in Fig. 11.12, it is necessary to take into account the different angles of incidence on the DOE. They result in a variation of the effective phase step height which generates a variation of the diffraction efficiency. In order to avoid signal degradation, it is necessary to optimize the DOE in order to yield homogeneous diffraction efficiencies. The use of the widths a and b of the phase steps (Fig. 11.14) as design freedom has been successfully demonstrated for this purpose [79].

In addition to the readout signal, the optical disc pick-up system has to yield information about the focusing as well as the tracking of the laser focus on the CD. For this purpose, the segmented detector system illustrated in Fig. 11.13 is advantageous. The focusing error is detected with the so-called spot size detection method. The DOE sections which focus on the two detector segments FE1 and FE2 incorporate a weak positive or negative focusing power, respectively. Thus, in the detector plane the focus is not perfectly sharp but rather yields two blurred focus circles. The wave incident on FE1, e.g., has not yet reached its focus, while the wave on FE2 has already passed its ideal focus spot. Therefore, perfect focusing on the optical disc is reached when both blurred spots on the detectors have equal diameter. Since the spot diameter is wider than the detector width, the size of the individual spots is measured via the detector signal. Comparison of the relative spot sizes yields an enhancement of the measurement precision of the focusing error. The tracking error is detected as the differential signal between detectors TE1 and TE2. If the beam is shifted, the DOE segments which deflect the beam onto these detectors are not illuminated with equal intensity. This generates the differential signal [77].

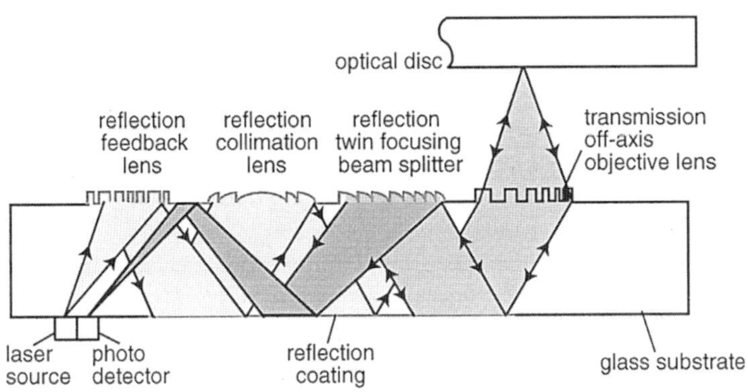

Fig. 11.15: Optical disc pick-up head integrated in planar optics.

The compact configuration of a disc pick-up using DOEs clearly demonstrates the potential of microoptics in optical storage systems. Even more compact disc pick-ups are possible using microoptical systems integration. A waveguide optical disc pick-up system has already been described in Chapter 6. A fully integrated planar free-space optical disc pick-up head has been suggested by Shiono et al. [80]. As illustrated in Fig. 11.15, multiplexed DOEs are again important components for this system. A laser diode is coupled into the substrate containing the necessary optical components as DOEs. Firstly, a reflection feedback lens is used for coupling a portion of the light back into the laser diode. This helps to stabilize the resonator and avoid wavelength variations. The off-axis collimator lens is optimized for collimation along the oblique optic axis (Chapter 5). The twin focusing beam splitter acts in an analogous way to the previously described optimized DOE. The zeroth order is used on the forward propagation to the optical disc. A transmissive off-axis diffractive lens focuses the beam onto the CD. On the backward path the beam is first collimated and coupled into the substrate to hit the twin focusing beam splitter. We now exploit the first diffraction orders of the segmented DOE. The implemented lens segments deflect the light and focus it onto the various photodiode segments for tracking and focusing error detection. Performance of an optimized off-axis transmission lens as well as a twin focusing beam splitter fabricated by e-beam writing has been tested successfully [80]. The diameter of the focus generated at the optical disc was 1 μm.

Resolution enhancement in optical pick-up heads

As mentioned above, it is desirable to increase the storage density of optical memories without increasing the numerical aperture to values much larger than 0.5. Therefore, a variety of techniques which allow one to achieve super-resolution have been suggested. A simple technique to achieve resolution enhancement is through the principle of synthetic aperture optics [81]. This is based on the observation that, by splitting the aperture of an optical imaging system, the extension of the central lobe of the diffraction pattern can be reduced. Fig. 11.16a shows a basic $4f$ imaging setup with an aperture splitting spatial filter. A simulation of the generated diffraction pattern in the image plane is shown in Fig. 11.16b which clearly shows a reduction of the lateral extension of the main central lobe for the split aperture. An optical pick-up system which exploits the potential of aperture splitting has been suggested by Yamanaka et al. [82]. They demonstrated resolution enhancement by a factor of 2 with respect to the conventional system. The disadvantage of aperture splitting is its low efficiency and the increased noise in the side lobes (Fig. 11.16b). In order to avoid crosstalk, an aperture stop has been used in front of the detector. A variety of alternative techniques for optical super-resolution has been suggested which might also be useful for optical data storage. Several of these techniques are based on the use of optimized diffraction gratings [83, 84]. The application of beam-shaping DOEs as described in the previous chapter can also prove helpful, e.g., to achieve enhanced depth resolution using non-diffracting Bessel beams.

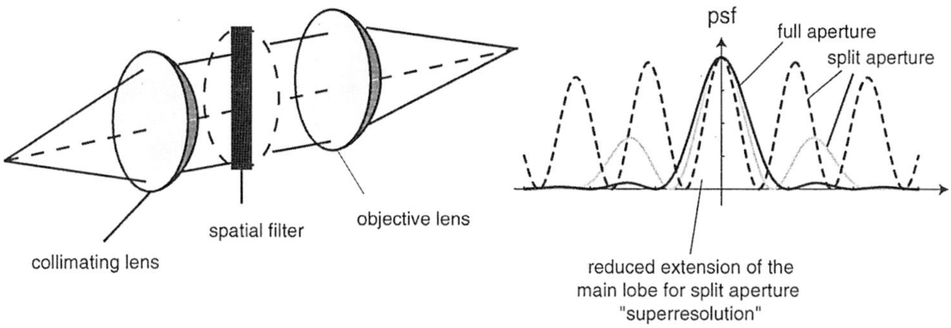

Fig. 11.16: The principle of resolution enhancement through aperture splitting; a) optical imaging system with split aperture; b) simulation of the the diffraction pattern of the aperture for different widths of the spatial filter.

Finally, we should like to discuss a further technique which can be exploited for high resolution imaging or sensing. Near-field imaging is a popular approach for resolution enhancement in optical microscopes. This technique is based on the fact that in the near field of an object the optical resolution is no longer limited by the wavelength, but rather by the lateral extension of the probe [85]. If a probe with an aperture much smaller than the wavelength is moved very close to the object it is possible to achieve super-resolution. The problems with this technique are twofold. Firstly, it is a technological challenge to fabricate the probe with apertures much smaller than the optical wavelength. For optical near-field microscopes fiber tips are often used with mirror pinholes of sub-wavelength aperture diameter [86]. The second problem related to near-field microscopy is the need to scan the probe over the object at very small distances.

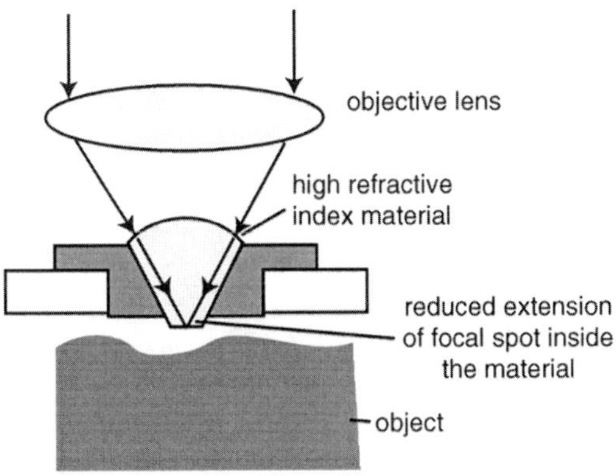

Fig. 11.17: The configuration of a solid immersion lens.

Near-field techniques have also been successfully demonstrated for application in optical data storage [87, 88]. Some of the technological problems of near-field microscopy are alleviated by using immersion techniques. In immersion microscopy the focal point of the objective lens lies in a medium with high refractive index. Since the wavelength scales as $\frac{1}{n}$, the focus size in the material can be significantly smaller than the vacuum wavelength ($\approx \frac{\lambda}{n}$). In combination with the scanning near-field technique, such a configuration also yields resolution enhancement. An example of this immersion technique is the solid immersion lens (SIL) [89]. As illustrated in Fig. 11.17, the SIL consists of an objective lens which generates a focus inside a high index material. In order to avoid refraction at the material interface the high index material has a spherical surface profile which is adjusted to the wavefront curvature of the focused beam. The "super-resolved" focus is formed at the lower surface of the high index material. Coupling between the object in the near field and the SIL takes place through the evanescent waves. The solid immersion lens has already been demonstrated for high density optical data storage. Focus diameters of several hundred nm have been achieved [90]. In the future, microoptical as well as lithographic techniques will help to overcome some of the remaining technological problems related to near-field optical data storage.

11.3.3 Volume optical memories

Stacked 2D optical storage systems

In this final section we shall focus on 3D optical data storage. As we have discussed above (Eq. 11.3.3), in order to increase the storage capacity C of optical memories it is interesting to increase the available storage area A or the volume multiplexing factor N_2. The conventional approach is to develop techniques for the efficient stacking of the layers of the storing material. This leads to magnetic tape or "juke box" memories, which can be considered as the first steps towards 3D data storage. Analogous optical approaches such as optical tape or optical disc stacks are also available in commercial systems. In fact, part of the increase in the storage capacity of the digital versatile disc (DVD) is due to the multiplexing in the third dimension. Both international standards which have been available commercially for about two years store information optically in two different layers of the disc ($N_2 = 2$) (Fig. 11.18). The remainder of the capacity increase is due to the step towards shorter wavelengths, which allows higher storage densities (Table 11.1).

In the multimedia compact disc we find one significant difference from pseudo 3D magnetic memory systems. Only one optical head with variable focal length is used for reading (or writing) the information stored in both layers at variable depths. The information in the lower level is read through the semi-transparent layer containing the upper data storage level. To this end, dual-focus heads have been developed using multiplexed diffractive optical elements [91]. Alternatively, variable focus optical heads could be used. An example for a variable focus optical head based on polarisation switching and a birefringent lens element has been

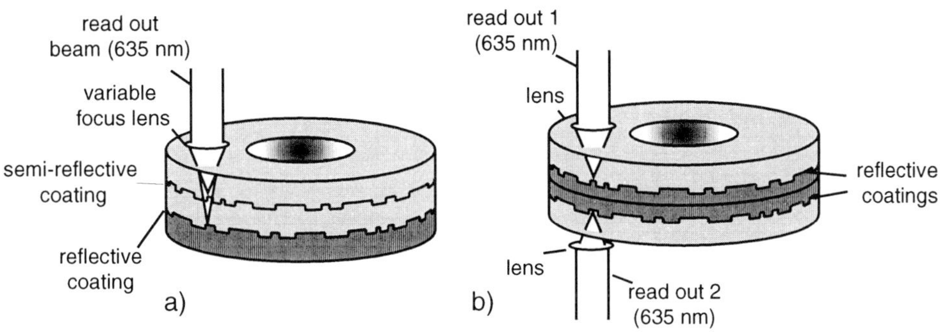

Fig. 11.18: Optical data storage in the digital versatile disc: a) multimedia compact disc (Sony/Philips); b) super density disc (Toshiba/Time Warner)

	CD	MMCD	SDD
diameter × thickness	120 mm × 1.2 mm	120 mm × 1.2 mm	120 mm × 1.2 mm
track pitch	1.6 μm	0.73 μm	0.84 μm
wavelength	780 nm	635 nm	635 nm
capacity	0.68 Gb	5/10 Gb	3.7/7.4 Gb

Tab. 11.1: Evolution of optical disc storage

tested by Wang et al. [92].

Two-photon optical recording

In the DVD approach, multilayer data storage is achieved by focusing the read or write beam through the partly transparent upper storage layer. This technique suffers from crosstalk. Since the laser beam is also partly reflected in layers which should not be read at a specific time, the signal-to-noise ratio in the signal is reduced. In order to keep the bit-error-rate sufficiently low, storage is only possible in a small number of layers which are separated sufficiently in depth. Precise control of the intensity in the focus of the read/write beam is required. For truly 3D data storage in a large number of layers ($N_2 \gg 1$), different techniques have to be used.

A promising approach to 3D storage is the so-called two-photon recording [93, 94]. Here, a positioning beam is used in combination with the data beam for providing enhanced discrimination between the storage layers (Fig. 11.19a). Two-photon storage has been successfully demonstrated in organic molecules such as spiropyrans. These molecules can be transformed into a modified state by photonic excitation. The diagram in Fig. 11.19b shows the energy levels of the molecule. The wavelengths of the data beam and the positioning beam are chosen such that an excitation is only possible if one photon of each beam is absorbed simultaneously. Due to the perpendicular incidence of the two beams it is possible to address one precise location in the 3D storing volume. Stimulated fluorescence of the excited molecules allows an analogous addressing scheme for memory readout. A possible source of crosstalk in this configuration is two-photon absorption initiated by a single incident beam of sufficient intensity. In Fig. 11.19 we consider a wavelength of 532 nm for the data beam and 1064 nm for the positioning beam. In this case two-photon interaction of the data beam could excite the molecules to a higher energy level (E_2). In order to minimize this source of crosstalk, the intensity of the data beam is reduced. At the same time, the intensity of the low energy positioning beam is increased proportionally. Thus, it is possible to achieve enhanced crosstalk suppression.

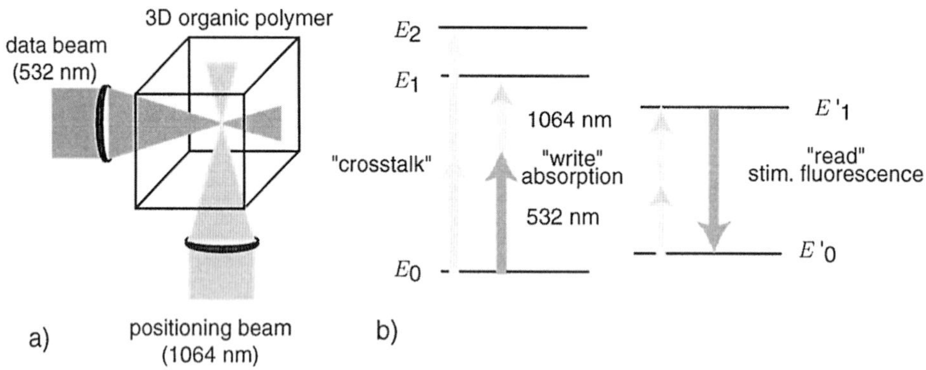

Fig. 11.19: 3D optical storage by two-photon recording: a) basic optical setup for writing and reading; b) energy levels of the recording medium.

The potential of this approach has been demonstrated in systems storing $N_2 = 100$ binary data planes each containing 100×100 bits in a 1 cm^3 memory cube of spirobenzopyran doped PMMA. In this case, one bit was stored in a volume of 30 μm $\times$ 30 μm $\times$ 80 μm corresponding to a storage density of $\approx 10^7 \frac{\text{bits}}{\text{cm}^3}$. Measured bit error rates were between 10^{-4} and 10^{-13}. Estimations show that storage densities as high as $10^{12} \frac{\text{bits}}{\text{cm}^3}$ can be expected for this approach [95]. Recently, optical disc-based two-photon optical memory systems have been demonstrated which might be promising candidates for future disc based fast access optical mass storage systems [94]. Microoptical approaches to packaging and integration will be necessary in order to develop compact and stable systems. For fast parallel data access, e.g., microlens arrays have been used in order to provide parallel access to

2D data planes.

Holographic memories

Another approach to 3D data storage is holography. This approach is fundamentally different from the previously described multiplexing techniques. In the two-photon memories independent access to the different layers in the storage volume is possible through the additional addressing beam. In volume holography the individual data planes are stored in the whole volume, rather than in one selected layer. These thick holograms are selective with respect to the wavelength or the angle of incidence of the reconstruction beam. This leads to multiplexing techniques which can be exploited for increasing the data storage capacity.

In Chapter 5 we discussed the application of thick holographic diffraction gratings as microoptical components. Holography, being a technique for recording complex optical wavefronts, can also be applied to optical data storage [96, 97, 98]. In order to understand the potential for storing the information in a 3D volume, we recall our analysis of thin and thick gratings. In Chapter 5 we defined thick diffraction gratings as gratings which generate ideally a single diffraction order which fulfils the Bragg condition. Coupled wave theory is a viable theoretical approach for analysing the diffraction in such thick holographic gratings. For diffraction gratings with finite thickness we found that the angular or wavelength selectivity of the gratings are useful criteria for deciding whether a grating is considered to be thin or thick. For optical data storage, the selectivity of thick holograms with respect to wavelength, angle of incidence or the phase of the reconstructing waves offers the potential to record multiple data planes in the storing volume (Fig. 11.20).

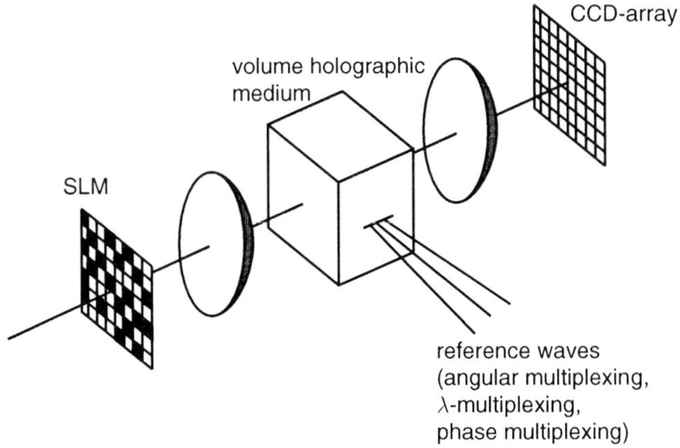

Fig. 11.20: Schematic optical setup for holographic data storage.

A variety of multiplexing schemes have been suggested for holographic volume data storage [99, 100, 101]. The most widely used techniques are wavelength, angular and phase multiplexing. In wavelength multiplexing, the wavelength is varied for the recording of each individual data plane. During the readout, the specific data planes can be addressed by the choice of the readout wavelength. For angular multiplexing, each hologram is recorded using a reference wave which is incident from a slightly different direction. In phase coding, the whole set of angular multiplexed reference waves is used for each of the holograms. However, the phase of each of the reference waves is varied deterministically from one recording step to the next. Optimum crosstalk suppression can be achieved if the phases of the reference beams vary between 0 and π and orthogonal phase addresses are chosen for each of the recorded images. Further enhancement of the crosstalk is possible if a uniform phase modulation is applied to the signal wave in order to reduce cross-correlation terms [102].

Using rigorous coupled wave analysis it is possible to find the angular selectivity $\Delta\theta_{\mathrm{R}}$ for the reconstruction of a hologram with thickness t [99]:

$$\Delta\theta_{\mathrm{R}} = \frac{8\lambda}{n\pi t} \frac{\cos(\theta_{\mathrm{S}})}{\sin(\theta_{\mathrm{R}} + \theta_{\mathrm{S}})} \tag{11.3.10}$$

Here θ_{S} and θ_{R} denote the angle of incidence of the signal and the reference beam, respectively, and n is the refractive index of the storing medium. If for the reference beam an angular range from θ_1 to θ_2 is used, the number N_θ of holograms which can be multiplexed in a volume of thickness t is:

$$N_\theta = 1 + \frac{n\pi t}{8\lambda} |\cos(\theta_1) - \cos(\theta_2)| \tag{11.3.11}$$

In an analogous way it is possible to calculate the possible number N_λ of wavelength-multiplexed data planes:

$$N_\lambda = 1 + \frac{nt}{2}\left(\frac{1}{\lambda_1} + \frac{1}{\lambda_2}\right) \tag{11.3.12}$$

Calculation of the overall capacity of the holographic memory is possible by replacing the multiplexing factor N_2 in Eq. 11.3.3 by the values calculated from Eq. 11.3.11 or 11.3.12 for angular or wavelength multiplexing.

Eqs 11.3.11 and 11.3.12 are only valid for the ideal case of diffraction of a plane wave at a thick volume grating. A general data plane is composed of a number of different plane wave components. This is the origin of crosstalk in angular multiplexed holograms. Similarly, the laser wavelength covers some spectral bandwidth, so that the storage density described by Eq. 11.3.12 cannot be reached without crosstalk. The phase-encoding of the reference beams is an approach which can significantly reduce crosstalk problems in multiplexed holograms. Currently, dynamic holographic materials such as photorefractive materials (e.g., $LiNbO_3$) are used for holographic optical memories. Due to the decay of the photorefractive phase changes in the material, which is supported by the writing of multiplexed holograms, it is necessary to refresh the holograms according to an optimized time schedule [103].

This brief discussion of the basics of holographic data storage suffices to point out the application areas for microoptics in holographic memory systems. Again, it will be necessary to fabricate integrated compact systems, which is possible only by using lithographically fabricated microoptical components. In today's laboratory experiments, microoptical components are already being used. For example, multiple beam splitters are useful for the generation of the reference beams in angular multiplexed holographic memories. Microlens arrays help to provide uniform illumination of spatial light modulators. The dispersion properties of diffractive optical components can provide reference beams for wavelength-multiplexing schemes.

11.4 Glossary

- **blocking and non-blocking interconnection networks**
  In complete interconnection networks it is possible to implement any arbitrary interconnection between the input and the output array. In non-blocking networks it is possible to implement any of these interconnections simultaneously. In blocking networks some interconnections are not possible while others are implemented.

- **digital versatile disc (DVD)**
  next stage in the evolution of optical disc storage, after the compact disc. The storage density and capacity of the system have been enhanced. The two major improvements are the step towards shorter wavelengths (635 nm compared to 780 nm), which allows better focusing and smaller tracking distances. Furthermore, spatial multiplexing is performed by storing information in two separate layers.

- **focusing error**
  High quality, low noise readout of an optical disc requires precise focusing of the optical beam onto the data pits on the optical disc. In order to be able to monitor the focus quality as well as the location of the focus, focusing and tracking error are measured during the readout. The focusing error is generally measured via the extension of the focus image on a split detector.

- **holographic optical memory**
  optical memory in which the data is stored holographically in the volume of the storing material (e.g., a photorefractive crystal). Large storage capacities can be achieved through multiplexing techniques.

- **multistage interconnection network**
  network providing switchable point-to-point interconnections between an input array and an output array. In order to be able to route a signal from any input to any output point in multistage networks, several planes (stages) with switching nodes are used.

- **Perfect Shuffle interconnect**
  scheme of fixed interconnections between an input and an output array. The interconnections are such that after propagating through the Perfect Shuffle input data from the upper half of the array is interlaced with input data from the lower half of the input array (Fig. 11.6a). The name stems from the shuffling of a pack of playing cards.

- **photorefractive effect**
  optically induced variation of the refractive index in specific materials, such as lithiumniobate ($LiNbO_3$). In the areas with intense incident light carriers are generated in the crystal which diffuse and are trapped at different locations; the resulting electric field inside the crystal leads to a modulation

of the refractive index, which allows one to record optical information holographically in the crystal. The photorefractive effect is reversible so that the memory is rewriteable.

- **switching node**
  switchable interconnection between a number of input channels and a number of output channels. Switching nodes are characterized by the number of input channels, the number of output channels and the number of simultaneously available connections. Integrated directional couplers are examples for (2,2,2) switching nodes, since they provide 2 inputs and 2 outputs, and both channels can be used simultaneously.

- **tracking error**
  error occurring upon readout of an optical disc if the focused laser beam is shifted laterally with respect to the data pits on the disc. The tracking error is generally measured as the differential signal of a split detector which receives light from different sections of the optical aperture.

- **two-photon recording**
  technique for optical data storage in the volume of the memory material rather than only at the surface. With this method, one specific point in the volume is addressed with two perpendicular optical beams. It is only possible to read or write information in the presence of both beams.

11.5 List of new symbols

AC:	autocorrelation
$a,\ b$:	parameters describing the width of the phase steps in a double pass DOE
C:	capacity of a storage system
DVD:	digital versatile disc
$\eta_\mathrm{N},\ \eta_\mathrm{S}$:	efficiencies of the noise and the signal beam in a double-pass DOE
$\tilde{F}(\nu_x,\ \nu_y)$	filter function in the spatial frequency domain; Fourier transform of $f(x,y)$
FE:	focusing error
$f(x,y)$:	filter function in the spatial domain
MIN:	multistage interconnection network
MMCD:	multimedia compact disc
N_λ, N_θ:	number of holograms which can be recorded in a holographic medium using wavelength or angular multiplexing
SDD:	super density disc
TE:	tracking error
t:	thickness of the storage volume of a 3D optical memory
θ_R; θS:	angle of incidence of a reference or a signal beam, respectively
$\Delta\theta$:	angular width of the main lobe of a Bragg-reflected wavefront, describing the angular selectivity
$\tilde{U}(\nu_x,\ \nu_y)$:	spatial frequency spectrum of a wavefront
$u(x,y)$:	complex amplitude distribution of a wavefront in the spatial domain
WORM:	write-once-read-many times data storage

11.6 Exercises

1. **Perfect Shuffle interconnection networks**
   A complete multistage interconnection network based on Perfect Shuffle interconnects consists of $\log_2 N$ stages.

   a) Make a drawing of the whole network for $N = 8$.

   b) Verify graphically that the network is complete, i.e., that it is possible to implement any arbitrary point-to-point interconnection between input and output array.

   c) Show that the Perfect Shuffle is not a non-blocking interconnection network (see Glossary), which means that it is not possible to implement arbitrary point-to-point interconnects simultaneously.

2. **Microlens arrays for arbitrary interconnects**
   Estimate the maximum number N of input channels (one-dimensional) that can be interconnected by microlens arrays in an arbitrary fashion.

   a) Assume a channel separation of $d = 125$ μm, a wavelength $\lambda = 0.633$ μm and a maximum propagation angle $\alpha_{max} = 45°$ with respect to the optical axis.

   b) How is this result affected if 4 phase level DOEs are to be used to provide the necessary deflections (minimum feature size 1 μm)?

3. **Storage density in different types of data storages**
   Calculate the storage density on

   a) a type-written sheet of paper (assume a page size of 210×300 mm^2 and about 5000 letters per page),

   b) an optical compact disc,

   c) a 16 Mb electronic CMOS memory chip with an area of ≈ 10 mm^2.

References

[1] A. W. Lohmann, "Optical information processing", Optik + INFO, Uttenreuth, Germany (1986).

[2] A. VanderLugt, "Optical Signal Processing", Wiley, New York (1992).

[3] B. Javidi and J. L. Horner (eds), "Real Time Optical Information Processing", Academic Press, Boston (1994).

[4] J. Jahns and S. H. Lee (eds), "Optical Computing Hardware", Academic Press, San Diego (1994).

[5] J. E. Midwinter (ed.), "Photonics in Switching I and II", Academic Press, San Diego (1993).

[6] H. S. Hinton, "An introduction to photonic switching fabrics", Plenum Press, New York (1993).

[7] A. B. Marchand, "Optical recording", Addison-Wesley, Reading (1990).

[8] D. Casasent (ed.), "Optical Data Processing", Springer Verlag, New York (1978).

[9] P. K. Das, "Optical Signal Processing", Springer Verlag, Berlin (1991).

[10] J. L. Horner (ed.), "Optical signal processing", Academic Press, San Diego (1987).

[11] F. T. S. Yu, "Optics and Information Theory", Wiley-Interscience, New York (1976).

[12] C. S. Weaver, J. W. Goodman, "A technique for optically convolving two functions", *Appl. Opt.* **5** (1966), 1248–1249.

[13] A. K. Ghosh, M. B. Lapis and D. Aossey, "Planar integration of a joint transform correlator", *Electr. Lett.* **27** (1991), 871–872.

[14] S. Reinhorn, Y. Amitai and A. A. Friesem, "Compact planar optical correlator", *Opt. Lett.* **22** (1997), 925–927.

[15] W. Eckert, J. Jahns, S. Sinzinger and V. Arrizón, "Design and fabrication of a compact planar-integrated correlator", LEOS 97 Annual Meeting, San Francisco (1997).

[16] J. W. Goodman, F. J. Leonberger, S. Kung and R. A. Athale, "Optical interconnections for VLSI systems", *Proc. IEEE* **72** (1984), 850–865.

[17] H. B. Bakoglu, "Circuits, interconnections, and packaging for VLSI", Addison-Wesley, Reading (1990).

[18] D. A. B. Miller and H. M. Ozaktas, "Limit to the bit-rate capacity of electrical interconnects from the aspect ratio of the system", *J. par. distr. comput.* **41** (1997), 42–52.

[19] J. Jahns, "Free-space optical digital computing and interconnection", in Progress in Optics **XXXVIII** (1998), E. Wolf (ed.), Elsevier, Amsterdam, 419–513.

[20] A. W. Lohmann, "What classical optics can do for the optical computer", *Appl. Opt.* **25** (1986), 1543–1549.

[21] M. R. Feldman, S. C. Esener, C. C. Guest and S. H. Lee, "Comparison between optical and electrical interconnects based on power and speed considerations", *Appl. Opt.* **27** (1988), 1742.

[22] N. Streibl, K.-H. Brenner, A. Huang, J. Jahns, J. Jewell, A. W. Lohmann, D. A. B. Miller, M. J. Murdocca, M. E. Prise and T. Sizer, "Digital Optics", *Proc. IEEE* **77** (1989), 1954–1969.

[23] D. A. B. Miller, "Optics for low-energy communication inside digital processors: quantum detectors, sources, and modulators as efficient impedance converters", *Opt. Lett.* **14** (1989), 146–148.

[24] J. W. Goodmann, "Fan-in and fan-out with optical interconnections", *Optica Acta* **32** (1985), 1489–1496.

[25] F. A. P. Tooley, "Challenges in optically interconnecting electronics", *IEEE J. Sel. Top. Quant. Electron.* **2** (1996), 3–13.

[26] D. M. Kuchta, D. Booth, W. Nation, C. DeCusatis and A. Muszynski, "Low cost 10 Gigabit/s optical interconnects for parallel processing", *IEEE Computer Society*, MPPOI98, Las Vegas (1998), 210–215.

[27] N. Streibl, R. Völkel, J. Schwider, P. Habel and N. Lindlein, "Parallel optoelectronic interconnections with high packing density through a light guiding plate using grating couplers and field lenses", *Opt. Comm.* **99** (1992), 167–171.

[28] T. Szymanski and H. S. Hinton, "A reconfigurable intelligent optical backplane for parallel computing and communications", *Appl. Opt.* **32** (1996), 1253–1268.

[29] A. G. Kirk et al., "Design and implementation of a two-stage optical power supply spot array generator for a modulator-based free-space interconnect", *EOS Meetings Digest* **3490**, Optics in Computing, Brugge (1998), 48–51.

[30] G. Verschaffelt et al., "Experimental demonstration of a multi-channel micro-optical bridge for multi-Gb/s, free-space intra MCM interconnects", *EOS Meetings Digest***3490**, Optics in Computing, Brugge (1998), 44–47.

[31] J. Jahns and B. Acklin, "Integrated Planar Optical Imaging System with High Interconnection Density", *Opt. Lett.* **18** (1993), 1594–1596.

[32] S. Sinzinger and J. Jahns, "Integrated microoptical imaging system with high interconnection capacity fabricated in planar optics", *Appl. Opt.* **36** (1997), 4729–4735.

[33] D. E. Smith, M. J. Murdocca and T. W. Stone, "Parallel optical interconnections", in Optical Computing Hardware, J. Jahns and S. H. Lee (eds), Academic Press, Boston (1994), 193–226.

[34] H. S. Hinton, J. R. Erickson, T. J. Cloonan and G. W. Richards, "Space-division switching", in Photonics in switching II, J. E. Midwinter (ed.), Academic Press, Boston (1993), 119–167.

[35] K.-H. Brenner and A. Huang, "Optical implementation of the perfect shuffle interconnection", *Appl. Opt.* **27** (1988), 135–137.

[36] S. Bian, K. Xu and J. Hong, "Optical perfect shuffle using Wollaston prisms", *Appl. Opt.* **30** (1991), 173–174.

[37] K.-H. Brenner and T. M. Merklein, "Implementation of an optical crossbar network based on directional switches", *Appl. Opt.* **31** (1992), 2446–2451.

[38] A. W. Lohmann, W. Stork and G. Stucke, "Optical perfect shuffle", *Appl. Opt.* **25** (1986), 1530–1531.

[39] J. Jahns, "Optical implementation of the Banyan network", *Opt. Comm.* **76** (1990), 321–324.

[40] T. C. Cloonan, "Free-space optical implementation of a feed forward crossbar interconnect", *Appl. Opt.* **29** (1990), 2006–2012.

[41] J. W. Goodman, A. R. Dias and L. M. Woody, "Fully parallel, high speed incoherent optical method for performing discrete Fourier transforms", *Opt. Lett.* **2** (1978), 1–3.

[42] A. A. Sawchuk, B. K. Jenkins, C. S. Raghavendra and A. Varma, "Optical crossbar interconnects", *Computer* **20** (1987), 50–60.

[43] M. Fukui and K. Kitayama, "Design considerations of the optical image crossbar switch", *Appl. Opt.* **31** (1992), 5542.

[44] C. Berger, N. Collings, R. Völkel, M. T. Gale and T. Hessler, "A microlens-array-based neural network application", *Pure Appl. Opt.* **6** (1997), 683–689.

[45] C. P. Barrett, P. Blair, G. S. Buller, D. T. Neilson, B. Robertson, E. C. Smith, M. R. Taghizadeh and A. C. Walker, "Components for the implementation of free-space optical crossbars", *Appl. Opt.* **35** (1996), 6934–6944.

[46] G. E. Lohman and A. W. Lohmann, "Optical interconnection network utilizing diffraction gratings", *Opt. Eng.* **27** (1988), 893–900.

[47] J. Jahns and M. J. Murdocca, "Crossover networks and their optical implementation", *Appl. Opt.* **27** (1988), 3155–3160.

[48] A. V. Krishnamoorthy, F. Xu, J. E. Ford and Y. Fainman, "Polarization-controlled multistage switch based on polarization selective holograms", *Appl. Opt.* **36** (1997), 997–1010.

[49] S. Kawai, "Free-space multistage optical interconnection networks using microlens arrays", *J. Lightwave. Techn.* **9** (1991), 1774–1779.

[50] J. Jahns and W. Däschner, "Optical cyclic shifter using diffractive lenslet arrays", *Opt. Comm.* **79** (1990), 407–410.

[51] B. Robertson, E. J. Restall, M. R. Taghizadeh and A. C. Walker, "Space-variant holographic optical elements in dichromated gelatin", *Appl. Opt.* **30** (1991), 2368–2375.

[52] U. Krackhardt, F. Sauer, W. Stork and N. Streibl, "Concept for an optical bus-type interconnection network", *Appl. Opt.* **31** (1992), 1730–1734.

[53] N. Davidson, A. A. Friesem and E. Hasman, "Realization of perfect shuffle and inverse perfect shuffle transforms with holographic elements", *Opt. Lett.* **31** (1992), 1810–1812.

[54] M. C. Hutley, P. Savander and M. Schrader, "The use of microlenses for making spatially variant optical interconnections", *Pure Appl. Opt.* **1** (1992), 337–346.

[55] R. K. Kostuk, J.-H. Yeh and M. Fink, "Distributed optical data bus for board-level interconnects", *Appl. Opt.* **32** (1993), 5010–5021.

[56] J. E. Morris and M. R. Feldman, "Reconfigurable optical interconnects by a spatial light modulator method", *Appl. Opt.* **33** (1994), 3683–3694.

[57] K. S. Urquart, P. Marchand, Y. Fainman and S. H. Lee, "Diffractive optics applied to free-space optical interconnects", *Appl. Opt.* **33** (1994), 3670–3682.

[58] Z. Zhou and T. J. Drabik, "Coplanar refractive-diffractive doublets for optoelectronic integrated systems", *Appl. Opt.* **34** (1995), 3048–3054.

[59] F. Sauer, J. Jahns, C. R. Nijander, A. Y. Feldblum and W. P. Townsend, "Refractive-diffractive micro-optics for permutation interconnects", *Opt. Eng.* **33** (1994), 1550–1560.

[60] J. Schwider, W. Stork, N. Streibl and R. Völkel, "Possibilities and limitations of space-variant holographic optical elements for switching networks and general interconnects", *Appl. Opt.* **31** (1992), 7403–7410.

[61] M. E. Prise, N. C. Craft, M. M. Downs, R. E. LaMarch, L. A. D'Asaro, L. M. F. Chirovsky and M. J. Murdocca, "Optical digital processor using arrays of symmetric self-electrooptic effect devices", *Appl. Opt.* **30** (1991), 2287.

[62] F. B. McCormick, F. A. P. Tooley, T. J. Cloonan, J. L. Brubaker, A. L. Lentine, R. L. Morrison, S. L. Hinterlong, M. J. Herron, S. L. Walker and J. M. Sasian, "Experimental investigation of a free-space optical switching network by using symmetric self-electro-optic devices", *Appl. Opt.* **31** (1992), 5431–5445.

[63] K.-H. Brenner, W. Eckert and C. Passon, "Demonstration of an optical pipeline adder and design concepts for its microintegration", *Opt. and Laser Techn.* **26** (1994), 229–237.

[64] S. H. Song and E.-H. Lee, "Planar optical configurations for crossover interconnects", *Opt. Lett.* **20** (1995), 617–619.

[65] S. H. Song, C. D. Carey, D. R. Selviah, J. E. Midwinter and E.-H. Lee, "Optical perfect-shuffle interconnection using a computer-generated hologram", *Appl. Opt.* **32** (1993), 5022–5025.

[66] S. H. Song, J.-S. Jeong and E.-H. Lee, "Beam-array combination with planar integrated optics for three-dimensional multistage interconnection networks", *Appl. Opt.* **36** (1997), 5728–5731.

[67] S. Reinhorn, Y. Amitai, A. A. Friesem, A. W. Lohmann and S. Gorodeisky, "Compact optical crossbar switch", *Appl. Opt.* **36** (1997), 1039–1044.

[68] J. Jahns and B. A. Brumback, "Integrated-optical split-and-shift module based on planar optics", *Opt. Comm.* **76** (1990), 318–320.

[69] J. Moisel and K.-H. Brenner, "Demonstration of a 3D integrated refractive microsystem", *Inst. phys. conf. series* **139**, International Conference on Optical Computing, Edinburgh (1994), 259–262.

[70] S. J. Walker and J. Jahns, "Optical clock distribution using integrated free-space optics", *Opt. Comm.* **90** (1992), 359–371.

[71] A. V. Krishnamoorthy and D. A. B. Miller, "Firehose architecture for free-space optically interconnected VLSI circuits", *J. par. distr. computing* **41** (1997), 109–114.

[72] A. Dickinson and M. E. Prise, "Free-space optical interconnection scheme", *Appl. Opt.* **29** (1990), 2001–2005.

[73] A. Huang, Y. Tsunoda, J. W. Goodman and S. Ishihara, "Optical computation using residue arithmetic", *Appl. Opt.* **18** (1979), 149.

[74] W. Erhardt and D. Fey, "Parallele digitale optische Recheneinheiten", Teubner Verlag, Stuttgart (1994).

[75] M. Mansuripur and G. Sincerbox, "Principles and techniques of optical data storage", *Proc. IEEE* **85** (1997), 1780–1796.

[76] C. F. Brucker, T. W. McDaniel and M. C. Gupta, "Optical data storage", in Handbook of Photonics, M. C. Gupta (ed.), CRC Press, Boca Raton (1997), 719–767.

[77] M. Kato, S. Kadowaki, Y. Komma and Y. Hori, "Recent advances in optical pick-up head with holographic optical elements", *Proc. SPIE* **1507**, Holographic Optics III: Principles and applications, The Hague (1991), 36–44.

[78] Y. Kimura, S. Sugama and Y. Ono, "Compact optical head using a holographic optical element for CD players", *Appl. Opt.* **27** (1988), 668–671.

[79] Y. Komma, S. Kadowaki, S. Nishino, Y. Hori and M. Kato, "A holographic optical element for an integrated optical head", *Opt. Rev.* **3** (1996), 251–257.

[80] T. Shiono and H. Ogawa, "Planar optical pick-up with diffractive micro-optics", *Appl. Opt.* **33** (1994), 7350–7355.

[81] J. W. Goodmann, "Synthetic aperture optics", in Progress in Optics **VIII** (1970), E. Wolf (ed), Elsevier, Amsterdam, 1–50.

[82] Y. Yamanaka, Y. Hirose, H. Fujii and K. Kubota, "High density recording by super-resolution in an optical disc memory system", *Appl. Opt.* **29** (1990), 3046–3051.

[83] D. Mendlovic and A. W. Lohmann, "Space-bandwidth product adaptation and its applications for super-resolution: fundamentals", *JOSA A* **14** (1997), 558–562.

[84] D. Mendlovic, A. W. Lohmann and Z. Zalevsky, "Space-bandwidth product adaptation and its application to super-resolution: examples", *JOSA A* **14** (1997), 563–567.

[85] D. W. Pohl, W. Denk and M. Lanz, "Optical stethoscopy: image recording with resolution $\lambda/20$", *Appl. Phys. Lett.* **44** (1984), 651.

[86] E. Betzig and J. K. Trautmann, "Near-field optics: microscopy, spectroscopy, and surface modification beyond the diffraction limit", *Science* **257** (1992), 189–195.

[87] E. Betzig, J. K. Trautmann, R. Wolfe, E. M. Gyorgy, P. L. Finn, M. H. Kryder and C.-H. Chang, "Near-field magneto-optics and high density data storage", *Appl. Phys. Lett.* **61** (1992), 142.

[88] B. D. Terris, H. J. Mamin and D. Rugar, "Near-field optical data storage", *Appl. Phys. Lett.* **68** (1996), 141–143.

[89] S. M. Mansfield and G. S. Kino, "Solid immersion microscopy", *Appl. Phys. Lett.* **57** (1990), 2615–2616.

[90] I. Ichimura, S. Hayashi and G. S. Kino, "High-density optical recording using a solid immersion lens", *Appl. Opt.* **36** (1997), 4339–4348.

[91] Y. Komma, S. Nishino and S. Mizuno, "Dual focus optical head for 0.6 mm and 1.2 mm discs", *Opt. Rev.* **1** (1994), 27–29.

[92] L. L. Wang, M. Hain and T. Tschudi, "Bifocal image system by using diffractive optical elements and liquid crystals", *EOS meetings digest* **12**, Diffractive Optics, Savonlinna (1997), 78–79.

[93] A. S. Dvornikov, S. E. Esener and P. M. Rentzepis, "Three-dimensional optical storage memory by means of two-photon interaction", in Optical Computing Hardware, J. Jahns and S. H. Lee (eds), Academic Press, Boston (1994), 287–324.

[94] F. B. McCormick, I. Cokgor, S. Dvornikov, J. Costa, N. Kim, H. Zhang, C. Chapman, I. Tomov, S. Esener and P. Rentzepis, "2-photon optical data storage", *EOS meetings digest* **3490**, Optics in Computing (OC98), Brugge (1998), 574–577.

[95] M. M. Wang, S. C. Esener, F. B. McCormick, I. Cokgor, A. S. Dvornikov and P. M. Rentzepis, "Experimental characterization of a two-photon memory", *Opt. Lett.* **22** (1997), 558–560.

[96] P. J. Van Heerden, "Theory of optical information storage in solids", *Appl. Opt.* **2** (1963), 393–400.

[97] E. N. Leith, A. Kozma, J. Upatnieks, J. Marks and N. Massey, "Holographic data storage in three-dimensional media", *Appl. Opt.* **5** (1966), 1303–1311.

[98] L. d'Auria, J. P. Huignard, C. Slezak and E. Spitz, "Experimental holographic read-write memory using 3D storage", *Appl. Opt.* **13** (1974), 808–818.

[99] H.-Y. S. Li and D. Psaltis, "Three-dimensional holographic optical discs", *Appl. Opt.* **33** (1994), 3764–3774.

[100] C. Denz, G. Pauliat, G. Roosen and T. Tschudi, "Volume hologram multiplexing using a deterministic phase encoding method", *Opt. Comm.* **85** (1991), 171–176.

[101] J. H. Hong, I. McMichael, T. Y. Chang, W. Christian and E. G. Paek, "Volume holographic memory systems: techniques and architectures", *Opt. Eng.* **34** (1995), 2193–2203.

[102] C. Denz, G. Pauliat, G. Roosen and T. Tschudi, "Potentialities and limitations of hologram multiplexing by using the phase-encoding technique", *Appl. Opt.* **31** (1992), 5700–5705.

[103] P. Yeh, C. Gu, C.-J. Cheng and K. Y. Hsu, "Hologram enhancement in photorefractive media", *Opt. Eng.* **34** (1995), 2204–2212.

12 Further application areas of microoptics

Throughout the discussions in this book, a variety of applications for microoptical components and systems have already been addressed. However, due to the abundance of application areas of microoptics, the list cannot be comprehensive. In this chapter we shall address several additional fields of application which are of actual interest in research or already in commercial use. We will discuss a few specific examples in some detail, while merely mentioning a variety of other important aspects. On reading this chapter, the reader will probably be able to find new possibilities for using microoptics.

Imaging is one of the most important tasks to perform in optical systems. In addition to the aspects of array imaging for optical interconnects (Chapters 7 and 11), in the first section of this chapter we wish to discuss the use of "microlens arrays for imaging" from a more general perspective. This helps us to understand the application of microlenses in systems such as photocopiers, fax machines and optical scanners. These applications have already been commercialized several decades ago. Imaging with microlens arrays may also be useful for unusual lithographic techniques. Finally, microlens arrays are useful for "beam steering", which is important in a variety of applications.

Another classical application area of optics is sensing. Compact integrated microoptical sensors are needed in areas such as medical or environmental monitoring. In the section on "microoptical sensors" we shall discuss the "Shack-Hartmann wavefront sensor" and "confocal sensing using microoptics" as two examples of free-space microoptical sensors. These systems have to compete with sensors based on integrated waveguide optics, some of which are addressed in Chapter 6.

In the final section, "microoptical components for optical design", we shall focus on the use of microoptics in optical design. We are going to address the correction of wavefront aberrations with hybrid diffractive/refractive elements.

12.1 Microlens arrays for imaging

With respect to the implementation of optical interconnects, we discussed in Chapter 7 imaging systems which are well suited for imaging discrete objects, e.g., laser arrays, detector arrays and smart pixel arrays. More specifically, we addressed the case of dilute arrays where the discrete points were not densely packed but spread over a large object plane. This section is devoted to another situation where the object is continuous or discrete but with densely packed object points. In this situation, microlenses can help in the construction of compact and cheaper imaging systems. The main problem related to conventional macroscopic imaging systems is the cost involved in high quality imaging lenses, which is increasing rapidly with the size of the image field. Secondly, the geometrical size of the system is becoming large, which makes alignment difficult and is the origin for problems with regard to system stability. In optical imaging systems, the field size Δx determines the diameter D of the imaging lenses and the length L of the imaging setup (focal length $F > D$) (Fig. 12.1a):

$$L > F > D > \Delta x \tag{12.1.1}$$

As we have seen in our discussion of the scaling behaviour of optical components (Chapter 2), for large lens systems it is becoming increasingly difficult to control wavefront aberrations. In combination with the technological challenges resulting from the processing of large lens surfaces, this results in an increase in cost for large imaging systems.

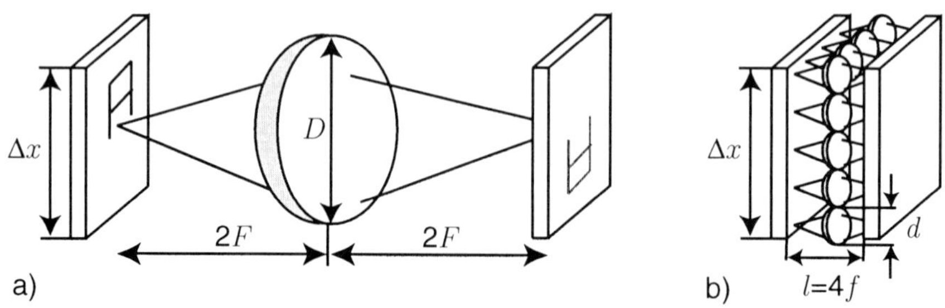

Fig. 12.1: a) Conventional macroscopic optical setup for imaging with no magnification; b) imaging setup using a microlens array

With microlens arrays it is possible to escape from this vicious circle [1]. Such arrays can be used to partition the image field and to image each of the parts separately through one of the lenses. Since microlens arrays can be fabricated in large arrays using lithographic fabrication and replication processes (Chapters 3–5), such systems can be significantly cheaper than systems with large lenses.

At the same time the systems are compact since the geometrical length l is now determined by the diameter d of the microlenses rather than by the diameter of the image field (Fig. 12.1b).

$$l > f > d \ll \Delta x \tag{12.1.2}$$

Reduction in the geometrical size of the imaging system is important for the packaging of the system. However, for the generation of high quality images with this setup several aspects have to be considered in designing the system:

- In the system shown in Fig. 12.1b, each of the microlenses generates an inverted image of a portion of the object plane. Unless the object has a very specific symmetry, this local inversion distorts the image. This can be avoided by using two microlens arrays forming an array of micro-telescopes, as illustrated in Fig. 12.2.

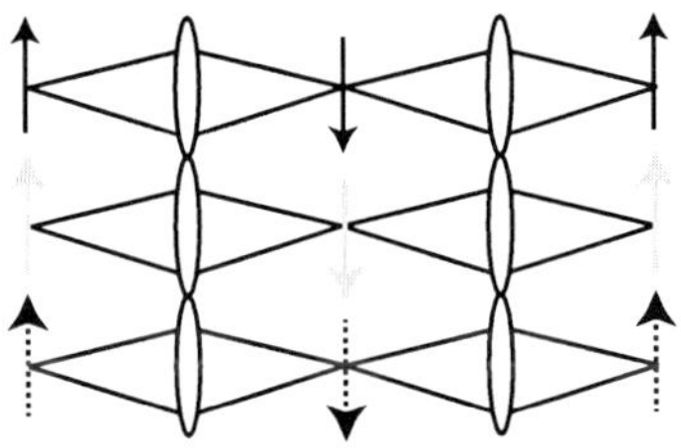

Fig. 12.2: Imaging using arrays of micro-telescopes which form erect images of each part of the object.

- Since the field of view of each microlens is not limited to the diameter of the lens, multiple images of each part of the object field are formed in the image plane. Geometrical considerations illustrate that these multiple images coincide for a strictly symmetrical configuration and form a so-called composite image. The relation to be fulfilled for the formation of the composite image is [1]:

$$z_C = z_O \tag{12.1.3}$$

However, as illustrated in Fig. 12.3a, this does not necessarily occur in the plane I, where sharp images are formed by the individual micro-telescopes. The distance z_I can be found by applying the basic imaging equation for each of the two imaging steps. If $z_I \neq z_C$, in the image plane multiple images of the object occur Fig. 12.3b. High quality imaging is possible in symmetrical setups, where the composite image plane coincides with the image plane of the microlenses. This symmetrical configuration leads to imaging systems

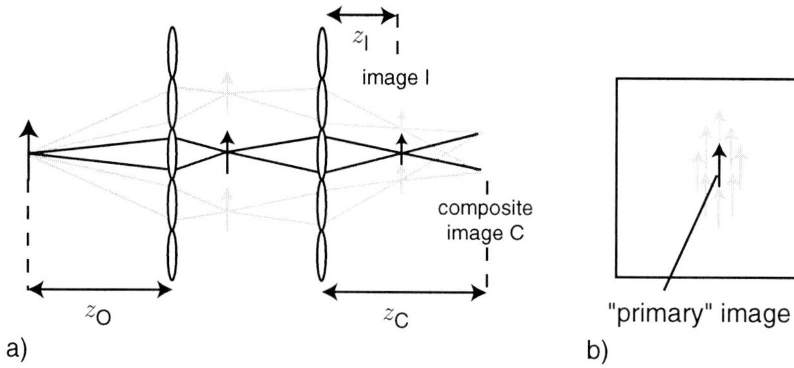

Fig. 12.3: Imaging using microlens arrays: formation of composite images. a) Location of the real image and composite image plane.
b) Light distribution (schematic) in the image plane.
The solid lines show the imaging of a section of the image field through the corresponding micro-telescope. The shaded lines indicate multiple imaging through adjacent channels.

which are free from antisymmetrical aberrations such as coma or distortion.

- In completely symmetrical microsystems, where the image plane I and the composite image plane C coincide, the only source of background noise stems from light transmitted through non-corresponding microlenses. This corresponds to the crosstalk discussed in Chapter 7 for microlens relay systems in optical interconnect systems. Field lenses help reducing this background noise due to crosstalk between the imaging channels [3].
- For the design of a microlens array-imaging system, it is necessary to consider the image field size of the microlenses as well as the fill factor of the arrays. Furthermore, for covering the whole image plane uniformly, it is required, that the lenses are capable of imaging field sizes larger than the lens area. Depending on the lens quality, this can be achieved at the expense of lateral resolution, if larger object and image distances are chosen [1].

Array-imaging systems as described above are already being used in fax machines, photocopiers and similar applications. Frequently, 1D arrays of GRIN SELFOC^{TM} lens arrays are used for the imaging [2]. In such systems it is important to achieve good uniformity in the intensity over the image plane. Aperture stops in the entrance and exit pupils as well as field lenses can be applied for increasing the uniformity. Combinations of refractive and diffractive microlens arrays with interesting imaging behaviour have been demonstrated [3]. The improvement of lithographic fabrication technology for microlens arrays with good imaging quality and high packaging density allows one to construct array-imaging systems for high resolution imaging. One of the frequently discussed new applications for such systems is microlens lithography. Here, microlens arrays are used for the imaging of a lithographic mask onto the wafer. Due to the small aperture, even diffraction-limited microlens imaging systems do not allow submicron resolution. Thus, they cannot compete with sophisticated lens systems designed, e.g., for high resolution lithography. Microlens array-imaging systems with a resolution of 2–5 μm have been demonstrated. The potential of microlens lithography lies in the possibility of using arrays with a large number of microlenses in imaging systems which can, nevertheless, be compact and cheap. They can be applied for projection lithography of large field sizes without the need of step and repeat or scanning systems. Due to the parallel arrangement of a large number of micro-imaging systems, the overall space-bandwidth product is very large.

The most critical feature of microlens arrays for imaging systems is the uniformity of the microlenses. The availability of cheap high quality microlens arrays will further increase the interest in array-imaging techniques such as integral photography or Gabor's "superlens" [4, 5]. Here, the array-imaging approach is generalized to the use of microlens arrays which differ in pitch or focal lengths. Significant potential might also be found in the combination with diffractive optical elements to form more general array optical elements [6].

12.2 Beam steering with microoptical components

The steering of a collimated laser beam is a fundamental function for a variety of optical systems. Typical applications, for example, can be found in optical scanners, optical data storage systems, laser printers, laser machining systems and in adaptive optics. High steering precision and speed, small optical aberrations as well as compact and lightweight systems are required for these applications. Traditionally, optical scanning systems consist of rotating mirrors or multiplexed holographic optical elements [7]. The use of microoptical elements in combination with micromechanical stages (MOEMS) has been proposed for this task [8, 9]. Flexible and mobile mirror structures can be used for the deflection of the reflected beam. However, in this case, beam steering requires rotational movements which are generally more difficult to control than lateral movements. Therefore, the use of pairs of lenses in a telescope-like arrangement has been suggested [10]. Lateral shifts of the lenses relative to each other provide agile beam deflection. Fig. 12.4a shows how beam steering is possible using two lenses which are shifted laterally. The phase functions of the shifted lenses are effectively combined to generate the phase wedge of a prism. The wedge angle of the prism depends on the relative shift. In order to be able to use micromechanical translation stages, the mass of the lenses has to be reduced as much as possible. This is one of the reasons why diffractive lenses are attractive for steering systems.

The deflection angle θ of the optical beam behind the telescope, decentered by Δx, is determined by:

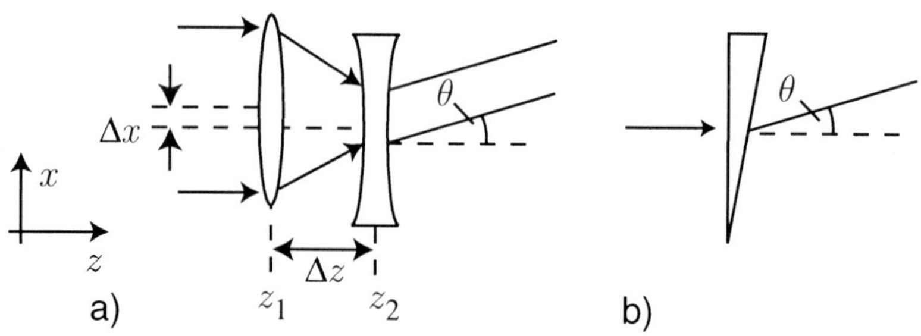

Fig. 12.4: a) Beam steering with shifted lenses in a telescope configuration with refractive lenses. b) Refractive prism resulting from an addition of the two shifted lenses.

$$\theta = -\arctan\frac{\Delta x}{f} = -\arctan\frac{\Delta x}{d(f/\#)} \tag{12.2.1}$$

where d is the diameter of the lenses with f-number $f/\#$. The negative sign indicates that the deflection takes place in the direction opposite to the lateral shift. For calculating the effect of the two lenses, we assume parabolic phase profiles:

$$t_1(x) = e^{-i\frac{\pi x^2}{\lambda f_1}} \quad \text{and} \quad t_2(x) = e^{i\frac{\pi(x-\Delta x)^2}{\lambda f_2}} \tag{12.2.2}$$

t_1 describes the transmission of the first lens element. The second lens element described by t_2 is shifted by Δx relative to the first lens. In order to yield a collimated beam in the output plane, the distance z_0 between the lenses has to fulfil the condition which is well known from the Galilean telescope:

$$\Delta z = f_1 - |f_2| \tag{12.2.3}$$

We assume that a uniform plane wave is incident onto the lens, which is described by the amplitude distribution $U(x, z_1 - 0) = 1$ (Fig. 12.4). Immediately behind the first lens ($z = z_1 + 0$), the wavefront has been modulated by the lens, $U(x, z_1 + 0) = U(x, z_1 - 0) \cdot t_1$. At a distance Δz behind the first lens, i.e., just in front of the second lens, the beam diameter is reduced and the beam curvature exactly matches the curvature of the second lens:

$$U(x, z_2 - 0) = U(x, z_1 - 0)\frac{f_1}{f_2}e^{-i\frac{\pi x^2}{\lambda f_2}} \tag{12.2.4}$$

Immediately behind the second lens the amplitude of the propagating wave is given by:

$$U(x, z_2 + 0) = \frac{f_1}{f_2}e^{i\frac{\pi \Delta x}{\lambda f_2}}e^{i\frac{2\pi x \Delta x}{\lambda f_2}} \tag{12.2.5}$$

The second exponential factor in Eq. 12.2.5 describes a linear phase shift across the aperture, which is proportional to the lateral shift Δx. This corresponds to the functionality of a beam-deflecting phase wedge.

According to Eq. 12.2.1, the deflection angle is inversely proportional to the diameter of the lenses. Small lens diameters allow large deflection angles for small translation distances. Therefore, microlens arrays have been suggested for steering systems yielding large deflection angles for translation distances of the order of the lenslet diameters (i.e., several hundred microns). Additional advantages are the design freedom for microlenses (especially when considering a diffractive implementation) and the potential for fabricating micro-opto-mechanical beam-steering systems based on microlens arrays [9].

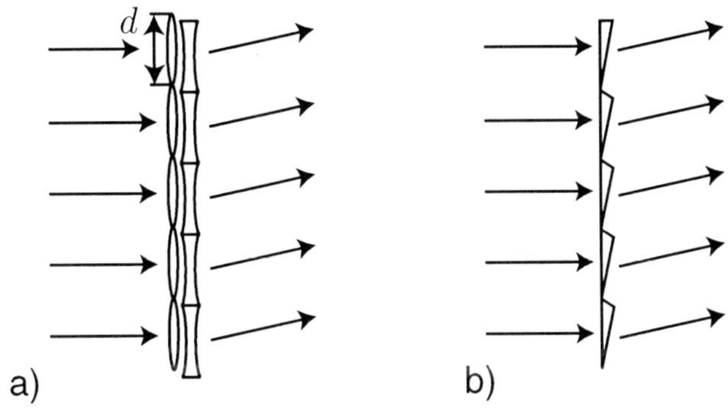

Fig. 12.5: Beam steering with microlens arrays; a) refractive implementation; b) corresponding blazed grating.

The periodic character of a microlens array causes some important effects on the behaviour of the beam-steering system. An investigation of the far-field diffraction pattern for the microlens array shows that optimum steering efficiency and crosstalk suppression is only achieved for discrete steering angles, corresponding to discrete translation shifts. The deflection is optimized if the steering angle coincides with the diffraction angles determined by the array factor of the microlens array. This leads to an angular resolution of:

$$\Delta\theta \approx \frac{\lambda}{d} \tag{12.2.6}$$

For intermediate angles, large amounts of light are diffracted into side orders [10]. In combination with Eq. 12.2.1 we find the best steering efficiency for discrete lateral shifts of multiples of δx where:

$$\delta x = d(f/\#)\tan(\Delta\theta) = d(f/\#)\tan\left(\frac{\lambda}{d}\right) \tag{12.2.7}$$

Another way of understanding this situation is by considering the resulting grating as a blazed grating. The angle of the prisms in each period depends on the lateral translation, while the period of the element remains constant. Optimum steering efficiency is achieved only if the blaze condition is fulfilled.

Diffractive implementation of microlens arrays with a quantized phase profile is especially interesting for beam-steering systems [10, 11]. It allows one to design the phase profile of the microlenses with a fill factor of 100%. In contrast to what one would expect intuitively, implementation of a beam-steering system using microlenses with quantized phase profiles yields better efficiency than a refractive arrangement. This is true independently of the fill factor of the individual microlens arrays. The reason for the good efficiency of the quantized steering system can be understood from Fig. 12.6. The efficiency of the refractive steering system is limited due to the reduced fill factor of the resulting blazed array. If the refractive lenses are shifted relative to each other, a linear phase slope results. However, at the edge of the period (for $x > d - n\delta x$) the overlap of the phase functions yields a second steeper slope with negative inclination. An incident beam is steered with an efficiency proportional to the fill factor of the desired phase slope in the array. Enhancement of the steering efficiency can be achieved if an additional lens array is introduced [12]. The total system then yields a confocal configuration of the three microlens arrays.

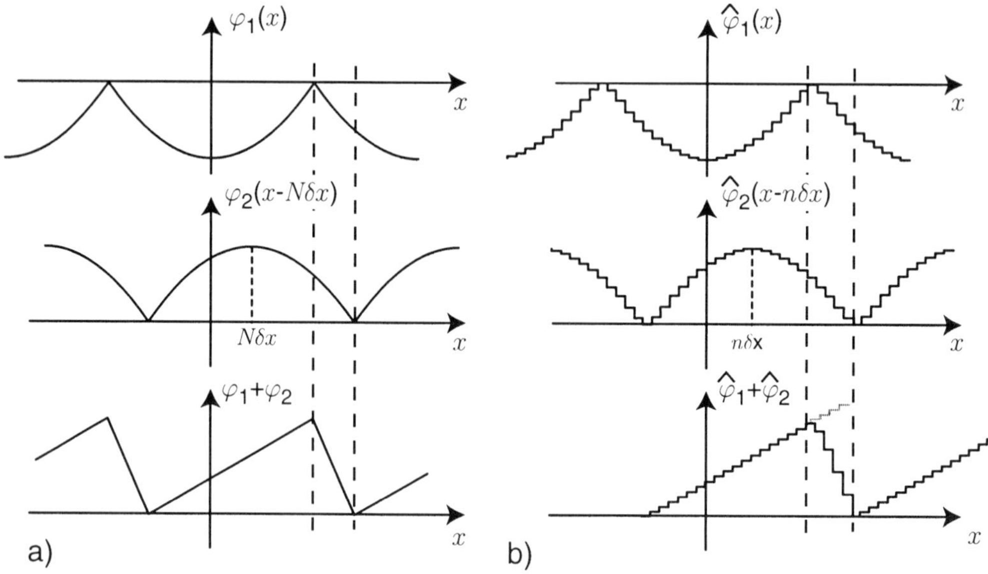

Fig. 12.6: Phase profiles of a section of the shifted microlens arrays: a) refractive lens profiles; b) quantized phase profiles.

The situation is different for the case of implementation with quantized phase profiles. As expected, the profile of the combined elements is also quantized. However, for the slope with negative inclination at the edges, the quantization has an important effect. Interestingly, the step height in this area is such that for the discrete steering shifts ($n\delta x$; n: integer), the profile differs by integral multiples of 2π from the profile desired for optimum steering (shaded line in Fig. 12.6) [11]. (In the following we shall denote the quantized phase functions by a "$\wedge$", i.e., $\hat{\varphi}$ is the quantized version of φ.)

For a relative lateral shift of $n\delta x$, we can calculate the resulting phase function $\hat{\Phi}(x) = \hat{\varphi_1}(x) + \hat{\varphi_2}(x - \delta x)$:

$$\hat{\Phi}(x) = \begin{cases} \dfrac{2\pi}{\lambda f}x^2 - \dfrac{2\pi}{\lambda f}(x - n\delta x)^2 = \text{const.} + \frac{4\pi}{\lambda f}xn\delta x & : \; x + n\delta x < \frac{d}{2} \\ \dfrac{2\pi}{\lambda f}x^2 - \dfrac{2\pi}{\lambda f}(x - n\delta x + d)^2 = \text{const.} + \frac{4\pi}{\lambda f}x(n\delta x - d) & : \; x > \frac{d}{2} - n\delta x \end{cases} \quad (12.2.8)$$

The two parts of Eq. 12.2.8 differ by a phase slope $\frac{4\pi}{\lambda f}xd$. Due to the periodicity of the lenslet array this corresponds to phase shifts of multiples of 2π. Using the paraxial approximation we can neglect such 2π phase jumps. In this case both parts of Eq. 12.2.8 represent an identical phase slope (shaded line in Fig. 12.6). One might also say that at the edges the continuous phase wedge is spatially undersampled, to yield precisely a quantized slope of "correct" inclination. Thus, as opposed to the refractive beam deflector, for the quantized lens arrays the aperture is filled completely by a quantized phase slope. The efficiency is given by the sinc-function which determines the efficiency of the quantized phase function (Chapter 5). For good quality and efficiency of the steering system, the uniformity of the microlens arrays is of significant importance [13]. This is yet another reason why lithographically fabricated microlens arrays are attractive for this application.

Microoptical steering systems which consist of arrays of decentered Kepler telescopes have also been analysed for beam steering and modulation [14]. For modulation, a spatial filter is introduced in the focal plane of the first array. The location of the focus is controlled by a shift of the microlenses in the input plane. If the spatial filter has a locally varying transmission, the beam can be modulated through this beam-steering process.

Alternative schemes for integrated steering systems use micro-electro-mechanically activated mirror structures [15] or active phase-modulating devices such as liquid-crystals or electro-optic materials. In liquid crystal device arrays, the implementation of variable lens profiles [16] as well as grating profiles for steering applications has been demonstrated [17, 18]. The implementation and cascading of optical phased array diffractive optical elements using electro-optic materials has been demonstrated recently [19].

12.3 Microoptical sensors

12.3.1 Shack-Hartmann wavefront sensor

Wavefront characterisation is an important task necessary for a variety of applications, e.g., adaptive optics for laser machining or astronomy [20, 21, 22]. While interferometric techniques yield high precision, they often require rather complex optical setups. Alternatively, microlens arrays can be applied to wavefront measurements in a so-called Shack-Hartmann configuration [23]. The basic measurement principle is fairly simple (Fig. 12.7). The wavefront of interest is incident on the microlens array and subdivided by the lens apertures (diameter: d). The positions of the foci generated by the microlenses determine the propagation direction of the respective section of the wavefront. The remaining task for determining the shape of the wavefront is to measure the location of the individual focus spots. The deviations Δx_i from the local optic axis is related to the local propagation direction (α_i: angles relative to the optic axis) of the wavefront:

$$\alpha_i \approx \tan(\alpha_i) = \frac{\Delta x_i}{d} \tag{12.3.1}$$

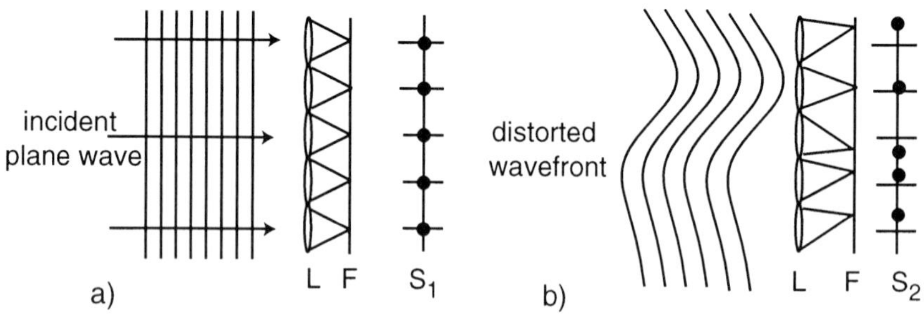

Fig. 12.7: Principle of wavefront sensing with the Shack-Hartmann sensor; a) ideal plane wave incident on the microlens array L, generating a regular pattern S_1 of spots in the focal plane F; b) distorted wavefront incident on the microlenses, generating an array of shifted foci S_2.

The propagation angle corresponds to the local direction of the wavevector $k_i = \frac{2\pi}{\lambda}\sin(\alpha_i)$. Just as in the ray-tracing design procedure, we can calculate the shape of the wavefront by partial integration (Chapter 2). This integration is possible with a variety of numerical methods, e.g., by polynomial fitting [24, 25].

In general, the resolution of the Shack-Hartmann sensor is determined by the wavelength (see Exercises). In combination with a demagnification of the wavefront onto the microlens array, however, it is possible to improve the phase resolution at the expense of lateral resolution. A critical issue for the accuracy of wavefront measurements with the Shack-Hartmann sensor is the precise measurement of the focus location. Numerical image-processing techniques help to reduce this source of error [21]. Another important aspect is again the uniformity of the microlens array. Here, a diffractive implementation offers advantages over refractive microlenses. On the other hand, the Shack-Hartmann technique is useful for determining deviations in the uniformity of microlens arrays.

12.3.2 Confocal sensing using microoptics

Fig. 12.8 shows a schematic diagram of a confocal imaging system. A point source is imaged onto an object by the lenses L_1 and L_2. The reflected or scattered light is collected by the objective lens L_2 and imaged via lens L_3 onto a point detector. Maximum intensity is measured at the detector if the object is in focus for both imaging systems. For a defocus distance Δz the detector intensity $I(\Delta z)$ drops sharply from the maximum value. Additionally, the detector intensity depends on the reflectivity and scattering behaviour of the object surface. By scanning the object it is possible to measure the surface structure and depth profile of the object. This principle is used, e.g., in microscopy, optical disc pick-ups (Chapter 11) and sensors. An analogous system is possible for use in transmission [26, 27, 28].

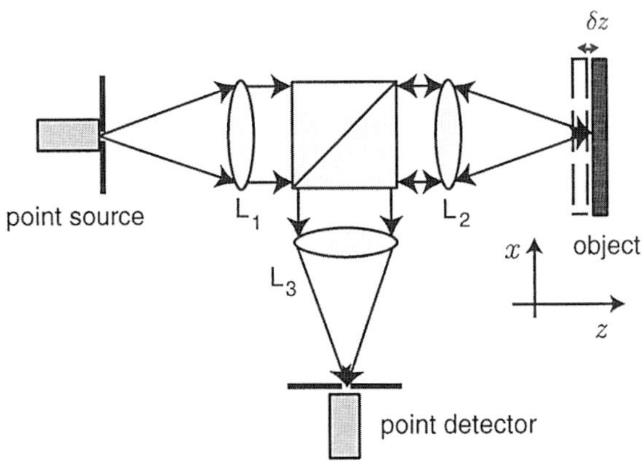

Fig. 12.8: Schematic of a confocal system for scanning microscopy or sensing applications. L_1–L_3: imaging lenses.

The main interest in the confocal configuration for imaging and sensing applications is initiated by the achievable resolution enhancement in an axial as well as a lateral direction [29]. Due to the two consecutive imaging steps, the point spread function of the confocal system is narrower than in the case of conventional microscopy with uniform illumination of the object. The 1D intensity distribution $I_D(x, z)$ in the detector plane can be approximated by [26]:

$$I_D(x,z) \approx \mathrm{sinc}^4\left[x\frac{2}{\lambda}\mathrm{NA}\right]\mathrm{sinc}^4\left[z\frac{2}{\lambda}\mathrm{NA}^2\right] \tag{12.3.2}$$

NA denotes the numerical aperture of the imaging systems and λ is the wavelength. x is a lateral coordinate and z is the axial coordinate in the imaging configuration. In Eq. 12.3.2, the intensity distribution is determined by the fourth power of sinc-functions. This describes the gain in lateral and axial resolution compared to a conventional imaging system, where the intensity distribution is approximated by the square of the sinc-functions (Fig. 12.9). The resolution enhancement in the axial direction corresponds to a reduced depth of focus. This leads to one of the main application areas for confocal sensors, e.g., in depth profiling and 3D image sensing. The gain in lateral and depth resolution is paid for by the need to scan the object in order to achieve a complete image of the object.

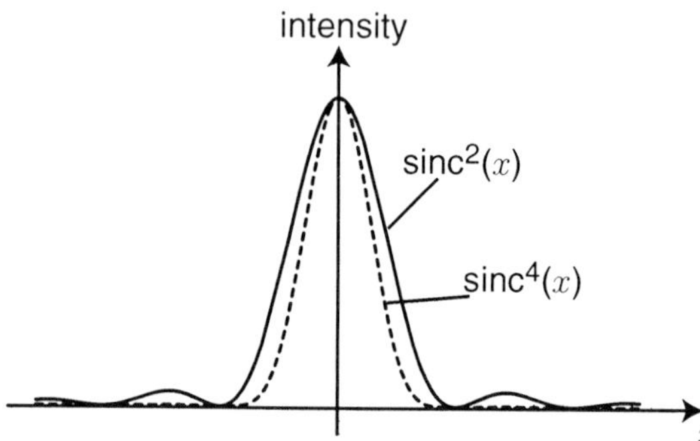

Fig. 12.9: Schematic of the square of the point spread function for conventional imaging and confocal imaging.

In the context of optical disc pick-ups we already discussed several possible application areas of microoptics in confocal sensors. We addressed the use of diffractive optics for obtaining integrated confocal pick-up heads as well as some possibilities for resolution enhancement. It is of specific interest to ease the scanning task. This is made possible using microlens arrays which split the aperture of the objective lens and project a large number of focused points onto the object. Such a confocal optical system has been suggested by Tiziani et al. (Fig. 12.10) [30, 31]. The setup is analogous to the conventional system except that lens L_3 is now used for forming an image of the microlens array on the CCD detector array. According to the concept of confocal imaging, a pinhole is used in the Fourier plane for spatial filtering of the signals from the individual microlenses. Due to the spatial multiplexing of confocal sensors using microlens arrays, the necessary scanning distance is significantly reduced.

Numerous microoptical techniques can be applied to confocal imaging, which help to improve the configuration of the system as well as the imaging and sens-

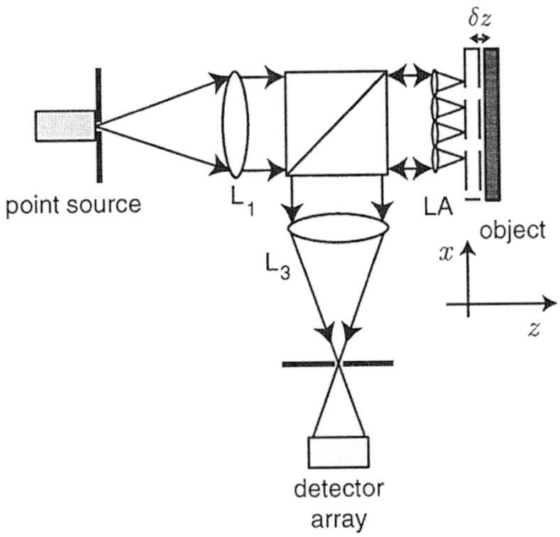

Fig. 12.10: Confocal imaging system using a microlens array (LA) instead of a single objective lens, as well as a detector array.

ing quality. Chromatic confocal microscopes have been suggested for wavelength-multiplexed confocal measurements [32, 33]. In particular, optimized diffractive optical elements can be useful in achieving specific chromatic behaviour of the imaging system (see following section) [34]. The spherical aberration introduced when focusing into the specimen is the main source of image distortion in confocal microscopy [35]. The design freedom of diffractive optics might be useful here for optimization of the image quality. Filtering and beam-shaping techniques can be applied for resolution enhancement in confocal sensors [36].

Due to the relatively simple optical configuration, confocal optical sensors are well suited for microoptical integration. Systems where a point source and a point detector are replaced by a scanning optical fiber have been suggested for applications in endoscopy [37, 38, 39]. One of the most critical issues for compact integration is the implementation of the beam splitter. Fiber-based systems often use fiber couplers for beam splitting. In free-space optics, double-pass diffractive optical elements, discussed in Chapter 11, can be applied. In side-looking confocal microscopy, imaging takes place along an oblique optic axis. Since the illumination and detection paths do not coincide in this case, no beam splitter is necessary [40]. This can be exploited for integration using planar microoptics [41]. Fully integrated confocal systems based on a combination of waveguide and free-space microoptics [42] as well as on planar optics [43] have been suggested. An extremely compact confocal system has been demonstrated recently by Dickensheets and Kino [44]. The optical system consists of monomode optical fibers for input and output as well as an optimized off-axis diffractive lens. The beam is scanned by two micromechanically fabricated mirrors along a folded optic axis. Except for the fused silica diffractive lens, the whole system is integrated in silicon with V-grooves for fixing the optical fibers. A lateral resolution of about 1 μm has been achieved with this system.

12.4 Microoptics for optical design

In our discussion of diffractive optical elements and their applications, we realized that one of their most important features is the design freedom provided by the lithographic fabrication. With the quantization rules introduced in Chapter 5, phase profiles with arbitrary shape can be implemented lithographically as a DOE. As opposed to refractive optics, the fabrication cost is not increased even for complex, "exotic" phase profiles. The only limitations for fabrication stem from the minimum feature size of the technology. In other words, the steepness of the phase slope must not be too large, in order to avoid features smaller than the minimum feature size.

For refractive optics, the situation is different. Very steep phase slopes are feasible with a variety of fabrication techniques such as diamond turning. Problems, however, arise for complex phase profiles with, e.g., aspherical shape. From these basic considerations we can already see the large potential of a combination of both approaches leading to hybrid diffractive/refractive optical elements (DROEs) [45]. The refractive part of these elements can involve large phase variations with steep slopes also being possible. The additional DOE applied to the refractive element can be designed, e.g., to correct the wavefront aberrations. In this section we shall discuss two applications of such DROEs, for chromatic correction and for athermalization.

12.4.1 Achromatic diffractive/refractive doublets

For many applications using broadband light sources (e.g., objective lenses for white-light photography, eyepieces for binoculars etc.), chromatic aberrations are the most important factor limiting the optical performance [46]. In refractive optical elements, chromatic aberrations stem from material dispersion. This material dispersion is generally described by the Abbe number ν:

$$\nu = \frac{[n(\lambda_{\mathrm{e}}) - 1]}{n(\lambda_{\mathrm{F}}) - n(\lambda_{\mathrm{C}})} \tag{12.4.1}$$

In the original definition of ν, λ_{e} denotes the wavelength of the green line in the mercury spectrum (λ_{e}= 546,1 nm), while λ_{F} and λ_{C} are the F and C lines in the Fraunhofer spectrum (λ_{C}= 656.3 nm and λ_{F}= 486.1 nm). These spectral lines have been chosen with respect to the spectral sensitivity of the human eye. For technical applications, the definition can be adapted to the desired wavelength region, so that λ_{e} is replaced by the center wavelength, and λ_{C} and λ_{F} mark the longest and the shortest wavelength in the spectral region of interest. Thus, ν is related to the specific spectral range. $\frac{1}{\nu}$ is often referred to, as the relative dispersion. Materials with normal dispersion have positive Abbe numbers. The dispersive effect of a lens element (focal length f) within the wavelength range $[\lambda_{\mathrm{F}}, \lambda_{\mathrm{C}}]$ is described

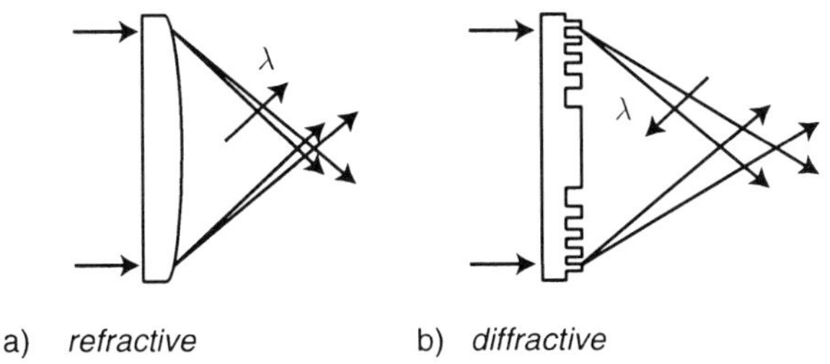

Fig. 12.11: Chromatic aberration of a) a refractive lens with normal dispersion resulting in an increasing focal length for increasing wavelength and b) a diffractive lens with negative dispersion, i.e., decreasing focal length for increasing wavelength.

by $f\nu_{\text{ref}}$. It is possible to combine two refractive lens elements with positive and negative focal lengths (f_1, f_2) to obtain an achromatic lens system. The element is achromatic if:

$$f_1\nu_1 + f_2\nu_2 = 0 \tag{12.4.2}$$

Except for the trivial solution ($f_1 = -f_2$), Eq. 12.4.2 can be fulfilled by a lens doublet made of two different materials. The resulting refractive achromatic doublets consist of a high power positive lens element made of a material with large Abbe number (e.g., BK7: $\nu_{\text{BK7}} = 64.12$) in combination with a low power negative lens in a low Abbe number material (e.g., F2: ν_{F2} =36.37) (Fig. 12.12a). Due to this combination of two lens elements which partly compensate each other in optical functionality, the refractive achromatic element is thick and clumsy [46].

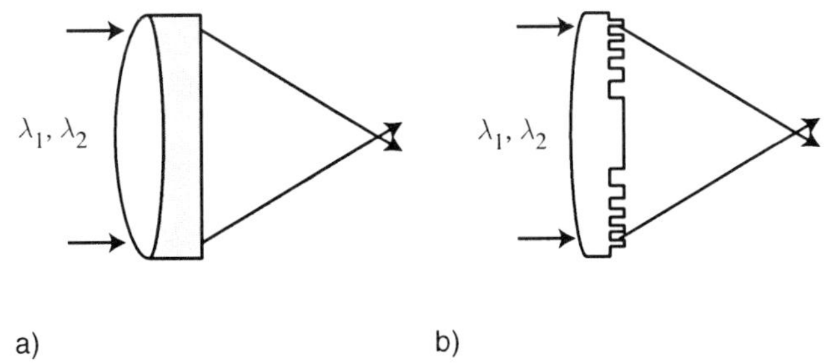

Fig. 12.12: Achromatic lens elements: a) refractive implementation consisting of positive and negative lens elements fabricated in different materials; b) hybrid diffractive/refractive implementation.

An alternative way for compensating chromatic dispersion in refractive elements is the use of diffractive optical elements (Fig. 12.12b) [47, 48, 49]. This is possible due to the negative optical dispersion observed for diffractive elements (Chapter 5).

By analogy to material dispersion, we can define an Abbe number for diffractive dispersion:

$$\nu_{\text{diff}} = \frac{\lambda_0}{\lambda_{\text{short}} - \lambda_{\text{long}}} \tag{12.4.3}$$

From this equation we find that the Abbe number of a diffractive element is always negative (visible region: $\nu_{\text{diff}}^{\text{vis}} = -3.45$). Using Eq. 12.4.2 it is possible to combine a positive refractive lens with a positive diffractive lens in order to achieve achromatic behaviour. For an achromatic diffractive/refractive lens with focal length f_{DROE} we have to combine a diffractive lens element with focal length f_{diff} and a refractive element with f_{ref}, where [47]:

$$f_{\text{diff}} = f_{\text{DROE}} \frac{\nu_{\text{diff}} - \nu}{\nu_{\text{diff}}} \qquad f_{\text{ref}} = f_{\text{DROE}} \frac{\nu - \nu_{\text{diff}}}{\nu} \tag{12.4.4}$$

A hybrid achromatic element can be thinner and, therefore, lighter than a refractive achromat. Since it is not necessary to compensate for the second element with negative power, the refractive element can be of relatively low optical power. This results in a reduction in wavefront aberrations. Due to the strong negative dispersion, a weak diffractive element is sufficient to compensate for the dispersion of a much stronger refractive element. Such a slow diffractive lens element can be fabricated with high quality and efficiency. Yet another advantage of hybrid achromats is due to the possible integration with the refractive element. Very often it is possible to fabricate the diffractive element on the surface of the refractive lens. Direct lithography has been demonstrated even on curved refractive surfaces [50]. Thus, no additional mechanical alignment of the two elements is necessary. The efficiency of DROEs is slightly lower than that of purely refractive achromatic elements. However, with state-of-the-art lithography it is possible to reach efficiencies greater than 90%, so that these problems are outweighted by the numerous advantages of achromatic DROEs. Apart from achromatisation, DROEs have also been applied to wavefront correction in eyepieces consisting of numerous lens elements [51].

12.4.2 Multi-order lenses

A second way of optimizing blazed diffractive optical elements for multiwavelength operation is possible varying the blaze depth [52]. This is a procedure well known from the design of echelette gratings for spectroscopy [53]. Dammann suggested using the phase depth of a diffraction grating as an additional design parameter, in order to optimize specific chromatic properties [54]. However, the application of this parameter for preventing chromatic errors of diffractive lenses for a discrete set of wavelengths has been suggested only recently. Again, the performance of the element can be understood as a combination of refractive and diffractive behaviour [55–59]. However, the refractive element is not added to the diffractive element as an additional element over the whole aperture. Instead, the blazed element can be considered as being composed of an array of refractive prism rings. In order to achieve the best performance, the wavefront diffracted at the array has to interfere constructively with the wave refracted at the individual array elements. Because of the different dependence of the diffractive and refractive performance on the wavelength tuning, such blazed elements show interesting features which can be exploited for the design of elements for multi-wavelength performance. Since reconstruction for each of the design wavelengths occurs in a different diffraction order, these elements are often called multi-order lenses [60].

In order to understand the spectral performance of a blazed DOE we consider the blazing procedure (Chapter 5). The quantization process is based on the insensitivity of a monochromatic wave to phase jumps of $\varphi = m2\pi$, where m is an integer. For the design wavelength λ_0 this condition is fulfilled in a transmission element of thickness t:

$$t = m\frac{\lambda_0}{n(\lambda_0) - 1}; \qquad m: \text{ integer} \tag{12.4.5}$$

Here m is the blaze order. An element of thickness t, however, also fulfils the blaze condition for other wavelengths λ_{blaze} (Fig. 12.13):

$$\lambda_{\text{blaze}} = t\frac{(n(\lambda_{\text{blaze}}) - 1)}{p} = \frac{m\lambda_0}{p}\frac{(n(\lambda_{\text{blaze}}) - 1)}{n(\lambda_0) - 1} \approx \frac{m\lambda_0}{p} \qquad m,\ p: \text{ integer} \tag{12.4.6}$$

Neglecting the dispersion of the lens material, we find that the blazed grating can be optimized for a number of wavelengths. These wavelengths have to be rational fractions of the design wavelength λ_0.

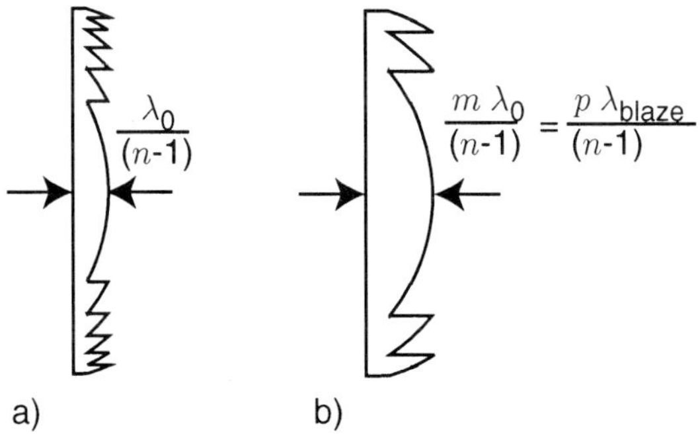

Fig. 12.13: Phase profile of a) a conventional blazed lens (design wavelength λ_0); b) a multi-order lens.

From these considerations we find that it is possible to use the blaze depth as a design parameter to achieve achromatic behaviour for a number of wavelengths (λ_1,

$\lambda_2, \ldots, \lambda_n$). For this purpose, we have to find a wavelength λ_0 which, divided by any of the wavelengths ($\lambda_1, \lambda_2, \ldots, \lambda_n$), yields an integral value (in other words we look for the lowest common multiple of the design wavelengths). This wavelength now becomes the design wavelength. The minimum necessary blaze depth of the multi-order element results, according to Eq. 12.4.5, from blazing the element into the first order for this design wavelength. It is therefore desirable to find the lowest design wavelength possible.

Use of the blaze depth as a design parameter for blazed DOEs can be understood as a concept for trading structural depth of the DOE, for achieving a desired chromatic behaviour. It is, however, important to beware of the difference between achromatisation with refractive or hybrid doublets and the use of DOEs with increased blaze depths. The dispersion for wavelengths other than the design wavelengths is moderate in the case of the doublet achromats, since one element compensates for the material dispersion of the other. In the case of multi-order diffractive lenses, the dispersion for intermediate wavelengths is increased significantly due to the reconstruction in higher diffraction orders. This introduces some image degradation when used with a continuous wavelength spectrum [61]. Multi-order lenses are, however, well suited for applications working with discrete wavelengths, such as multiplexed optical interconnects [62].

12.4.3 Athermalization with hybrid elements

Another important application of hybrid diffractive/refractive elements is athermalization. This is important in the case of harsh environmental conditions which might occur in military, aerospace or high power applications. For designing a DROE which is insensitive to thermal variations, it is necessary to find a parameter which like the Abbe number for the case of the chromatic errors, determines the thermal sensitivity of the individual elements. For this purpose, an opto-thermal coefficient can be defined. For refractive lenses, the influence of temperature on the optical performance is due to an effective index change under temperature variation as well as due to the thermal expansion of the lens material (thermal expansion coefficient: τ). The opto-thermal coefficient ξ_{ref} is defined as [45, 63]:

$$\xi_{\mathrm{ref}} = \frac{1}{f}\frac{df}{dT} = \tau - \frac{1}{n-1}\frac{dn}{dT} \tag{12.4.7}$$

For diffractive lenses, the effect of temperature change mainly stems from a variation of the spacing between the Fresnel zones. Effects of an index change on the diffraction efficiency can be neglected. For the opto-thermal coefficient ξ_{diff} of a diffractive lens we find [64, 65, 66]:

$$\xi_{\mathrm{diff}} = \frac{1}{f}\frac{df}{dT} \approx 2\tau \tag{12.4.8}$$

The variation (Δf) of the focal length under temperature change (ΔT) is calculated from:

$$\Delta f = f\xi\Delta T \qquad (12.4.9)$$

Again, we find differences in the behaviour of diffractive and refractive lenses which can be exploited for element optimization. Typical values for some of the most common materials for optical elements can be found in Tab. 12.1 [45]. ξ_{ref} can be positive or negative depending on the magnitude of the thermal expansion coefficient τ and the variation $\frac{dn}{dT}$ of the refractive index under temperature variation. ξ_{diff} is dominated by the thermal expansion coefficient which always assumes positive values. For polymer materials in particular, which are interesting for the fabrication of inexpensive and lightweight optical elements, we find large opto-thermal expansion coefficients. Germanium, which is interesting for infrared applications, exhibits a similarly high temperature dependence.

Material	ξ_{ref} in $10^{-4}\frac{\%}{^\circ\text{C}}$	ξ_{diff} in $\frac{10^{-4}\%}{^\circ\text{C}}$
BK7	0.98	13.89
SF11	-10.35	11.29
fused silica	-21.10	1.10
acrylic	315.00	129.00
polycarbonate	246.00	131.00
Ge	-123.76	12.20

Tab. 12.1: Opto-thermal expansion coefficients for several materials.

By analogy with the fabrication of achromatic lens doublets, it is possible to use two lens elements made of two different materials for compensating the temperature dependence of the element. However, the opto-thermal coefficients are different for diffractive and refractive lenses made of the same material. Therefore, it is also possible to design DROEs made of a single material, which are insensitive to temperature variations. For a DROE with total focal length f we find a net opto-thermal coefficient ξ of:

$$\xi = \xi_{\text{ref}}\frac{f}{f_{\text{ref}}} + \xi_{\text{diff}}\frac{f}{f_{\text{diff}}} \qquad (12.4.10)$$

where f_{ref} and f_{diff} are the focal lengths of the refractive and diffractive parts of the hybrid element. From Eq. 12.4.10 it is possible to determine the relative contributions of a diffractive and a refractive element to the performance of the hybrid athermat. If both ξ_{ref} and ξ_{diff} are positive, for fabricating an athermal element, it is necessary to combine a positive and a negative lens element which partly compensate each other in optical power. Athermal elements made of materials with opposite signs of ξ_{ref} and ξ_{diff}, on the other hand, result from two positive or negative lens elements, so that their optical power is added. For the optimization of mounted optical systems, it is necessary to take into consideration the thermal expansion coefficient of the mount material and adjust the lens performance for optimized system performance. An optimized acrylic DROE with focal length $f = 100$ mm, mounted with the detector in an aluminum mount, has been realized in practice. The focal lengths of the refractive and diffractive parts were $f_{\mathrm{ref}} = 111$ mm and $f_{\mathrm{diff}} = 55.6$ mm [65].

For many applications, it is necessary to athermalize systems which perform under broadband illumination. Simultaneous athermalization and achromatisation is possible in a triplet lens system. Alternatively, the thermal expansion coefficient of the mount material can be used as a degree of freedom in this design process [45, 67].

In this section we have focused on the optimization of optical elements with the help of microoptical structures. We outlined some of the basic design considerations necessary for the optimization of these hybrid or doublet elements. The optimization of multi-element optical systems is a more complex task with significantly more degrees of freedom. A variety of optical systems has been suggested and demonstrated for optimized performance [68–72] .

12.5 Glossary

- **Abbe number**
  number which describes the average material dispersion over a specific spectral range. In the original definition it was applied to the visible region. It is straightforward to extend the definition to different spectral regions and also to dispersion which does not depend on material dispersion, e.g., Abbe number for diffraction.

- **adaptive optics**
  effort to develop optical systems which adjust themselves to changing environmental conditions. This technique is often applied in astronomy, to optimize systems under changing atmospheric conditions. The first task in adaptive optics is to detect wavefront variations (e.g., with a Shack-Hartmann sensor) and then to change the optical system for optimized performance.

- **athermat**
  (optical) element which is insensitive to temperature variations.

- **composite image plane**
  In imaging with microlens arrays multiple images of the object are formed in the image plane due to the periodic arrangement of the microlenses. The composite image plane is located at the distance behind the second microlens array where these multiple images coincide (Fig. 12.2). For microlenses with equal focal lengths, this is the case in a strictly symmetrical setup. For sharp undistorted image formation with microlens arrays, it is necessary that the composite image plane coincides with the conventional image plane for the individual microlenses.

- **Galilean telescope**
  telescope formed by a long focal length positive lens (focal length: f_1) and a short focal length negative lens (f_2) at a distance $l = f_1 - |f_2|$. The magnification of the telescope is determined by the ratio $\frac{f_1}{|f_2|}$. The second popular telescope configuration is the so-called **Kepler telescope** which consists of two positive lenses.

- **microlens lithography**
  modification of projection lithography (Chapter 3), where microlens arrays are used for the imaging of the mask pattern onto the substrate. This approach has advantages with respect to the cost of the imaging system as well as the possible size of the image field.

12.6 List of new symbols

DROE:	diffractive/refractive optical element
L_1, L_2, L_3:	lenses in a confocal system
LA:	microlens arrays
λ_e, λ_C, λ_F:	wavelengths corresponding to specific spectral lines
λ_{long}, λ_{short}:	wavelengths determining a specific spectral region
λ_0:	design wavelength for a blazed optical element
λ_{blaze} :	number of wavelengths which fulfil the blaze condition in higher orders
λ_1, λ_2, ..., λ_n:	design wavelengths for a multi-order element
ν, ν_{ref}, ν_{diff} :	Abbe numbers: general definition, for refractive and diffractive elements
t_1, t_2:	transmission functions of microlenses
θ:	steering angle in a beam-steering system
U_0, U_1, U_2, U_3:	optical wavefronts at different planes in an optical system,
Δx:	relative lateral shift of two microlenses in a beam-steering system
ξ, ξ_{ref}, ξ_{dif}:	opto-thermal expansion coefficients: net coefficient for a compound element, coefficients for the refractive and diffractive elements
z_C:	composite image distance for imaging with microlens arrays
z_I:	conventional image distance
z_O:	object distance

12.7 Exercises

1. **Shack-Hartmann wavefront sensor**
   Which parameters of the microlens arrays determine the lateral and depth resolution of the microoptical wavefront sensor?

2. **Diffractive lenses and axial scanning by wavelength tuning**
   Due to the large chromatic length aberration of diffractive lenses, in diffractive imaging systems it is possible to perform axial scanning by wavelength detuning. Calculate the wavelength shift necessary to shift the focus of a diffractive lens by 1 μm. Assume a focal length of the lens of f = 10 mm at λ= 633 nm.

3. **Multiorder diffractive lenses**
   a) What is the profiling depth necessary for a blazed lens which is achromatized for the wavelengths λ_1 = 360 nm and λ_2 = 600 nm? Assume a transmission element with refractive index n= 1.457. What are the corresponding reconstruction orders?

   b) What would the necessary thickness be if achromatic behaviour is required for two additional wavelengths λ_3 = 440 nm, and λ_4 = 520 nm?

References

[1] R. H. Anderson, "Close-up imaging of documents and displays with lens arrays", *Appl. Opt.* **18** (1979), 477–484.

[2] J. D. Rees, "Non-Gaussian imaging properties of GRIN fiber lens arrays", *Appl. Opt.* **21** (1982), 1009–1012.

[3] R. Völkel, H.-P. Herzig, P. Nussbaum and R. Dändliker, "Imaging properties of microlens array systems", *EOS Meetings Digest* **9**, Free-Space Micro-Optical Systems, Engelberg (1996), 62–63.

[4] M. C. Hutley and R. F. Stevens, "The formation of integral images by afocal pairs of lens arrays", *IOP Short Meetings Series* **30**, Microlens Arrays, Teddington (1991), 147–154.

[5] N. Davies, M. McCormick and M. Brewin, "Design and analysis of an image transfer system using microlens arrays", *Opt. Eng.* **33** (1994), 3624–3633.

[6] W. Shaomin and L. Ronchi, "Principles and design of optical arrays", in Progress in Optics **XXV** (1988), E. Wolf (ed.), Elsevier, Amsterdam, 279–348.

[7] R. V. Pole, H. W. Werlich and R. J. Krusche, "Holographic light deflection", *Appl. Opt.* **17** (1978), 3294–3297.

[8] M. E. Motamedi, S. Park, A. Wang, M. S. Dadkhah, A. P. Andrews, H. O. Marcy, M. Khoshnevisan, A. E. Chiou, R. J. Huhn, C. F. Sell and J. G. Smits, "Development of micro-electro-mechanical optical scanners", *Opt. Eng.* **36** (1997), 1346–1353.

[9] S. Glöckner, R. Göring and T. Possner, "Micro-opto-mechanical beam deflectors", *Opt. Eng.* **36** (1997), 1339–1345.

[10] W. Glostos and M. Holz, "Agile beam steering using binary optics microlens arrays", *Opt. Eng.* **29** (1990), 1392–1397.

[11] M. W. Farn, "Agile beam steering using phased arraylike binary optics", *Appl. Opt.* **33** (1994), 5151–5158.

[12] E. A. Wilson, "Analysis of beam steering with decentered microlens arrays", *Opt. Eng.* **32** (1993), 2665–2670.

[13] S. Glöckner and R. Göring, "Investigation of statistical variations between lenslets of microlens arrays", *Appl. Opt.* **36** (1997), 4438–4445.

[14] S. Glöckner and R. Göring, "Analysis of a micro-optical light modulator", *Appl. Opt.* **36** (1997), 1467–1471.

[15] Y.-A. Peter, E. Rochat and H. P. Herzig, "Micro-opto-mechanical systems: applications in pulsed fiber lasers and optical switching", *Proc. SPIE* **3226**, Microelectronic structures and MEMS for optical processing III, Austin, TX, USA (1997), 102–111.

[16] S. Masuda, S. Takahashi, T. Nose, S. Sato and H. Ito, "Liquid-crystal microlens with a beam-steering function", *Appl. Opt.* **36** (1997), 4772–4778.

[17] G. D. Love, J. V. Major and A. Purvis, "Liquid-crystal prisms for tip-tilt adaptive optics", *Opt. Lett.* **19** (1994), 1170–1172.

[18] D. P. Resler, D. S. Hobbs, R. C. Sharp, L. J. Friedman and T. A. Dorschner, "High-efficiency liquid-crystal optical phased-array beam steering", *Opt. Lett.* **21** (1996), 689–691.

[19] J. A. Thomas and Y. Fainman, "Optimal cascade operation of optical phased-array beam deflectors", *Appl. Opt.* **37** (1998), 6196–6212.

[20] N. Lindlein, J. Pfund and J. Schwider, "Wave aberration measurement using microlens arrays", *EOS Meetings Digest* **9**, Free-Space Micro-Optical Systems, Engelberg (1996), 14–15.

[21] G. Y. Yoon, T. Jitsuno, M. Nakatsuka and S. Nakai, "Shack-Hartmann wave-front measurement with a large f-number plastic microlens array", *Appl. Opt.* **35** (1996), 188–192.

[22] R. G. Lane and M. Tallon, "Wave-front reconstruction using a Shack-Hartmann sensor", *Appl. Opt.* **31** (1992), 6902–6908.

[23] J. Ghozeil, "Hartmann and other screen tests", in Optical Shop Testing, D. Malacara (ed.), John Wiley & Sons, New York (1992), 367–426.

[24] W. H. Southwell, "Wave-front estimation from wave-front slope measurements", *JOSA* **70** (1980), 998–1006.

[25] J. Primot, G. Rousset and J. C. Fontanella, "Deconvolution from wave-front sensing: a new technique for compensating turbulence-degraded images", *JOSA A* **7** (1990), 1598–1608.

[26] R. H. Webb, "Confocal microscopes", *Optics & Photonics News* (July 1991), 8–13.

[27] E. A. Ash (ed.), "Scanned Image Microscopy", Academic Press, London (1980).

[28] T. Wilson and C. Sheppard, "Theory and practice of scanning optical microscopy", Academic Press, London (1984).

[29] D. K. Hamilton, T. Wilson and C. J. R. Sheppard, "Experimental observation of the depth-discrimination properties of scanning microscopes", *Opt. Lett.* **6** (1981), 625–626.

[30] H. J. Tiziani and H. M. Uhde, "Three-dimensional image sensing by chromatic confocal microscopy", *Appl. Opt.* **33** (1994), 1838–1843.

[31] H. J. Tiziani, R. Achi, R. N. Krämer and L. Wiegers, "Theoretical analysis of confocal microscopy with microlenses", *Appl. Opt.* **35** (1996), 120–125.

[32] H. J. Tiziani, R. Achi and R. N. Krämer, "Chromatic confocal microscopy with microlenses", *J. Mod Opt.* **43** (1996), 155–163.

[33] S. L. Dobson, P.-C. Sun and Y. Fainman, "Diffractive lenses for chromatic confocal imaging", *Appl. Opt.* **36** (1997), 4744–4748.

[34] S. Sinzinger, V. Arrizón and J. Jahns, "Confocal imaging with diffractive optics and broadband light sources", *Proc. SPIE* **3002**, Light-Emitting Diodes: Research, Manufacturing, and Applications, San Jose (1997), 186–189.

[35] C. J. R. Sheppard and M. Gu, "Aberration compensation in confocal microscopy", *Appl. Opt.* **30** (1991), 3563.

[36] M. Martínez-Corral, P. Andrés, E. Silvestre and C. Barreiro, "Improvement of three-dimensional resolution in confocal microscopy", *Proc. SPIE* **2730**, Second Iberoamerican Meeting on Optics, Guanajuato, México (1995), 634–638.

[37] L. Giniūnas, R. Juškaitis and S. V. Shatalin, "Endoscopes with optical sectioning capability", *Appl. Opt.* **32** (1993), 2888–2890.

[38] T. Wilson, "Image formation in two-mode fiber-based confocal microscopes", *JOSA A* **10** (1993), 1535–1543.

[39] N. P. Rea, T. Wilson and R. Juškaitis, "Semiconductor confocal and interference microscope", *Opt. Comm.* **125** (1996), 158–167.

[40] T. Ode, S. Komiya, Y. Yamauchi and D. Awamura, "Side looking confocal microscope", *Proc. SPIE* **1319**, Optics in complex systems (1990), 462–463.

[41] S. Sinzinger and J. Jahns, "Planar optical confocal microscope for imaging and sensing", *EOS Meetings Digest* **9**, Free-Space Micro-Optical Systems, Engelberg (1996), 46–47.

[42] S. Sheard, T. Suhara and H. Nishihara, "Integrated-optic implementation of a confocal scanning optical microscope", *IEEE J. Lightwave Techn.* **11** (1993), 1400–1403.

[43] T. Shiono and H. Ogawa, "Planar optical pick-up with diffractive micro-optics", *Appl. Opt.* **33** (1994), 7350–7355.

[44] D. L. Dickensheets and G. S. Kino, "Micromachined scanning confocal optical microscope", *Opt. Lett.* **21** (1996), 764–766.

[45] G. P. Behrmann and J. N. Mait, "Hybrid (refractive/diffractive) optics", in Microoptics: elements, systems, and applications, H.-P. Herzig (ed.), Taylor & Francis, London, UK (1997), 259–292.

[46] G. M. Morris and K. J. McIntyre, "Optical system design with diffractive optics", in Diffractive Optics for industrial and commercial applications, J. Turunen and F. Wyrowski (eds), Akademie Verlag, Berlin (1997), 81–104.

[47] T. Stone and N. George, "Hybrid diffractive-refractive lenses and achromats", *Appl. Opt.* **27** (1988), 2960–2971.

[48] C. Londoño and P. P. Clark, "Modelling diffraction efficiency effects when designing hybrid diffractive lens systems", *Appl. Opt.* **31** (1992), 2248–2252.

[49] N. Davidson, A. A. Friesem and E. Hasman, "Analytic design of hybrid diffractive-refractive achromats", *Appl. Opt.* **32** (1993), 4770–4774.

[50] E. B. Kley, "Lithographic techniques in diffractive optics", *EOS Meetings Digest* **12**, Diffractive Optics, Savonlinna (1997), 72–73.

[51] M. D. Missig and G. M. Morris, "Diffractive optics applied to eyepiece design", *Appl. Opt.* **34** (1995), 2452–2461.

[52] N. Davidson, R. Duer, A. A. Friesem and E. Hasman, "Blazed holographic gratings for polychromatic and multidirectional incidence of light", *JOSA A* **9** (1992), 1196–1199.

[53] M. C. Hutley, "Diffraction Gratings", Academic Press, London (1982).

[54] H. Dammann, "Colour separation gratings", *Appl. Opt.* **17** (1978), 2273–2279.

[55] H. Dammann, "From prism-prism via grating-prism to grating-grating compounds", *Appl. Opt.* **19** (1980), 2276–2278.

[56] S. Sinzinger and M. Testorf, "The transition between diffractive and refractive micro-optical components", *Appl. Opt.* **34** (1995), 5970–5976.

[57] M. Rossi, R. E. Kunz and H. P. Herzig, "Refractive and diffractive properties of planar micro-optical elements", *Appl. Opt.* **34** (1995), 5996–6007.

[58] M. Kovatchev and R. Ilieva, "Diffractive, refractive optics or anything more? Comparative analysis and trends of development", *J. Mod. Opt.* **43** (1996), 1535–1541.

[59] T. R. M. Sales and G. M. Morris, "Diffractive refractive behaviour of kinoform lenses", *Appl. Opt.* **36** (1997), 253–257.

[60] D. Faklis and M. Morris, "Spectral properties of multiorder diffractive lenses", *Appl. Opt.* **34** (1995), 2462–2468.

[61] D. W. Sweeney and G. E. Sommargren, "Harmonic diffractive lenses", *Appl. Opt.* **34** (1995), 2469.

[62] E. Quertemont and R. Chevallier, "Multi-order diffractive optical elements for achromatic interconnections at 1.3 μm and 1.55 μm", *Opt. Comm.* **140** (1997), 191–194.

[63] T. H. Jamieson, "Thermal effects in optical systems", *Opt. Eng.* **20** (1981), 156–160.

[64] J. Jahns, Y. H. Lee, C. A. Burrus and J. Jewell, "Optical interconnects using top-surface-emitting microlasers and planar optics", *Appl. Opt.* **31** (1992), 592–597.

[65] G. P. Behrmann and J. P. Bowen, "Influence of temperature on diffractive lens performance", *Appl. Opt.* **32** (1993), 2483–2489.

[66] C. Londoño, W. T. Plummer and P. P. Clark, "Athermalization of a single-component lens with diffractive optics", *Appl. Opt.* **32** (1993), 2295–2302.

[67] I. Friedman, "Thermo-optical analysis of two long focal length areal reconnaissance lenses", *Opt. Eng.* **20** (1981), 161–165.

[68] G. M. Morris, "Diffraction theory for an achromatic Fourier transformation", *Appl. Opt.* **20** (1981), 2017–2025.

[69] B. Packross, R. Eschbach and O. Bryngdhal, "Achromatisation of the self-imaging (Talbot) effect", *Opt. Comm.* **50** (1984), 205–209.

[70] J. Lancis, E. Tajahuerce, P. Andrés, V. Climent and Tepichín, "Single-zone-plate achromatic Fresnel-transform setup: pattern tunability", *Opt. Comm.* **136** (1997), 297–305.

[71] J. Schwider, "Achromatic design of holographic optical interconnects", *Opt. Eng.* **35** (1996), 826–831.

[72] B. Lunitz and J. Jahns, "Tolerant design of a planar optical clock distribution system", *Opt. Comm.* **134** (1997), 281–288.

Conclusion

With this book, we hope to have shown that microoptics is an interesting field to study and a technology of great commercial potential. Some 30 years after the demonstration of the first lithographically fabricated optical elements, the technology is in a relatively mature state. Several areas for using microoptics were described in the previous chapters. Others, not so much in the spotlight at the moment, might emerge in the future.

One of the main driving forces for the development of technology is the desire for bandwidth in areas like computing, communications and data storage. This is becoming a strong motivation for using photons rather than electrons. Microoptics is bound to be the technological foundation of photonics.

At the moment, however, an issue that still remains to be solved is cost. Some aspects of reducing cost, such as techniques for replication and integration, have already been discussed. Another aspect is the access to microoptics technology. For a potential user, obtaining custom-made components (such as a lenslet array or beam splitter grating) can be expensive. This problem, in turn, limits the growth of the field unless suitable measures are taken. As a solution, the use of fabrication networks or foundries has been discussed and partially implemented. The realization of joint shuttle runs for multiple users automatically necessitates standardization of the components. Standardization means that parameters such as the size, pitch and the wavelengths used are assigned only certain values. This allows manufacturers to offer off-the-shelf components, which in turn means faster access for customers and lower prices.

We started in Chapter 1 by pointing out that microoptics shares the same technology base with microelectronics. It is important to note that this does not only imply lithographic fabrication itself but also the design and modeling of the optics. It can be foreseen that, as in electronics, computer-aided design tools for layout, system simulation and data generation will become more and more part of the fabrication procedure.

Despite the similarities between microoptics and microelectronics, the development of microoptics will probably be different from the development of microelectronics, not as revolutionary, but evolutionary. Given further progress in technology and the ingenuity of scientists and engineers, microoptics will gradually be used in more and more applications.

Finally, we would like to end this book with a quote from Ernest Hemingway's "Death in the Afternoon":

"However, if I had waited long enough I probably would never have written anything at all since there is a tendency when you really begin to learn something about a thing not to want to write about it but rather to keep on learning about it always and at no time, unless you are very egotistical, which, of course, accounts for many books, will you be able to say: now I know all about this and will write about it. Certainly I do not say that now; every year I know there is more to learn, but I know some things which may be interesting now, and I may be away from the bullfights for a long time and I might as well write what I know about them now."

Abbreviations

AIL:	array illuminator
AOM:	acousto-optical modulator
AOD:	acousto-optical deflector
C4:	controlled collapse chip connection; flip-chip bonding
CAIBE:	chemically assisted ion-beam etching
CD:	compact disc
CGH:	computer-generated hologram
CMOS:	complementary metaloxide semiconductor
CVD:	chemical vapor deposition
DCG:	dichromatic gelatine
DMD™;	deflecting mirror device, micromirror projector (trademark of Texas Instruments)
DMSM:	diffractive mode-selective mirror
DOE:	diffractive optical element
DROE:	diffractive/refractive optical element
DVD:	digital versatile disc
epsf:	electron point spread function
FE:	focusing error in a disc pick-up
FSO:	free-space optics
FWHM:	full-width half maximum
FZP:	Fresnel zone plate
GRIN:	gradient-index
HEBS glass:	high-energy beam sensitive glass
HeCd:	helium-cadmium
HeNe:	helium-neon
HOE:	holographic optical element
IGWO:	integrated waveguide optics
IFSO:	integrated free-space optics
IFTA:	iterative Fourier transform algorithm
KDPH:	kinoform detour-phase hologram
LIDAR:	light detecting and ranging; laser radar
LIGA:	Lithography, Galvanoformung (galvanic forming) und Abformung (moulding)
MBE:	molecular beam epitaxy

MEM:	micro-electro-mechanical system
MIN:	multistage interconnection network
MMCD:	multimedia compact disc
MOCVD:	metal-organic CVD
MOEM:	micro-opto-electro-mechanical system
MQW:	multiple quantum well
MSM:	mode-selective mirror
MST:	micro systems technology
NA:	numerical aperture
OPD:	optical path difference
PECVD:	plasma enhanced CVD
PML^{TM}:	planar microlens array (trademark of Nippon Sheet Glass, Inc.)
PMMA:	polymethyl methacrylate
psf:	point spread function
PVD:	physical vapor deposition
QCSE:	quantum-confined Stark effect
REM:	raster electron microscope
RIE:	reactive ion etching
rms:	root mean square
ROE:	refractive optical element
rpm:	rotations per minute
SAW:	surface acoustic wave
SBP:	space-bandwidth product
SDD:	super density disc
SEED:	self-electro-optics-effect device; often used with the prefixes: D-: diode; S-: symmetric; L-: logic; FET-: field effect transistor
SELFOC^{TM}:	GRIN rod lenses (trademark of Nippon Sheet Glass, Inc.)
SIL:	solid immersion lens
SNR:	signal-to-noise ratio
SPA:	smart pixel array
TE:	tracking error in a disc pick-up
UV:	ultra-violet
VCSEL:	vertical cavity surface emitting laser
VLSI:	very large scale integration
WDM:	wavelength division multiplexing
WORM:	write-once-read-many times

Solutions to exercises

Chapter 2

1. **The second Gaussian moment of the point spread function**

   G_{psf} is defined as:

$$G_{\text{psf}} = \int x^2 \left|\text{psf}\right|^2 dx = \int x^2 \left|p(x)\right|^2 dx.$$

From Fourier mathematics we know:

$$\int \left|g(x)\right|^2 dx = \int \left|\tilde{g}(\nu_x)\right|^2 \, d\nu_x$$

and

$$\tilde{g}(\nu_x) = \frac{d\tilde{p}(\nu_x)}{d\nu_x} \qquad \longrightarrow \qquad g(x) = 2\pi i x p(x)$$

Here $\tilde{g}(\nu_x)$ is the Fourier transform of $g(x)$.

We use these relations to express G_{psf} as:

$$(2\pi)^2 \int x^2 \left|p(x)\right|^2 dx = \int \left|\frac{\tilde{p}(\nu_x)}{\nu_x}\right|^2 \, d\nu_x$$

$$\longrightarrow \qquad G_{\text{psf}} = \frac{1}{(2\pi)^2} \int \left|\frac{d\tilde{p}(\nu_x)}{\nu_x}\right|^2 \, d\nu_x$$

with $\tilde{p}(\nu_x) = A(\nu_x) e^{ik\Psi(\nu_x)}$ it follows that:

$$\begin{aligned} G_{\text{psf}} &= \frac{1}{4\pi^2} \int \left| \frac{dA(\nu_x)}{d\nu_x} \cdot e^{ik\Psi(\nu_x)} + A(\nu_x) \cdot ik \, \frac{d\Psi(\nu_x)}{d\nu_x} \cdot e^{ik\Psi(\nu_x)} \right|^2 \, d\nu_x \\ &= \frac{1}{4\pi^2} \left[\underbrace{\overline{\frac{A(\nu_x)}{d\nu_x}}^2}_{\text{waveoptical}} + \underbrace{A(\nu_x)^2 \, k^2 \, \frac{d\Psi(\nu_x)}{d\nu_x}^2}_{\text{rayoptical}} \right] \end{aligned}$$

2. **Rayleigh criterion and geometrical aberrations**

We are considering the presence of only primary spherical aberration. At the Rayleigh limit we get:

$$|\Psi(x)| = |\mathcal{A}|\ x^4 \leq \frac{\lambda}{4}$$

$$\longrightarrow \quad |\mathcal{A}| \leq \frac{\lambda}{4\ x^4}$$

$\mathcal{A}$ is the aberration coefficient. This amount of wavefront aberration $\Psi(x)$ causes ray aberrations $\xi(x)$ where:

$$\xi(x) = f \cdot \frac{\partial \Psi(x)}{\partial x} = |\mathcal{A}| \cdot f \cdot 4x^3 \leq \frac{\lambda}{4\ x^4} \cdot f \cdot 4x^3$$

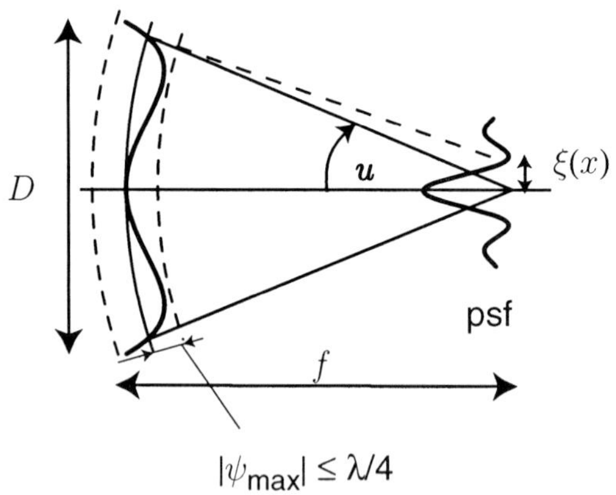

Fig. 1: Wavefront and ray aberrations.

The ray aberrations $\xi(x)$ reach a maximum for maximum values of x, i.e., at the edge of the lens aperture ($x_{\max} = \frac{D}{2} = \sin(u) f$):

$$\xi(x) \leq \frac{\lambda}{\sin(u)}$$

If we compare this to the size of the diffraction-limited psf we find:

$$\Delta x = 2f\frac{\lambda}{D} = 2f \cdot \frac{\lambda}{2 \cdot f \cdot \sin(u)} = \frac{\lambda}{\sin(u)} = \xi(x)$$

Thus, at the Rayleigh limit the ray aberrations caused by primary (spherical) aberrations are just covered by the extension of the diffraction-limited psf.

3. **Strehl ratio and rms wavefront aberrations**

$$\mathcal{S} = \left| \iint e^{i\frac{2\pi}{\lambda}\Psi(x,y)} dxdy \right|^2$$

Taylor expansion:

$$\begin{aligned}
\mathcal{S} &= \left| \iint \left[1 + ik\Psi(x,y) + \frac{1}{2}(ik\Psi(x,y))^2 + \text{higher order terms} \right] dxdy \right|^2 \\
&= \left| 1 + ik\overline{\Psi(x,y)} - \frac{k^2}{2}\overline{\Psi(x,y)^2} + \ldots \right|^2
\end{aligned}$$

From this we obtain:

$$\mathcal{S} = 1 - k^2\,\overline{\Psi(x,y)}^2 - k^2\,\overline{\Psi(x,y)^2} + \frac{k^4}{4}\,\overline{\Psi(x,y)^2}^2 + \ldots$$

By neglecting the higher order terms, which are quickly converging to zero, we find:

$$\mathcal{S} \approx 1 + k^2\,\overline{\Psi(x,y)}^2 - k^2\,\overline{\Psi(x,y)^2} = 1 - \Psi^2_{\text{rms}}(x,y)$$

Chapter 3

1. **Photoresist**
A photoresist is a polymer material which is sensitive to optical or high energy radiation. By analogy to photographic films, we distinguish between positive and negative resists. In positive resists the solubility of the photoresist material is increased upon exposure. In a developing process the exposed areas can be removed from the coating. In negative resists the effect of exposure is a reduction of the solubility. After development the exposed areas remain on the substrate, while non-exposed parts are removed.

   Photoresist materials are important for lithographic processing since they are very resistant to a large variety of chemicals. They can be used as protective coating for many etching processes.

2. **Resolution limits in lithography**
Optical lithography is limited by diffraction. In e-beam lithography the limiting factors are scattering effects of the electrons in the photoresist and the substrate (proximity effect).

3. **Lift-off processing**
The lift-off process allows the structuring of resistant coatings. The chemicals involved need not dissolve the coating materials. The material is removed through dissolving the underlying photoresist layer. This is especially important, if the chemicals necessary to etch the coating could be harmful for the substrate material.

   Problems with the lift-off process occur for structures with high aspect ratios and very thick coatings. In this case the resist developer may not be able to reach and dissolve the resist layer.

4. **Reactive ion etching**
Ion etching is performed with ions generated in the reaction chamber in the vicinity of the substrate. For ion-beam etching the ion source is located in a separate chamber. The ions are then accelerated by an electric field in the direction of the main chamber. This separation of ion source and reaction chamber has the effect that a collimated ion beam is available for the etching process. This yields very high anisotropy in the etching process.

   In both reactive ion etching and reactive ion-beam etching the ratio between (anisotropic) physical etching and (isotropic) chemical etching processes can be manipulated over a wide range through control of the gas composition. The etch rates for different materials (selectivity) can vary correspondingly. This explains the significance of these processes for lithographic processing.

Chapter 4

1. **Technology for microlens fabrication**

a) From Eqs 4.1.1 and 4.1.4 we know that the focal length f is related to the lens sag h by the quadratic equation (D: lens diameter):

$$h^2 - 2f(n-1)h + \frac{D^2}{4} = 0$$

The solutions for this equation are:

$$h_{1,2} = \frac{2f(n-1) \pm \sqrt{4f^2(n-1)^2 - D^2}}{2}$$

For the two lenses the minimum lens sags are:

$$\begin{aligned} h_{200\,\mu\mathrm{m}} &= 17.22\ \mu\mathrm{m} \\ h_{3000\,\mu\mathrm{m}} &= 556.4\ \mu\mathrm{m} \end{aligned}$$

The second mathematical solutions of the quadratic equation for both cases results in sags, which are larger than the corresponding diameter. Physically these are no feasible solutions.

b) Due to the moderate diameter and sag, the 200 μm diameter lenses can be fabricated with a variety of different technologies. The fabrication by dispensing droplets of a curable polymer seems the best approach. Using the multidroplet approach, sufficiently large lenses with precisely defined focal lengths are possible.

Fabrication with the reflow technique is also feasible. Sufficiently thick coatings are challenging in this case. The numerical aperture of reflow lenses tends to be larger than the one required here. The use of base layers or preshaped resist profiles allows one to achieve the required parameters. Similar statements apply for the mass-transport approach in semiconductor materials. Profiling techniques, such as analog lithography using direct-writing or HEBS-glass grey-scale masks, will give difficulties achieving the necessary profiling depth. In combination with etching techniques, e.g., CAIBE, the parameters can be achieved. Ion-exchange technology in glass is a further technique useful for lenses with this size and numerical aperture. Even GRIN rod lenses might be considered if individual lenses are required rather than lens arrays.

The fabrication of "larger" microlenses with the relatively long focal length is more challenging. For many technological approaches a large diameter as well as a long focal length are difficult to achieve. Analog lithographic approaches are impracticable for profiling depths larger than several 10s of micrometers. If at all capable of the fabrication, reflow and droplet approaches would both need significant optimization procedures. Preshaping and local reflow through ion-, or e-beam polishing can lead to a solution.

The dimensions of the microlens are such that optimized micromechanical techniques are recommended for the fabrication. Single-point diamond turning is the most promising approach. If necessary an optimization of the surface quality can be achieved by ion polishing or thermal reflow.

2. **Deflection by microprisms**

Fig. 2 shows the deflection of the incident light beam by the prism. From Snell's law we may write:

$$
\begin{aligned}
n_1 \sin(\epsilon) &= n_0 \sin(\alpha + \epsilon) \\
&= n_0(\sin(\alpha)\cos(\epsilon) + \sin(\epsilon)\cos(\alpha))
\end{aligned}
$$

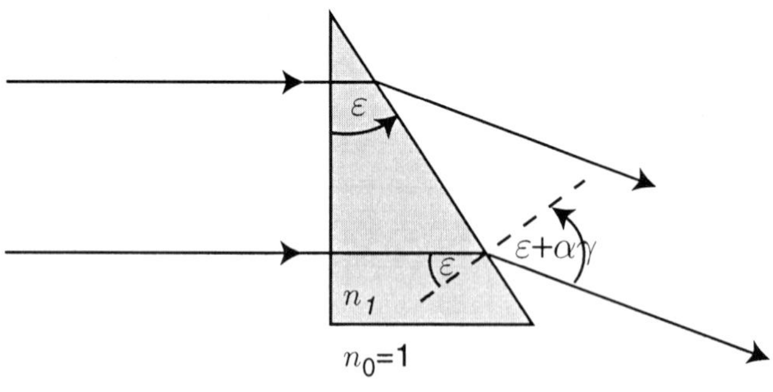

Fig. 2: Light deflection by a microprism.

For the desired deflection angle of $\alpha = 10°$, we find wedge angles of:

$$
\begin{aligned}
& & \epsilon &= \arctan\left(\frac{\sin(\alpha)}{\frac{n_1}{n_0} - \cos(\alpha)}\right) \\
n_1 = 1.6 &\rightarrow & \epsilon &= 20.1° \\
n_1 = 3.5 &\rightarrow & \epsilon &= 3.9°
\end{aligned}
$$

For a prism with a lateral extension w_{prism} the wedge angle is given by:

$$\tan(\epsilon) = \frac{h_{\text{prism}}}{w_{\text{prism}}}$$

Since the prisms have to be at least as large as the beam diameter we find for the minimum depth of the prisms:

$$\begin{aligned} n = 1.46 \quad &\rightarrow \quad h_{1.46} &= 73.1\mu m \\ n = 3.5 \quad &\rightarrow \quad h_{3.5} &= 13.8\mu m \end{aligned}$$

We note that the fabrication of such prisms in GaAs is feasible with lithographic techniques. For a fused silica prism, however, the profiling depth is much larger. This significantly increases the fabrication problems.

3. **Thermal reflow and mass transport mechanisms**

The basic processing steps for the reflow process and the mass transport are the same. In fact, the melting of the photoresist is a kind of mass transport phenomenon. After a preform is fabricated in photoresist or, e.g., GaP the substrates are heated. This heating initiates the mass transport in both cases.

The physical effects which cause the mass movement are quite different. The photoresist material melts and due to surface tension assumes the spherical shape of a droplet. In semiconductor compounds, mass transport takes place at much higher temperatures, which are, however, still below the melting temperatures of the substrate. In order to avoid chemical decomposition of the compound, a suitable atmosphere has to be generated in the reaction chamber, e.g., in a sealed ampoule. In this atmosphere the material still decomposes, but after a diffusion process chemically recombines at different locations on the substratc. As in the melting process the decomposition and recombination is supported by surface energy. The shape which forms in the mass transport process is similar to the shape formed if the material were melted.

Chapter 5

1. **Etching depth for the fabrication of multilevel phase gratings**

   For the fabrication of an ideal blazed grating (blazed into the first diffraction order) the maximum phase jump at the edge of each period is $\Delta\varphi_{\max} = 2\pi$. The thickness of the grating necessary for such a phase step depends on whether the grating is used in reflection or in transmission. Due to the phase quantization the linear phase slope is approximated by a regular staircase profile. The step height is determined as $2\pi/N$ where N is the number of steps. Using a logarithmic mask sequence three mask layers are needed for the fabrication of an 8 phase level grating. With each masking step the phase depth is reduced to 0.5 of the previous phase depth.

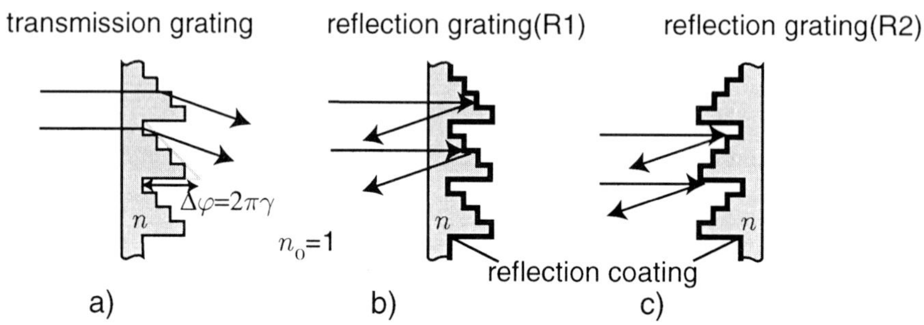

Fig. 3: Illustration of the etching depths necessary for the different illumination configurations.

a) According to the figure (left), in order to achieve a phase step of $\Delta\varphi_{\max}$ in a transmission element we need an etching depth of:

$$t^{\mathrm{T}} = \frac{\Delta\varphi_{\max}}{2\pi}\frac{\lambda}{(n-1)}$$

Thus for the blazed transmission grating with N phase steps we calculate the etching depths from:

$$t_N^{\mathrm{T}} = \frac{\Delta\varphi_{\max}}{2\pi 2^N}\frac{\lambda}{(n-1)} = \frac{1}{2^N}\frac{\lambda}{(n-1)}$$

b) As illustrated in the two right figures there are two possible ways to use the DOE in reflection. Thus, two different sets of optimum etching depths result:

$$t_N^{\mathrm{R1}} = \frac{1}{2^{N+1}}\frac{\lambda}{n}$$

$$t_N^{\mathrm{R2}} = \frac{1}{2^{N+1}}\lambda$$

For the individual etching steps one obtains:

# of etching steps N	1	2	3
transmission grating t^{T}	0.692 μm	0.346 μm	0.173 μm
reflection in substrate t^{R1}	0.217 μm	0.109 μm	0.054 μm
hline reflection outside the substrate t^{R2}	0.316 μm	0.158 μm	0.079 μm

2. **The validity of the paraxial approximation for the design of diffractive lenses**

   According to Eq. 5.3.48 we have :

$$w_{\min} = \frac{2\lambda f/\#}{N} \quad \longrightarrow \quad f/\# = \frac{w_{\min} N}{2\lambda}$$

   Using Eq. 5.3.41 we can calculate the minimum focal length and maximum radius for which the lenses with the calculated f-numbers still belong to the paraxial regime:

$$f < (f/\#)^4 \cdot 32 \cdot \lambda 2r > 32\lambda f/\#$$

# of phase levels N	2	4	8
$f/\#$	1.6	3.2	6.4
focal length f_{min}	132.8 μm	2.12 mm	34 mm
diameter $2r_{\mathrm{max}}$	83 μm	664 μm	5.31 mm

3. **Effect of fabrication errors on the performance of kinoform gratings**

   a) One period of the phase grating resulting from the lateral misalignment δs of the N-th mask relative to the $(N-1)$st mask layer can be described by using (Fig. 4):

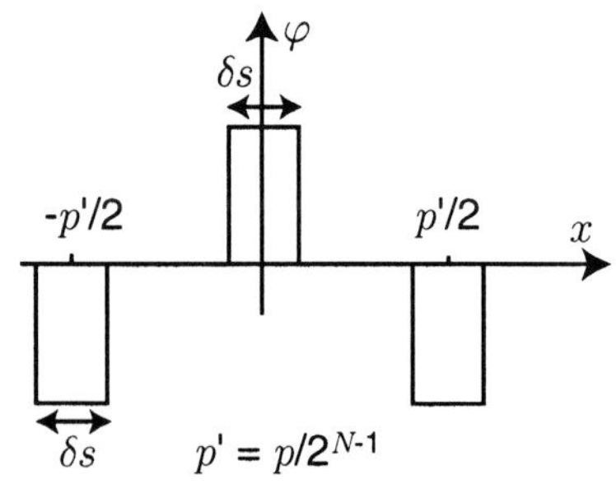

Fig. 4: The phase grating introduced by a lateral misalignment of the N-th mask layer.

$$f_{\text{error}}(x) = \text{rect}\left(\frac{x}{\delta s}\right) e^{i\frac{2\pi}{2N}} + \text{rect}\left(\frac{x - \frac{p'}{4}}{\frac{p'}{2} - \delta s}\right) + $$
$$+ \text{rect}\left(\frac{x - p'/2}{\delta s}\right) e^{-i\frac{2\pi}{2N}} + \text{rect}\left(\frac{x - \frac{3p'}{4}}{\frac{p'}{2} - \delta s}\right)$$

Here p' is the period of the N-th mask for which the etching depth is $\varphi = 2\pi/2^N$.

b) The diffraction amplitudes resulting from this phase grating are calculated from the Fourier transform:

$$\begin{aligned}
A_n &= e^{i\frac{2\pi}{2N}} \frac{1}{p'} \int_{-\frac{\delta s}{2}}^{\frac{\delta s}{2}} e^{-2\pi i n \frac{x}{p'}} dx + \frac{1}{p'} \int_{\frac{\delta s}{2}}^{\frac{p'}{2} - \frac{\delta s}{2}} e^{-2\pi i n \frac{x}{p'}} dx + \\
&+ e^{-i\frac{2\pi}{2N}} \frac{1}{p'} \int_{\frac{p'}{2} - \frac{\delta s}{2}}^{\frac{p'}{2} + \frac{\delta s}{2}} e^{-2\pi i n \frac{x}{p'}} dx + \frac{1}{p'} \int_{\frac{p'}{2} + \frac{\delta s}{2}}^{p' - \frac{\delta s}{2}} e^{-2\pi i n \frac{x}{p'}} dx = \\
&= \frac{\delta s}{p'} \,\text{sinc}\left(n \frac{\delta s}{p'}\right) \left(e^{i\frac{2\pi}{2N}} + e^{i(\pi n - \frac{2\pi}{2N})}\right) + \\
&+ \left(\frac{1}{2} - \frac{\delta s}{p}\right) \text{sinc}\left(n\left(\frac{1}{2} - \frac{\delta s}{p}\right)\right) \left(e^{i - \pi\frac{n}{2}} + e^{-i\pi\frac{n3}{2}}\right)
\end{aligned}$$

Fig. 5 shows the intensity loss in the first diffraction order A_1^2 of this error function, as a function of the relative misalignment $\delta s/p'$ for the second and third mask layer.

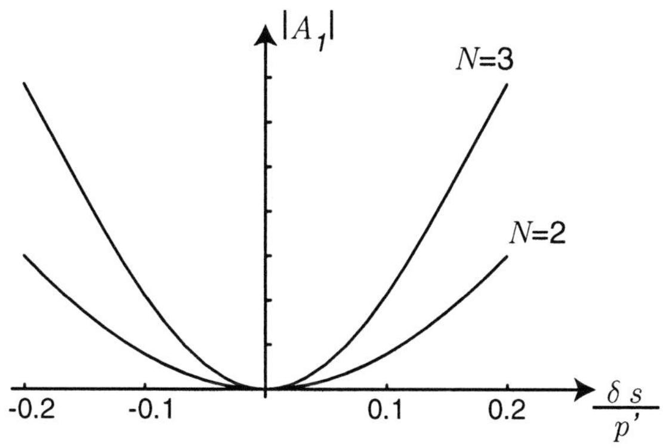

Fig. 5: First diffraction order of the error function introduced by lateral misalignment of different lithographic masks.

4. Efficiency of holographically recorded blazed gratings

a)

$$g_1(x) = \frac{1}{2}[1 + \cos(2\pi\frac{x}{p})] = \frac{1}{2} + \frac{1}{4}\left(e^{2\pi i\frac{x}{p}} + e^{2\pi i\frac{x}{p}}\right)$$

For the diffraction amplitudes we calculate the Fourier transform:

$$\begin{aligned}\tilde{g}(\nu_x) &= \int_{-\infty}^{\infty} g(x)e^{-2\pi i n\nu_x x}dx = \\ &= \frac{1}{2}\delta(\nu_x) + \frac{1}{4}\delta\left(\nu_x - \frac{1}{p}\right) + \frac{1}{4}\delta\left(\nu_x + \frac{1}{p}\right)\end{aligned}$$

Thus, a cosinusoidal amplitude grating has 3 diffraction orders with intensities $|A_0|^2 = 0.25$ and $|A_{\pm 1}|^2 = 0.0625$.

b)

$$g_2(x) = e^{i\phi\cos(2\pi\frac{x}{p})}$$

$$\begin{aligned}A_n &= \frac{1}{p}\int_{-\frac{p}{2}}^{\frac{p}{2}} e^{i\phi\cos(2\pi\frac{x}{p})}e^{2\pi i n\frac{x}{p}}dx \\ &= \frac{1}{p}\int_{0}^{p} e^{i\phi\cos(2\pi\frac{x}{p})}e^{2\pi i n\frac{x}{p}}dx\end{aligned}$$

$$= \frac{1}{2\pi} \int_0^{2\pi} e^{i\phi \cos x} e^{inx} dx = i^{-n} J_{-n}(\phi) = i^n J_n(\phi)$$

Thus, the Fourier transform $\tilde{g}(\nu)$ can be written as:

$$\tilde{g}(\nu_x) = 2\pi \sum_{n=-\infty}^{\infty} i^n J_n(\phi) \delta \left(\nu_x - n\frac{1}{p} \right)$$

The cosinusoidal phase grating has an infinite number of nonzero diffraction orders. Nevertheless, since no light is absorbed by the grating, the diffraction efficiency in the first order is generally higher than for the cosinusoidal amplitude grating. For a modulation depth of $\phi = \pi/2$ we obtain for the maximum intensity in the first order: $|A_1|^2 = 0.339$.

Chapter 6

1. **The numerical aperture of a symmetrical waveguide**
In the ray diagram for a mode to be guided there has to be total internal reflection. Thus, the propagation angle of the ray inside the waveguide has to fulfil the condition:

$$\theta_m \leq \frac{\pi}{2} - \alpha_c$$

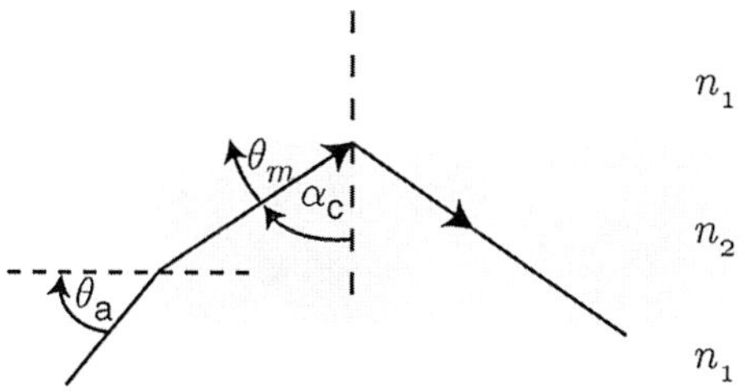

Fig. 6: The aperture angle of a symmetrical waveguide.

The propagation direction of this beam outside the dielectric waveguiding layer determines the numerical aperture:

$$\text{NA} = \sin(\theta_a) = n_2 \sin\left(\frac{\pi}{2} - \alpha_c\right)$$

Using Eq. 6.1.1 we thus get:

$$\text{NA} = n_2 \cos(\alpha_c) = n_2\sqrt{1 - \sin^2(\alpha_c)} = \sqrt{n_2^2 - n_1^2}$$

2. **Number of modes in slab waveguides:**
According to Eq. 6.1.6:

$$M > 2\sqrt{n_2^2 - n_1^2}\,\frac{d}{\lambda_0}$$

For the number of modes in both cases we obtain:

$$M_{1\,\mu\text{m}} = 4; \quad M_{3\,\text{mm}} = 10045$$

From Eq. 6.1.4 we find the angular separation between two modes:

$$\Delta\theta_{1\,\mu\text{m}} = 0.32°; \quad \Delta\theta_{3\,\text{mm}} = 10^{-4}°$$

While for the 1 μm waveguide we find few clearly separated modes, a very large number of quasi-continuously distributed modes propagate in the 3 mm dielectric slab. Propagation in the thick slab corresponds to unguided free-space optics.

3. **Cut-off wavelength**
In a monomode waveguide only a single mode can be guided. From Eq. 6.1.6 we see that the number of modes which propagate in the waveguide is inversely proportional to the wavelength of the incident light. Thus for each waveguide a cut-off wavelength exists. For wavelengths larger than the cut-off wavelength λ_{co} the waveguide performs as a monomode waveguide, i.e., has only one propagating mode. For the $d = 1\ \mu$m waveguide of the previous problem we find:

$$\lambda_{co} = 1.45\ \mu\text{m}$$

4. **Optical tunneling**
Optical tunneling is the effect that light energy can be coupled through a dielectric barrier between two waveguiding structures. This is possible via the exponentially decaying evanescent tails of the waves which travel outside the waveguide. The evanescent waves result as solutions of the wave equation and the boundary conditions (i.e., continuity and finite energy content). Evanescent waves occur, e.g., during total internal reflection of a lightwave in a prism. If another optical component, e.g., another prism or waveguide, is located close enough to allow an overlap of the evanescent tails of the modes, light is coupled through the gap into the neighbouring component. At the same time the energy in the reflected beam is reduced. Therefore, optical tunneling is sometimes also referred to as "frustrated total internal reflection". Prism and grating couplers exploit optical tunneling for coupling a light wave into a waveguide.

Chapter 7

1. **Divergence of Gaussian beams**
According to Eq. 7.2.11, in order to transmit 99% of the light through the lens aperture, the waist w_0 of the Gaussian beams has to fulfil the condition:

$$w_0 = \frac{d_{\text{lens}}}{1.52}$$

We assume that the waist of the Gaussian beam lies in the plane of the microlens. Using Eqs 7.2.2 and 7.2.3, we find the diameter at a distance z behind the beam waist:

$$w(z) = w_0 \left[1 + \left(\frac{z\lambda}{\pi w_0^2}\right)^2\right]^{\frac{1}{2}}$$

Thus, we find the Gaussian beam diameters at a distance $z = 10$ cm. The light efficiency upon transmission through a second lens can be calculated from Eqs 7.2.9 and 7.2.10:

$$\frac{P}{P_{\text{max}}} = 1 - e^{-2\frac{R^2}{w_0^2}}$$

microlens diameter	Gaussian beam width at $z = 0$	Gaussian beam width at $z = 10$ cm	clipping efficiency
50 μm	32.9 μm	613.3 μm	1.3 %
100 μm	65.79 μm	313.25 μm	36.8 %
250 μm	164.5 μm	205.1 μm	94.9 %
500 μm	328.95 μm	334.65 μm	98.85 %
1000 μm	657.9 μm	658 μm	99 %
10000 μm	6578.95 μm	6578.95 μm	99 %

2. **Microchannel imaging**
In microchannel imaging the interconnection distance is limited by crosstalk due to diffraction at the microlens aperture. A light beam transmitted through a microlens (diameter: d_{lens}) is diffracted at an angle α:

$$\sin(\alpha) = \frac{\lambda}{d_{\text{lens}}}$$

For a rectangular lens aperture, the beam profile in the focus of the lens is described by a sinc function. At a distance δz behind the lens the central lobe of the beam has a width Δx of:

$$2\Delta x = 2\Delta z \sin(\alpha) = 2\lambda \frac{\Delta z}{d_{\text{lens}}}$$

In order to avoid crosstalk we demand the central lobe to be smaller than the diameter of the subsequent microlens:

$$2\Delta x \leq d_{\text{lens}}$$

Thus, we conclude:

$$\Delta z \leq \frac{d_{\text{lens}}^2}{2\lambda}$$

3. **Hybrid imaging**
   Hybrid imaging splits the imaging task in a favourable way between microlens arrays and imaging lenses. The microlens arrays provide the numerical aperture for collimating or focusing with high resolution. Thus, the imaging lenses only have to resolve the microlens apertures and can be much slower than necessary for resolving the source or detector windows. In contrast to microchannel imaging, the intermediate $4F$ imaging step allows large distances between the input and output microlens arrays.

Chapter 8

1. **Heat dissipation in an array of VCSELs**
The emitted optical power of 1 mW per laser diode corresponds to 15% of the total power consumption. Thus, for the dissipated heat results:

$$P_{\text{diss}} = 1 \text{ mW } \frac{100}{15} = 6.67 \text{ mW}$$

per diode. For the total heating power P_{total} generated by the whole array, we obtain:

$$P_{\text{total}} = 16^2 \; P_{\text{diss}} = 1.71 \; W$$

The total area covered by the array, however, only amounts to 16 mm^2. In order to be able to use air cooling it would be necessary to spread the dissipated heat over an area A of:

$$A = \frac{P_{\text{total}}}{0.1 \text{ W/cm}^2} = 17.1 \text{ cm}^2$$

2. **Modulator vs active devices**
In a modulator device an incident beam of light is modulated, e.g., in intensity. This is opposed to actively emitting devices such as VCSELs, where a new (modulated) light beam is generated. In SEEDs the reflectivity or absorption characteristic of the device is changed. From a systems perspective it has to be taken into account that optical readout beams are necessary to read the state of the device.

3. **Doping and hetero-superlattices**
Superlattices are stacks of multiple layers of semiconductor materials with different electronic band structures. These differences in the electronic bands can be generated either through different doping (doping-superlattice) or through different composition of the semiconductor material (hetero or composition-superlattice).

4. **VCSELs vs edge-emitting laser diodes**
Due to the orientation of the laser cavity vertical to the planar substrate surface, the light emission of VCSELs naturally takes place perpendicularly to the substrate. For edge-emitting laser diodes deflecting mirrors or gratings are necessary to achieve this. For integration with optical systems vertical emission is desirable since the planar substrate surfaces of light sources and optical components can be aligned more easily (e.g., with planar bonding techniques). Vertical resonators are easily fabricated in large, densely packed 2D arrays. This is not possible for edge emitting laser diodes for geometrical reasons. 2D data planes are adjusted to the 3D nature of free-space (micro)optical systems.

   The beam shape of the light emitted by the laser cavity is determined by the shape of the cavity. Since the VCSEL cavities are symmetrical, so is the emitted beam profile. Side-emitting laser diodes, on the other hand, have rectangular cavities, which results in astigmatic beams.

Because of the microcavities, the quantum efficiency of VCSELs can be very high, which yields high output power and little heat dissipation compared to other light sources. The planar structure also allows one to access the laser arrays, e.g., for cooling purposes.

Chapter 9

1. **Compression ratio of aperture dividing array illuminators**
The spacing Δx between the individual beams resulting from aperture division is determined by the diameter of the individual apertures of the lenslets or prisms. The most important influence on the compression ratio thus stems from the diameter of the generated foci or beamlets. In the case of Fig. 9.8a), the size of the foci generated by the lenslet array is determined by the numerical aperture. For the compression ratio we find:

$$\mathcal{D}_a = \frac{D^2}{2f}$$

For the telescopelet arrays we know that the ratio of the beam diameters is determined by the ratio of the focal lengths:

$$\mathcal{D}_{b/c} = \left| \frac{f_1}{f_2} \right|$$

In the Brewster telescope arrangement, the prism angles determine the deflection and thus the anamorphotic beam compression.

2. **Optimization of a Dammann grating**

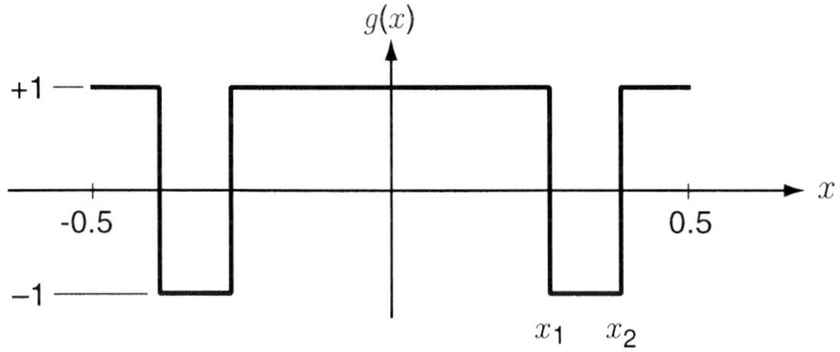

Fig. 7: The phase profile of one period of a Dammann grating with two transition points x_1 and x_2.

Fig. 7 shows one period of the Dammann grating with $N = 2$ transition points x_1 and x_2. With such a grating it is possible to optimize $2N + 1 = 5$ diffraction orders for equal intensity.

For an optimization of the grating we write for the diffraction orders according to Eq. 9.3.5:

$$A_0 = 4(x_1 - x_2) - 1$$

$$A_{\pm 1} = \frac{2}{\pi}\left[\sin(2\pi x_1) - \sin(2\pi x_2)\right]$$

$$A_{\pm 2} = \frac{1}{\pi}\left[\sin(4\pi x_1) - \sin(4\pi x_2)\right]$$

For the optimization we look for the minima (zero transitions) of the cost function:

$$C(x_1, x_2) = \frac{1}{2N+1} \sum_{m=-N}^{N} |I_m - \bar{I}|$$

where

$$\bar{I} = \frac{1}{2N+1} \sum_{m=-N}^{N} I_m$$

Fig. 8 shows the cost function C depending on the variables x_1 and x_2:

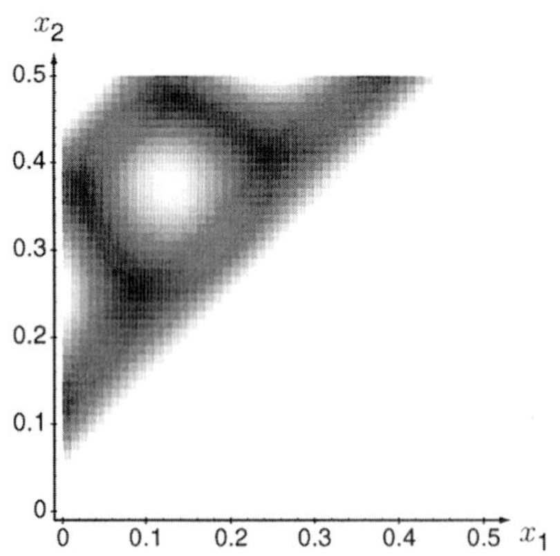

Fig. 8: Grey-scale plot of the cost function C depending on x_1 and x_2. The larger the value of C, the brighter the pixel.

We find $4 = 2^N$ zero transitions for $C(x_1, x_2)$. These solutions of the optimization problem are the following locations for phase transitions:

$$\begin{aligned}
x_1 &= 0.0854, & x_2 &= 0.2577, & \eta &= 0.482\\
x_1 &= 0.0193, & x_2 &= 0.3677, & \eta &= 0.774\\
x_1 &= 0.2423, & x_2 &= 0.4146, & \eta &= 0.482\\
x_1 &= 0.1323, & x_2 &= 0.4807, & \eta &= 0.774
\end{aligned}$$

Here η denotes the diffraction efficiency which is calculated from $\eta = (2N + 1)\bar{I}$. Only two of these solutions are independent. The third solution results from the first solution by a cyclic shift. So does the fourth solution from the second.

3. **Compression ratio of Fourier array generators**

The compression ratio of array illuminators is defined as the ratio between the extension of the individual diffraction order and their spacing. Due to the Fourier relationship between the component plane and the diffraction pattern we know that the extension of the component determines the size of the diffraction order. If we assume a grating consisting of Z periods with period length p, we find for the extension δx of the diffraction order:

$$\delta x \propto \frac{\lambda f}{Zp}$$

where f is the focal length of the optical Fourier transforming system.

At the same time, the spacing Δx between the diffraction orders is determined by:

$$\Delta x = \frac{\lambda f}{p}$$

We thus find for the compression ratio $\mathcal{D}$:

$$\mathcal{D} = \frac{\Delta x}{\delta x} \propto Z$$

The compression ratio in Fourier array illuminators is determined by the number of illuminated periods of the component.

Chapter 10

1. **Propagation of Gaussian beams; design of an astigmatic element for laser diode collimation**
The propagation of the Gaussian beam is described by a Fresnel transform:

$$u(x,y,z) = \int\int u(x',y') e^{\frac{i\pi}{\lambda z}[(x-x')^2+(y-y')^2]} dx'dy'$$

With Eq. 10.2.1 one obtains:

$$u(x,y,z) = \int\int A_0 e^{-\left(\frac{x'^2}{2\sigma_{x,0}^2}+\frac{y'^2}{2\sigma_{y,0}^2}\right)} e^{\frac{i\pi}{\lambda z}[(x-x')^2+(y-y')^2]} dx'dy'$$

This integral is separable in x and y. We find:

$$u(x,z) = e^{i\frac{\pi x^2}{\lambda z}} \int e^{-\pi x'^2\left(\frac{1}{2\pi\sigma_{x,0}^2}-\frac{i}{\lambda z}\right)} e^{2\pi\frac{x'x}{\lambda z}} dx'$$

We note that $u(x,z)$ is described by a Fourier transform of a Gaussian distribution with complex width. Solving this integral we find:

$$u(x,z) = e^{i\frac{\pi x^2}{\lambda z}} e^{-\pi\frac{x^2}{\lambda^2 z^2}\left(\frac{1}{2\pi\sigma_{x,0}}-\frac{i}{\lambda z}\right)^{-1}}$$

In order to calculate the width of the corresponding Gaussian function we compare this result with Eq. 10.2.1:

$$\frac{x^2}{2\sigma_x} = \frac{\pi x^2}{\lambda^2 z^2}\left(\frac{1}{2\pi\sigma_{x,0}} - \frac{i}{\lambda z}\right)^{-1}$$

From this we find:

$$\frac{1}{\sigma_x^2(z)} = \frac{4\pi^2\sigma_{x,0}}{\lambda^2 z^2 + 4\pi^2\sigma_{x,0}^4} + i\frac{8\pi^3\sigma_{x,0}^4}{4\pi^2\sigma_{x,0}^4\lambda z + \lambda^3 z^3}$$

For $\sigma_y(z)$ an analogous relation results. The distance z_0 can be derived as the distance where the real values of σ_x and σ_y are equal. We find the condition:

$$(\lambda^2 z_0^2 + 4\pi^2\sigma_{x,0}^4)\sigma_{y,0}^2 = (\lambda^2 z_0^2 + 4\pi^2\sigma_{y,0}^4)\sigma_{x,0}^2$$

For z_0 one obtains:

$$z_0 = \frac{2\pi}{\lambda}\sigma_{x,0}\sigma_{y,0}$$

2. **Bessel beams**
   According to Eq. 10.3.6:

$$w = 0.766\frac{\lambda}{\sin(\theta)} = 2 \cdot 0.766\lambda f/\# = 29.1\ \mu\text{m}$$

$$z_{\max} = \frac{D}{2}\sqrt{\frac{1}{\sin^2(\theta)} - 1} = \frac{D}{2}\sqrt{4(f/\#)^2 - 1} = 30.1\ \text{cm}$$

3. **Super-Gaussian beam shape**
   Fig. 9 shows the amplitude distribution of super-Gaussian beams with the parameter n varying from $n = 1, 2, 3, 4, 5, 6, 10$.

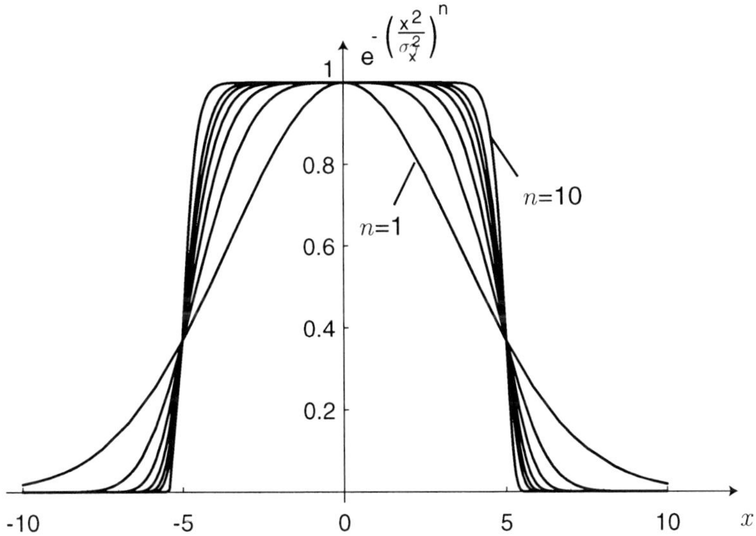

Fig. 9: Super-Gaussian distributions for different parameters n.

Numerical integration of the distribution shows that for $n = 5$ about 98% of the beam intensity is contained in the rectangular area with width 2σ.

Chapter 11

1. **Perfect Shuffle interconnection networks**

a) and b)

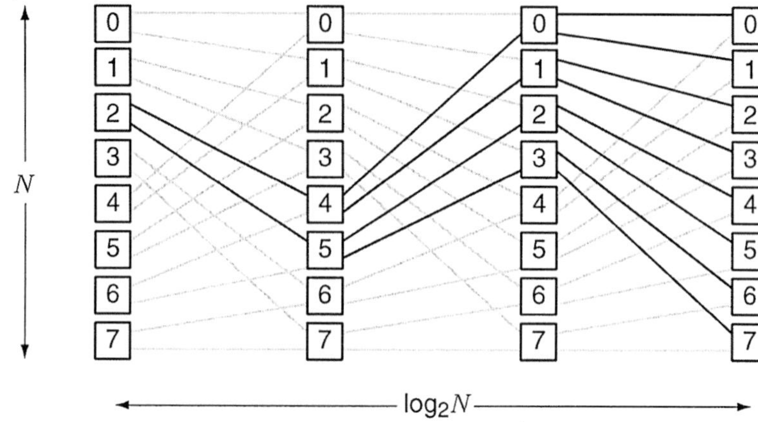

Fig. 10: Perfect shuffle interconnection network.

Fig. 10 shows a complete Perfect Shuffle interconnection network. For the input channel No. 2 all possible interconnections are indicated, which shows that arbitrary interconnects can be implemented.

c)

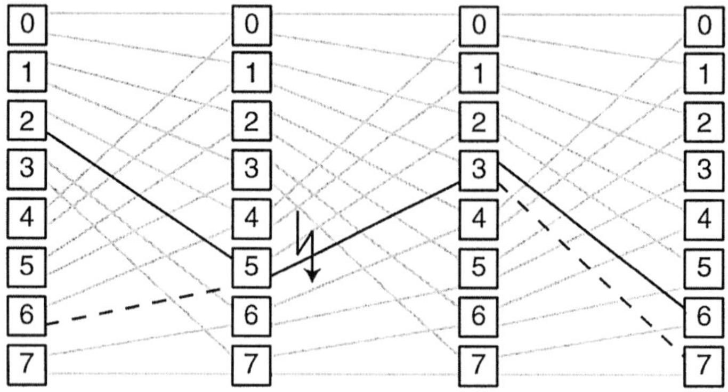

Fig. 11: Blocking in perfect shuffle interconnection networks.

The interconnections indicated in Fig. 11 show that the perfect shuffle is not a non-blocking network. If we need to implement connections between input 2 and output 6 as well as between input 6 and output channel 7, this is not possible simultaneously, since both connections need the same data channel in the second shuffle stage.

2. Microlens arrays for arbitrary interconnects

a) The maximum possible interconnection length which has to be supported by the microlenses interconnects the lower-most channel in the input with the upper-most channel in the output (Fig. 11.9a). With the maximum propagation angle α we can calculate for this interconnection a length of:

$$l_{\max} = \frac{Nd}{\sin(\alpha)}$$

On the other hand we know that in the microchannel system the interconnection distance has to be limited to values smaller than $l_{\max} < \frac{d^2}{\lambda}$. This is necessary to avoid crosstalk between the data channels (Chapter 7). Thus, for the maximum number of channels we find:

$$N = \frac{d}{\lambda} \sin(\alpha) = 139$$

b) 4 phase level diffractive optical elements may have a minimum period of $p = 4\ \mu$m. Thus, we obtain a maximum propagation angle of:

$$\sin(\alpha) = \frac{\lambda}{p}$$

In this case, we find for the total number of channels which can be arbitrarily interconnected:

$$N = 31$$

3. Storage density in different types of data storages

	paper	CD	CMOS chip
area	63000 mm^2	10000 mm^2	10 mm^2
storage capacity	5000 $\times$ 8 bit = 40000 b	$\approx$ 1 Gb	16 Mb
density	0.63 bit/mm^2	$\approx 10^5$ bit/mm^2	1.6 10^5 bit/mm^2

Tab. 1: Comparison of the storage density in different media.

Chapter 12

1. **Shack-Hartmann wavefront sensor**
The lateral resolution is determined by the diameter d of the microlenses.

depth resolution: the angular resolution $\delta\alpha$ of the lenslets can be estimated by means of:

$$\delta\alpha \sim \frac{\lambda}{d}$$

The angle α by which the wavefront is inclined over the lens aperture can be calculated from:

$$\tan(\alpha) = \frac{\delta t}{d}$$

Thus we find that the depth resolution for a Shack-Hartmann sensor is proportional to the wavelength:

$$\delta t \sim \lambda$$

An increase in resolution is possible with reduction optics, which reduces the diameter of the tested wavefront. This virtually creates an increased diameter d of the microlenses.

2. **Diffractive lenses and axial scanning by wavelength tuning**
According to Eq. 5.3.52 in Chapter 5, the wavelength dependence of a diffractive lens is determined by:

$$\frac{\partial f}{\partial \lambda} = \frac{f}{\lambda}$$

Thus, we find that we can achieve a shift of $\delta f = 1\ \mu$m in the focal length for a wavelength shift of $\delta\lambda = 63.3$ nm. It is important to realize that lenses with relatively long focal lengths need to be used if long focal length shifts are to be achieved with a small wavelength tuning range.

3. **Multi-order diffractive lenses**
a) Design wavelengths: $\lambda_1 = 360$ nm and $\lambda_2 = 600$nm:
The smallest wavelength which yields integral numbers if divided by any of those two wavelengths (lowest common multiple) is:

$$\lambda_0 = 1800 \text{ nm}$$

The corresponding profiling depth according to Eq. 12.4.5 is:

$$t_1 = \frac{\lambda_0}{n-1} = 3939 \text{ nm}$$

In a blazed element of thickness t_1, the 5th diffraction order for λ_1 coincides with the 3rd diffraction order for λ_2.

b) The smallest design wavelength for a multiorder lens designed for the wavelengths λ_1–λ_4 results as:

$$\lambda_0 = 257400 \text{ nm}$$

Thus, the necessary thickness of the multi-order element is:

$$t_2 = 563238 \text{ nm}$$

In this case the coinciding diffraction orders are:
$p_{360\,\text{nm}} = 715$, $p_{440\,\text{nm}} = 585$, $p_{520\,\text{nm}} = 495$, $p_{600\,\text{nm}} = 429$.

This result clearly demonstrates the limits for the design of multi-order lenses. If the difference in the design wavelengths decreases, the necessary blaze depth increases rapidly.

Index